THE KELLEY STATISTICAL TABLES

Revised 1948

LONDON : GEOFFREY CUMBERLEGE

Oxford University Press

THE KELLEY
STATISTICAL TABLES

Revised 1948

BY TRUMAN LEE KELLEY
HARVARD UNIVERSITY

CAMBRIDGE, MASSACHUSETTS
HARVARD UNIVERSITY PRESS
1948

PREFACE TO 1938 EDITION

Many years ago my inspiring teacher, Henry Lewis Rietz, observed that statistical procedures derived from the normal distribution were born of a higher realm than other procedures. I disbelieved this with a religious fervor. Though, as time has passed, I have espoused curvilinear regression and skew distributions with gusto, I have found myself frequently slipping, for the data would not support me, and linear relationships and nearly normal distributions have in my experience as a psychologist cropped up with a frequency which has chided and mocked me. I still reserve judgment as to the place of birth of the normal distribution, but that its sphere of usefulness is extended in connection with biological and psychological phenomena I no longer have the slightest doubt. This constitutes the chief justification for this book and the heavy labor that has been involved in the computation of the deviates and ordinates of the normal distribution.

T. L. K.

Cambridge, Mass.
February, 1938

PREFACE TO 1948 EDITION

The practice of providing tabled entries to eight figures has been continued, but sundry improvements have been made to assure an actual utility to eight figures.

The normal distribution functions provided in the first edition have been supplemented by the addition at the feet of columns of E'' and E''' values, which indicate the maximum interpolation error for the region in question when linear and quadric interpolation are employed.

Precise results from tables having sufficient decimal places in the tabled values, whatever the tabled function, can in general be obtained by (a) linear interpolation in a table having arguments at small intervals, or by (b) quadric, cubic, or higher-order interpolation in a table with coarse argument intervals. In favor of (a) is ease of use and against it is cost of construction and bulk of a table having arguments at small intervals. In favor of (b) is low cost and small bulk and against it is the greater labor of interpolation of higher order than linear. Some compromise between (a) and (b) has always been necessary. The (b) procedure is greatly simplified if tables of Lagrangian interpolation coefficients are available. The first edition tables of interpolation coefficients for direct interpolation are here replaced by more detailed tables, and inverse three- and four-point interpolation is described. These tables constitute generally useful devices in connection with tables of all sorts where precision greater than that given by linear interpolation is demanded. In general, the use of direct interpolation coefficients as here tabled in connection with a 3 × 8 table (3-place arguments and 8-place consequents) will be more accurate than linear interpolation in a 5 × 8 table, which is, of course, 100 times as bulky.

A statistic of universal importance is P, the probability of a situation, deviating from the hypothesis as much as does the observed situation, arising as a matter of chance. The calculation of P depends upon the form of the distribution yielded by chance. It has long been known that a large number of chance distributions have the normal form and it has been

more recently established that many other chance distributions
fall in the class of variance ratio, or F, distributions. To
table the common F-distributions with the refinement in which
the normal distribution has been tabled is prohibitive, but
happily a transformation is available which normalizes F-distributions so that the detailed table of the normal distribution may be used to get highly precise (i.e., to the third
decimal place) P values from F. The most involved step in
this meritorious transformation is the extraction of a cube
root. A complete 3×8 table of cube roots is given herein.

The variety of occasions in which natural logarithms are
needed is served by an 8-place table for arguments from 1.00
to 10.00.

The chief objectives of this work are: (1) To provide certain
8-place tables of general utility, but not duplicating the
operations readily handled upon a computing machine, (2) to
provide interpolation coefficients which enable the ready use
of tables with a precision approximating the maximum precision
inherent in the tabled entries, (3) to provide a detailed table
of the normal distribution, and (4) to provide tables and transformation equations enabling the use of the fully tabled normal
distribution to evaluate the areas in t (Student's t), z (Fisher's
z from r), χ^2 (Pearson's), and F (variance ratio) distributions.

Truman Lee Kelley

Cambridge, Massachusetts
January, 1948

CONTENTS

CONTENTS

THE KELLEY STATISTICAL TABLES

Revised 1948

THE KELLEY STATISTICAL TABLES
Revised 1948

INTRODUCTORY SECTION: BEARING UPON ALL THE TABLES AND INCLUDING INVERSE INTERPOLATION PROCEDURES

The need of a table of deviates and ordinates corresponding to proportionate tail areas of a normal distribution was met in part by the Kelley-Wood Table* and Table I of *The Kelley Statistical Tables* (1938). Table I herewith is Table I of the 1938 edition, with the addition throughout of maximum inverse two-point and direct two- and three-point interpolation errors.

The $\sqrt{pq}$, $\sqrt{1-p^2}$, and $\sqrt{1-q^2}$ columns of Table I have been incorporated for good measure. These square roots are commonly needed in probability and correlation work, and since the arguments, values from .0000 to 1.0000 by .0001's were available, it seemed a happy opportunity to use them. Of course, the labor cost of these columns has been but a trifle in comparison with that of the deviate and ordinate columns.

Table VI gives probability values for different values of χ^2 and different d.o.f. (degrees of freedom). The argument $\chi / \sqrt{n}$ has been chosen as it yields a convenient spacing of values for interpolation purposes, especially when the number of d.o.f., n, is large. An alternative method for obtaining P from χ^2 is that for getting P from F, a variance ratio, for χ^2 with n d.o.f. is an $F_{n\infty}$. Table VI yields an answer expeditiously, but when interpolation is necessary it generally is not quite as accurate as the P from F method described in Section VIII.

Tables II, III, IV, and V provide Lagrangian interpolation coefficients, which the writer believes will generally be found to be more simple to use than first, second, third, fourth, fifth, or sixth order differences, even when these are published. When not published and when a computing machine is available, the economy of this method of interpolation over Everett's, or any other higher order difference formula, is great.

INVERSE INTERPOLATION: The numerical values for the maximal inverse two- and three-point interpolation errors of argument columns, are based upon the following method of inverse inter-

* Truman L. Kelley, STATISTICAL METHOD, 1923.

polation:

Let four consecutive equally spaced arguments with interval i be a_{-1}, a_0, a_1, a_2 and let the tabled values and differences be as indicated in Table A herewith.

TABLE A. ARGUMENT, CONSEQUENT, AND DIFFERENCE NOTATION

ARGUMENTS	CONSEQUENTS		DIFFERENCES	
a_{-1}	t_{-1}			
		Δ^{I}_{-1}		
a_0	t_0		Δ^{II}_{-1}	
		Δ^{I}_{0}		Δ^{III}_{-1}
a_1	t_1		Δ^{II}_{0}	
		Δ^{I}_{1}		
a_2	t_2			

Corresponding to a is a value of the function t. a is some proportionate distance p between a_0 and a_1 and t is some proportionate distance p' between t_0 and t_1. The problem of inverse interpolation is to find p knowing p'. The linear approximation to p is designated p^{-ii}, the quadric p^{-iii}, and the cubic p^{-iv}. The simplest solution maintains when a is a linear function of t, for then $p^{-ii} = p'$ and a is given by [1]. We let $p' = (t-t_0)/\Delta^{I}_{0}$ and the value of a as given by inverse two-point interpolation, designated a^{-ii} is

$$a^{-ii} = a_0 + i\,p' \quad \dots\dots\dots\dots \quad [1]$$

Though [1] is axiomatic we derive it herewith to illustrate the principles followed in obtaining more complex inverse interpolation formulas. The general expression of a as a linear function of t is $a^{-ii} = k + b\,t$. Imposing the condition that this straight line pass through the two points (a_0, t_0) and (a_1, t_1), we have

$$a_0 = k + b\,t_0 \quad \text{and} \quad a_1 = k + b\,t_1$$

from which, since $i = a_1 - a_0$, we obtain $k = a_0 - \dfrac{i\,t_0}{\Delta^{I}_{0}}$ and $b = \dfrac{i}{\Delta^{I}_{0}}$, thus giving equation [1].

We can in a comparable manner pass a quadric through the three points (a_0, t_0), (a_1, t_1), (a_2, t_2). It proves convenient for three- and four-point interpolation to write t in terms of Δ^{I}_{0} and to make other substitutions as follows:

$$p' = \frac{t-t_0}{\Delta^{I}_{0}}; \qquad T_0 = \frac{t_0 - t_0}{\Delta^{I}_{0}} = 0; \qquad T_1 = \frac{t_1 - t_0}{\Delta^{I}_{0}} = 1;$$

$$T_2 = \frac{t_2 - t_0}{\Delta_0^I} = 2 + \Delta,$$ in which Δ is a small quantity if a deviates but slightly from being a linear function of t

Otherwise expressed,

$$\Delta = \frac{\Delta_0^{II}}{\Delta_0^I} \qquad \ldots \ldots \ldots \ldots \ldots \ldots [2]$$

$$T_{-1} = \frac{t_{-1} - t_0}{\Delta_0^I} = -1 + \Delta - 3\Delta^2 + \delta,$$ in which δ is a small quantity if a deviates but slightly from being a quadric function of t. Also Δ^2 is ordinarily small with reference to Δ

Otherwise expressed,

$$\delta = \frac{-\Delta_{-1}^{III}}{\Delta_0^I} + 3\Delta^2 \qquad \ldots \ldots \ldots \ldots [3]$$

Passing the quadric $a^{-iii} = k + bT + cT^2$ through the three points (a_0, T_0), (a_1, T_1), (a_2, T_2) permits solving for k, b, and c and obtaining equation [4], wherein $q' = 1 - p'$. We first obtain p^{-iii}:

$$p^{-iii} = p' + \frac{\Delta}{2 + 3\Delta + \Delta^2} p'q'$$

$$a^{-iii} = a_0 + ip' + i \frac{\Delta p'q'}{2 + 3\Delta + \Delta^2}$$

or approximately

$$a^{-iii} = a_0 + ip' + \frac{1}{2} i \Delta p'q' (1 - \frac{3\Delta}{2}) \qquad \begin{array}{c} \text{Inverse 3-point interpolation} \\ (.5 < p' < 1.0) \end{array} \qquad [4]$$

If p' is greater than .5 the tabled values to use are t_0, t_1, and t_2, while if less than .5 interpolate in the opposite direction using t_1, t_0, and t_{-1}.

A close approximation to the error in the linear inverse interpolation answer [1] is

$$e^{-ii} = a^{-iii} - a^{-ii} = \frac{1}{2} i \Delta p'q' (1 - \frac{3\Delta}{2}) = \frac{1}{2} i \Delta p'q' \qquad \ldots \ldots \ldots [5]$$

(holding when terms of order Δ^2, δ, etc. are negligible)

The maximum error is found when $p' = .5$. We designate the absolute value of this maximum error in linear or two-point inverse interpolation E^{-ii} and have recorded it in connection with the principal arguments of this book of tables.

$$E^{-ii} = .125 \, i \, |\Delta| \qquad \ldots \ldots \ldots [6]$$

Following the principles of derivation involved in [4], [5], and [6] yields formula [7] for inverse four-point interpolation, formula [8] for the error in inverse three-point interpolation, and formula [9] for the maximal error.

$$a^{-iv} = a_0 + ip' + ip'q'\left(\frac{1}{2}\Delta - \frac{3}{4}\Delta^2\right) + ip'q'(1+q')\frac{\delta}{6} \quad \ldots \ldots \quad [7]$$

$$e^{-iii} = a^{-iv} - a^{-iii} = \frac{p'q'(1+q')}{6}\delta \quad \begin{array}{l}\text{(Holding when terms of order} \\ \Delta^3, \Delta\delta, \delta^2, \text{ etc., are negli-} \\ \text{gible.}\end{array} \quad [8]$$

$$E^{-iii} = .0625\, i\, |\delta| \qquad \text{Maximum 3-point inverse interpolation error} \qquad [9]$$

Attention is called to the italicized rules given in Section I covering the inverse interpolation error consequent to the rounding-off error in tabled entries.

More precise methods of high order inverse interpolation than here given are presented in Davis* and Salzer†.

Formulas for three-, four-, five-, six-, and seven-point interpolation are given by Salzer. The first three of them are given herewith in the notation of this treatment. Quantities r, s, t, and u, defined by the following equations, and differing in their definitions depending upon the number of interpolation points used, are introduced in formula [21] to yield the desired inverse interpolated value.

Let $\dfrac{1}{d_3} = t_1 - t_{-1}$

Let $\dfrac{1}{d_4} = -t_2 + 6t_1 - 3t_0 - 2t_{-1}$

Let $\dfrac{1}{d_5} = -2t_2 + 16t_1 - 16t_{-1} - 2t_{-2}$

Then r, s, t, and u are defined as follows:

For three-point inverse interpolation,—

$$r = 2\, d_3\, (t_p - t_0) \quad \ldots \ldots \ldots \ldots \ldots \quad [10]$$

$$s = d_3\, (t_1 - 2t_0 + t_{-1}) \quad \ldots \ldots \ldots \ldots \quad [11]$$

$$t = u = 0 \quad \ldots \ldots \ldots \ldots \ldots \ldots \quad [12]$$

*H. T. Davis, TABLES OF THE HIGHER MATHEMATICAL FUNCTIONS, Vol. 1, 1933-1935, pages 80-83.

†Herbert E. Salzer, "Tables of the Coefficients for Inverse Interpolation with Central Differences," JOURNAL MATHEMATICS AND PHYSICS, Vol. 22, No. 4, December, 1943, pages 210-224.

——, "A New Formula for Inverse Interpolation," BULLETIN AMERICAN MATHEMATICAL SOCIETY, Vol. 50, No. 8, August, 1944, pages 513-516.

TABLE I 7

For four-point inverse interpolation,—

$$r = 6 d_4 (t_p - t_0) \quad \ldots \ldots \ldots \ldots \ldots \ldots [13]$$

$$s = 3 d_4 (t_1 - 2t_0 + t_{-1}) \quad \ldots \ldots \ldots \ldots [14]$$

$$t = d_4 (t_2 - 3t_1 + 3t_0 - t_{-1}) \quad \ldots \ldots \ldots [15]$$

$$u = 0 \quad \ldots \ldots \ldots \ldots \ldots \ldots \ldots \ldots [16]$$

For five-point inverse interpolation,—

$$r = 24 d_5 (t_p - t_0) \quad \ldots \ldots \ldots \ldots \ldots [17]$$

$$s = d_5 (-t_2 + 16t_1 - 30t_0 + 16t_{-1} - t_{-2}) \quad \ldots \ldots [18]$$

$$t = 2 d_5 (t_2 - 2t_1 + 2t_{-1} - t_{-2}) \quad \ldots \ldots \ldots [19]$$

$$u = d_5 (t_2 - 4t_1 + 6t_0 - 4t_{-1} + t_{-2}) \quad \ldots \ldots [20]$$

When, for any order of inverse interpolation, the values of r, s, t, and u are introduced into [21] the desired proportionate distance p is obtained.

$$p = r - r^2 s + r^3 (2s^2 - t) + r^4 (-5s^3 + 5st + u)$$
$$+ r^5 (14s^4 - 21s^2 t + 3t^2 + 6su)$$
$$+ r^6 (-42s^5 + 84s^3 t - 28st^2 - 28s^2 u + 7tu) + \ldots \quad [21]$$

It will be noted that when this formula is used to get several values between the same t_0 and t_1 the only change in variable in the right hand member is in r.

SECTION I. BEARING UPON TABLE I.

The early standard tables of normal distribution functions* used the deviate as the argument and an area as the consequent, or tabled measure. Table I herewith interchanges these. It is based on a unit normal distribution which is defined by the equation

$$z = \frac{1}{\sqrt{2\pi}} e^{\frac{-x^2}{2}}$$

and the integral $\quad p = \displaystyle\int_{-\infty}^{x} z \, dx$

*James Burgess, "On the Definite Integral $\frac{2}{\pi} \int_{0}^{t} e^{-t^2} dt$ with Extended Tables of Values," TRANS. ROYAL SOCIETY OF EDINBURGH, Vol. 39, Part 2, 1897-1898, pp. 257-321.

    Karl Pearson, TABLES FOR BIOMETRICIANS AND STATISTICIANS, 1914 (in which are included Sheppard's Tables).

This unit normal distribution has an area of 1.00, a mean of zero, and a standard deviation of 1.00. The computation of z for a given value of x is relatively simple, but the computation of x for a given value of p is laborious. Many methods were tried and that finally adopted involved the following steps:

(a) Interpolation using higher order differences in Kondo and Elderton's table*. Differences as high as the twelfth order were tried but they proved uneconomical, so that, generally speaking, differences beyond the fifth were not used in this step.

(b) Refinement of values thus obtained by computation, using formula [23], of all values which were marginal, that is, in the neighborhood of 50 in the ninth and tenth decimal places, since it has been the aim to have the table accurate to the last published figure. For values of p near .5000 few recomputations were necessary, but for values of p above .9700 from 20 to 50 per cent of recomputation was needed. As a check on the accuracy of interpolation, the intervals used overlapped so that two values of x were obtained, the difference between them giving the order of accuracy of the interpolation method employed. Also, to test this accuracy, the high order interpolation error formulas† of Sections II-V were employed. The interpolated values were computed on a Moon-Hopkins machine, printing the last nine figures (decimal places six to fourteen inclusive) and guaranteeing accuracy, as each subsequent term depended upon those preceding. The first five decimal places, which change but little, were computed mentally. It is believed that if an inaccuracy is present in the last published figure, it is only of size consequent to a carry-over from the eleventh or higher decimal place.

(c) Occasional computation by Schlömilch's formula‡ [22]. This formula proved of limited use because it is only serviceable for a few values at the upper end of the table.

(d) Computation by formula [23] of certain intermediate values; for example, values for p ending in 5 in the fourth decimal place, and then the use of these together with Kondo and Elderton values to get interpolated values.

(e) More precise computation for $p > .9950$, using Laplace's continued fractions‡ and quadric inverse interpolation.

---

*T. Kondo and E. M. Elderton, "Abscissae, Ordinates, and Ratios, $z/\frac{1}{2}(1 \pm a)$, $\frac{1}{2}(1 \pm a)/z$, to Ten Significant Figures of the Normal Curve to Each Permille of Frequency," BIOMETRIKA, Vol. 22, 1930-1931. pages 368-376.

†Edward V. Huntington, "Tables of Lagrangian Coefficients for Interpolating Without Differences," PROCEEDINGS AMERICAN ACADEMY OF ARTS AND SCIENCES, Vol. 63, No. 11, March, 1929.

‡E. R. Enlow, "Quadrature of the Normal Curve," ANNALS OF MATHEMATICAL STATISTICS, Vol. 5, June, 1934, pages 136-146.

‡Burgess, op. cit.

TABLE I   9

$$p = \frac{1}{2} + \frac{1}{2}\left\{ 1 - \frac{2e^{-\frac{x^2}{2}}}{x\sqrt{2\pi}}\left[1 - \frac{1}{x^2+2} + \frac{1}{(x^2+2)(x^2+4)} \right.\right.$$

$$- \frac{5}{(x^2+2)(x^2+4)(x^2+6)} + \frac{9}{(x^2+2)(x^2+4)(x^2+6)(x^2+8)}$$

$$- \frac{129}{(x^2+2)(x^2+4)(x^2+6)(x^2+8)(x^2+10)}$$

$$\left.\left. + \frac{57}{(x^2+2)(x^2+4)\ldots(x^2+12)} \right] \right\} \quad [22]$$

FORMULA GIVING AREAS FROM $-\infty$ TO X UNDER A UNIT NORMAL DISTRIBUTION

$$p = \int_{-\infty}^{x} \frac{1}{\sqrt{2\pi}}\, e^{-\frac{x^2}{2}}\, dx$$

$$p = \frac{1}{2} + x(.39894,22804,014)\left[1 - \frac{1}{6}x^2 + \frac{P}{6.66666,66666,67}x^2 - \frac{P}{8.4}x^2 \right.$$

$$+ \frac{P}{10.28571,42857,1}x^2 - \frac{P}{12.22222,22222,2}x^2 + \frac{P}{14.18181,81818,2}x^2$$

$$- \frac{P}{16.15384,61538,5}x^2 + \frac{P}{18.13333,33333,3}x^2 - \frac{P}{20,11764,70588,2}x^2$$

$$+ \frac{P}{22.10526,31578,9}x^2 - \frac{P}{24,09523,80952,4}x^2 + \frac{P}{26.08695,65217,4}x^2$$

$$- \frac{P}{28.08000,00000,0}x^2 + \frac{P}{30.07407,40740,7}x^2 - \frac{P}{32.06896,55172,4}x^2$$

$$+ \frac{P}{34.06451,61290,3}x^2 - \frac{P}{36,06060,60606,1}x^2 + \frac{P}{38,05714,28571,4}x^2$$

$$- \frac{P}{40.05405,40540,5}x^2 + \frac{P}{42,05128,20512,8}x^2 - \frac{P}{44,04878,04878,0}x^2$$

$$+ \frac{P}{46.04651,16279,1}x^2 - \frac{P}{48,04444,44444,4}x^2 + \frac{P}{50.04255,3191}x^2$$

$$- \frac{P}{52.04081,6326}x^2 + \frac{P}{54,03921,5686}x^2 - \frac{P}{56,03773,5849}x^2$$

$$+ \frac{P}{58.03636}x^2 - \frac{P}{60.03509}x^2 + \frac{P}{62.03}x^2 + \text{etc.}\left. \right] [23]$$

P is not a constant. It is, in any term, the absolute value of the immediately preceding term.

The size of the interval in Kondo and Elderton's table is ten times that of Table I herewith.

It is possible to determine differences for small intervals, knowing them for large intervals, if it is sound to take differences of certain high order as equal to zero. If δ^I, δ^{II}, δ^{III}, et cetera, are the first, second, third, et cetera, differences for intervals j times that for which the differences are Δ^I, Δ^{II}, Δ^{III}, et cetera, we can express the δ's in terms of the Δ's. The following formulas give the relationships and in these C_a^j is the number of the combinations of j things a at a time. In each of these formulas the form is obvious so they may be terminated with any term desired.

$$\delta^I = C_1^j \Delta^I + C_2^j \Delta^{II} + C_3^j \Delta^{III} + C_4^j \Delta^{IV} + \text{etc.} \quad \ldots \ldots \ldots \quad [24]$$

$$\delta^{II} = (C_2^{2j} - 2C_2^j)\Delta^{II} + (C_3^{2j} - 2C_3^j)\Delta^{III} + (C_4^{2j} - 2C_4^{2j})\Delta^{IV} + \text{etc.} \quad [25]$$

$$\delta^{III} = (C_3^{3j} - 3C_3^{2j} + 3C_3^j)\Delta^{III} + (C_4^{3j} - 3C_4^{2j} + 3C_4^j)\Delta^{IV}$$
$$+ (C_5^{3j} - 3C_5^{2j} + 3C_5^j)\Delta^{V} + \text{etc.} \quad \ldots \ldots \ldots \ldots \quad [26]$$

$$\delta^{IV} = (C_4^{4j} - 4C_4^{3j} + 6C_4^{2j} - 4C_4^j)\Delta^{IV} + (C_5^{4j} - 4C_5^{3j} + 6C_5^{2j} - 4C_5^j)\Delta^{V}$$
$$+ (C_6^{4j} - 4C_6^{3j} + 6C_6^{2j} - 4C_6^j)\Delta^{VI} + \text{etc.} \quad \ldots \ldots \ldots \quad [27]$$

$$\delta^{V} = (C_5^{5j} - 5C_5^{4j} + 10C_5^{3j} - 10C_5^{2j} + 5C_5^j)\Delta^{V}$$
$$+ (C_6^{5j} - 5C_6^{4j} + 10C_6^{3j} - 10C_6^{2j} + 5C_6^j)\Delta^{VI}$$
$$+ (C_7^{5j} - 5C_7^{4j} + 10C_7^{3j} - 10C_7^{2j} + 5C_7^j)\Delta^{VII} + \text{etc.} \quad \ldots \quad [28]$$

$$\delta^{VI} = (C_6^{6j} - 6C_6^{5j} + 15C_6^{4j} - 20C_6^{3j} + 15C_6^{2j} - 6C_6^j)\Delta^{VI}$$
$$+ (C_7^{6j} - 6C_7^{5j} + 15C_7^{4j} - 20C_7^{3j} + 15C_7^{2j} - 6C_7^j)\Delta^{VII}$$
$$+ (C_8^{6j} - 6C_8^{5j} + 15C_8^{4j} - 20C_8^{3j} + 15C_8^{2j} - 6C_8^j)\Delta^{VIII} + \text{etc.} \quad [29]$$

$$\delta^{VII} = (C_7^{7j} - 7C_7^{6j} + 21C_7^{5j} - 35C_7^{4j} + 35C_7^{3j} - 21C_7^{2j} + 7C_7^j)\Delta^{VII}$$
$$+ (C_8^{7j} - 7C_8^{6j} + 21C_8^{5j} - 35C_8^{4j} + 35C_8^{3j} - 21C_8^{2j} + 7C_8^j)\Delta^{VII} + \text{etc.} \quad [30]$$

$$\delta^{VIII} = (C_8^{8j} - 8C_8^{7j} + 28C_8^{6j} - 56C_8^{5j} + 70C_8^{4j} - 56C_8^{3j} + 28C_8^{2j} - 8C_8^j)\Delta^{VIII} + \text{etc.} \quad [31]$$

If we now let j be fractional and write $i = \dfrac{1}{j}$ and substitute in these formulas, we obtain, after considerable algebraic expansion

TABLE I 11

and simplification, the following formulas giving δ's for small intervals, knowing Δ's for intervals i times as large. The formulas as written assume that Δ's of ninth and higher order are negligibly small. The subscript zero attaching to each has been omitted.

$$\delta^{I} = \frac{\Delta^{I}}{i} - \frac{1}{2i^{2}}(i-1)\Delta^{II} + \frac{1}{3! \, i^{3}}(i-1)(2i-1)\Delta^{III}$$

$$- \frac{1}{4! \, i^{4}}(i-1)(2i-1)(3i-1)\Delta^{IV}$$

$$+ \frac{1}{5! \, i^{5}}(i-1)(2i-1)(3i-1)(4i-1)\Delta^{V}$$

$$- \frac{1}{6! \, i^{6}}(i-1)(2i-1)(3i-1)(4i-1)(5i-1)\Delta^{VI}$$

$$+ \frac{1}{7! \, i^{7}}(i-1)(2i-1)(3i-1)(4i-1)(5i-1)(6i-1)\Delta^{VII}$$

$$- \frac{1}{8! \, i^{8}}(i-1)(2i-1)(3i-1)(4i-1)(5i-1)(6i-1)(7i-1)\Delta^{VIII} \quad [32]$$

$$\delta^{II} = \frac{1}{i^{2}}\Delta^{II} - \frac{(i-1)}{i^{3}}\Delta^{III} + \frac{1}{12i^{4}}(i-1)(11i-7)\Delta^{IV}$$

$$- \frac{1}{12i^{5}}(i-1)(2i-1)(5i-3)\Delta^{V}$$

$$+ \frac{2}{6! \, i^{6}}(i-1)(2i-1)(137i^{2}-132i+31)\Delta^{VI}$$

$$+ \frac{1}{5! \, i^{7}}(i-1)(2i-1)^{2}(3i-1)(7i-3)\Delta^{VII}$$

$$+ \frac{1}{7! \, 4i^{8}}(i-1)(2i-1)(3i-1)(2178i^{3}-2573i^{2}+1002i-127)\Delta^{VIII} \quad [33]$$

$$\delta^{III} = \frac{1}{i^{3}}\Delta^{III} - \frac{3}{2i^{4}}(i-1)\Delta^{IV} + \frac{1}{4i^{5}}(i-1)(7i-5)\Delta^{V}$$

$$- \frac{1}{8i^{6}}(i-1)(3i-2)(5i-3)\Delta^{VI}$$

$$+ \frac{1}{5! \, i^{7}}(i-1)(2i-1)(116i^{2}-141i+43)\Delta^{VII}$$

$$- \frac{1}{5! \, 4i^{8}}(i-1)(7i-3)(2i-1)(67i^{2}-78i+23)\Delta^{VIII} \quad [34]$$

$$\delta^{IV} = \frac{1}{i^4}\Delta^{IV} - \frac{2}{i^5}(i-1)\Delta^{V} + \frac{1}{6i^6}(i-1)(17i-13)\Delta^{VI}$$

$$- \frac{1}{6i^7}(i-1)(3i-2)(7i-5)\Delta^{VII}$$

$$+ \frac{1}{5!\,2i^8}(i-1)(967i^3 - 1833i^2 + 1157i - 243)\Delta^{VIII} \qquad [35]$$

$$\delta^{V} = \frac{1}{i^5}\Delta^{V} - \frac{5}{2i^6}(i-1)\Delta^{VI} + \frac{1}{6i^7}5(i-1)(5i-4)\Delta^{VII}$$

$$- \frac{5}{4!\,i^8}(i-1)(4i-3)(7i-5)\Delta^{VIII} \quad \ldots \ldots \ldots \quad [36]$$

$$\delta^{VI} = \frac{1}{i^6}\Delta^{VI} - \frac{3(i-1)}{i^7}\Delta^{VII} + \frac{1}{4i^8}(i-1)(23i-19)\Delta^{VIII} \quad \ldots \ldots \quad [37]$$

$$\delta^{VII} = \frac{\Delta^{VII}}{i^7} + \frac{-7}{2i^8}(i-1)\Delta^{VIII} \quad \ldots \ldots \ldots \ldots \ldots \quad [38]$$

$$\delta^{VIII} = \frac{1}{i^8}\Delta^{VIII} \quad \ldots \ldots \ldots \ldots \ldots \ldots \ldots \quad [39]$$

In these formulas δ^{I}, δ^{II}, δ^{III}, etc., are the first, second, third, etc., differences for the small interval, and Δ^{I}, Δ^{II}, Δ^{III}, etc., the differences for intervals i times as large. As employed, i was usually 10, occasionally 5, and for a short stretch 2.5. When for the upper portion of the table this last interval proved too large, an interpolation method was no longer followed and each successive value computed by formula [23], this finally requiring for the higher values some thirty 15-decimal-place terms for each successive approximation. The number of such approximations to get a single value varied from two or three to ten or twelve, dependent upon the excellence of the initial starting value and upon sundry vicissitudes of computation. The first computations in such successive approximations were made upon a ten-bank Monroe machine with a cumulator dial, thus obviating the labor and chance for error due to copying. And later computations were made upon a thirteen-bank Monroe, and still later ones on this thirteen-bank machine so supplemented by slide rule as to give fifteen-place accuracy. This extended computation was found to be necessary in view of the rapid change in x corresponding to a change in p.

The $\sqrt{pq}$, $\sqrt{1 - p^2}$, and $\sqrt{1 - q^2}$ columns offered little dif-

TABLE I   13

ficulty of computation.

The uses of Table I are so numerous and they will be so obvious to statistical workers that illustrations in the main are superfluous. We will, however, note procedures involving inverse interpolation and the accuracy of direct interpolation of high parabolic degree.

If the proportion, p, to the left of the point of dichotomy, x, of a normal distribution is desired, the x column becomes the argument and the p column the consequent. For example, let p be desired for $x = .6$. By linear inverse interpolation we have

$$p' = \frac{.6000\ 0000 - .5998\ 5931}{.6001\ 5941 - .5998\ 5931} = \frac{.0001\ 4069}{.0003\ 0010} = .4688\ 104$$

Since the fifth significant figure, namely 9, of the dividend is subject to a rounding-off error, the quotient is subject to such an error in the fifth significant figure and no method of interpolation can produce a result not subject to this rounding-off error in the fifth significant figure of p', which corresponds to the ninth decimal place in a. We thus have

$$a^{-ii} = a_0 + ip' = .7257 + .0000\ 4688\ 1 = .7257\ 4688\ 1$$

with a maximal error as given by [6] and recorded at the foot of column p of .0000 0000 2.

We find $i = .0001$ and Δ, given by [2], $= -.0002\ 00$, so three-point inverse interpolation by [4] yields

$$a^{-iii} = .7257\ 4688\ 1 + .5(-.0001)(-.0002\ 00)(.4688\ 1)(.5311\ 9)(1.0003\ 00)$$
$$= .7257\ 4688\ 3$$

Since the rounding-off error in the tabled measures (eighth decimal place) carried over to an error in the fifth significant figure in p', that is to the ninth decimal place in a, the error in a^{-iii} is in this ninth decimal place in spite of E^{-iii}, as given by [9], indicating an error of but .0000 0000 02. We may write the rules:

In connection with a^{-ii} whichever of the two errors, (1) that deriving from the rounding-off error in the original tabled values, or (2) that given by [6], is the greater is the error to be attached to a^{-ii}.

In connection with a^{-iii} whichever of the two errors, (1) that deriving from the rounding-off error in the original tabled values, or (2) that given by [9], is the greater is the error to be attached to a^{-iii}.

The accuracy of direct interpolation of any order may be determined by computing E^{ii}, E^{iii}, etc., as given in Table C, but in no case can the accuracy exceed that of the basic tabled entries, which may have an error as great as ½ in the last figure. For practical purposes, if an interpolation procedure yields an answer with a maximal error not greater than 1 in the place given by the last published figure in the table, it may be considered entirely adequate.

We designate direct two-, three-, etc., point interpolation errors e^{ii}, e^{iii}, etc., and designate their maximal values E^{ii}, E^{iii}, etc. The formulas of Table C have been used to compute the maximal errors recorded at the feet of the columns of Table I. Where a single E^{ii}, or E^{iii}, value is recorded it applies specifically to the region near the middle of the column, i.e., the region half-way down the page.

A scanning of the upper bounds of error, E^{ii}, shows that when p is the argument the error in x is less than 2 in the last, or eighth decimal place of the table for the region $p < .8$; that it is less than 2 in the seventh decimal place for $p < .95$; etc.

Three-point interpolation yields an x accurate to within 2 in the eighth decimal place for the region $p < .99$.

In the extreme tail region, when $p = .99935$, eight point interpolation yields an x answer in which the upper bound of error is 2 in the sixth decimal place.

The inverse error bounds that are recorded, E^{-ii} and E^{-iii}, apply when x is the argument and p or q is the consequent. Linear inverse interpolation yields a p which is correct to within 2 in the ninth decimal place for $x < .50$ and to within 2 in the eighth place for $x < 1.75$, and three-point inverse interpolation yields a p correct to within 2 in the eighth place for $x < 2.9$.

Many uses of the functions tabled in columns 4-6 will be obvious to the statistician. A number of unusual uses, not served by other tables, may arise; for example, if the sine of an angle corresponding to a given cosine is desired, this important relationship for four-decimal-place arguments is immediately given in the pairs of columns, p and $\sqrt{1-p^2}$ and q and $\sqrt{1-q^2}$. If p or q is the value of a correlation coefficient, $\sqrt{1-p^2}$ and $\sqrt{1-q^2}$ are the corresponding alienation coefficients. The function $\sqrt{1-r^2}$ so universally needed in simple and multiple correlation work, is provided in the $\sqrt{1-p^2}$ and $\sqrt{1-q^2}$ entries. The common functions $\sqrt{Npq}$ and $\sqrt{pq/N}$ are readily gotten since Table I provides pq and Table VII provides $\sqrt{N}$.

SECTION II-V: BEARING UPON TABLES II, III, IV, AND V
Giving three-, four-, six-, and eight-point interpolation coefficients

The equations of parabolas which pass through one, two, three, . . . nine tabled values which pertain to equally spaced arguments are given in Table B.

TABLE B
FORMULAS FOR PARABOLIC INTERPOLATION OF 0, 1, 2, 3, 4, 5, 6, 7, 8 DEGREES

$$t^{i} = t_0 \quad \ldots \ldots \ldots \ldots \ldots \ldots \quad [40]$$

$$t^{ii} = (1-p)t_0 + pt_1 \quad \ldots \ldots \ldots \text{(Two-point interpolation)} \quad [41]$$

$$t^{iii} = p(1-p^2)\left[\frac{-1}{2(1+p)}t_{-1} + \frac{1}{p}t_0 + \frac{1}{2(1-p)}t_1\right] \quad \begin{array}{c}\text{(Three-point}\\ \text{interpolation)}\end{array} \quad [42]$$

$$t^{iv} = \frac{1}{6}p(1-p^2)(2-p)\left[-\frac{1}{1+p}t_{-1} + \frac{3}{p}t_0 + \frac{3}{1-p}t_1 - \frac{1}{2-p}t_2\right] \quad \ldots \ldots \quad [43]$$

$$t^{v} = \frac{1}{24}p(1-p^2)(4-p^2)\left[\frac{1}{2+p}t_{-2} - \frac{4}{1+p}t_{-1} + \frac{6}{p}t_0 + \frac{4}{1-p}t_1 - \frac{1}{2-p}t_2\right] \quad [44]$$

$$t^{vi} = \frac{1}{120}p(1-p^2)(4-p^2)(3-p)\left[\frac{1}{2+p}t_{-2} - \frac{5}{1+p}t_{-1} + \frac{10}{p}t_0\right.$$

$$\left. + \frac{10}{1-p}t_1 - \frac{5}{2-p}t_2 + \frac{1}{3-p}t_3\right] \quad \ldots \ldots \ldots \quad [45]$$

$$t^{vii} = \frac{1}{720}p(1-p^2)(4-p^2)(9-p^2)\left[-\frac{1}{3+p}t_{-3} + \frac{6}{2+p}t_{-2} - \frac{15}{1+p}t_{-1}\right.$$

$$\left. + \frac{20}{p}t_0 + \frac{15}{1-p}t_1 - \frac{6}{2-p}t_2 + \frac{1}{3-p}t_3\right] \quad \ldots \ldots \ldots : \ldots \quad [46]$$

$$t^{viii} = \frac{1}{5040}p(1-p^2)(4-p^2)(9-p^2)(4-p)\left[-\frac{1}{3+p}t_{-3} + \frac{7}{2+p}t_{-2}\right.$$

$$\left. - \frac{21}{1+p}t_{-1} + \frac{35}{p}t_0 + \frac{35}{1-p}t_1 - \frac{21}{2-p}t_2 + \frac{7}{3-p}t_3 - \frac{1}{4-p}t_4\right] \quad [47]$$

$$t^{ix} = \frac{1}{40320}p(1-p^2)(4-p^2)(9-p^2)(16-p^2)\left[\frac{1}{4+p}t_{-4} - \frac{8}{3+p}t_{-3}\right.$$

$$\left. + \frac{28}{2+p}t_{-2} - \frac{56}{1+p}t_{-1} + \frac{70}{p}t_0 + \frac{56}{1-p}t_1 - \frac{28}{2-p}t_2\right.$$

$$\left. + \frac{8}{3-p}t_3 - \frac{1}{4-p}t_4\right] \quad \text{(Nine-point interpolation)} \quad [48]$$

In each instance the value sought is that between t_0 and t_1 which corresponds to an argument p proportion of the distance from a_0 to a_1. When this value is determined from the two neighboring points it is designated t^i, when from the three nearest points t^{iii}, when from the nearest four t^{iv}, etc. The solutions, t^i, t^{ii}, t^{iii}, etc., are succeedingly better and better estimates of the true value of t which corresponds to the argument $a_0 + ip$ (in which i is the size of the interval, i.e., $i = a_1 - a_0 = a_0 - a_{-1} = a_2 - a_1 =$ etc.).

Linear interpolation by equation [41] calls for such simple functions of p that there is no need of tabling them, but tabling the analogous coefficients when a larger number of interpolation points are used is a time saving device. Equations [42] to [48] are Lagrangian formulas for three-point (quadric) to nine-point (octic) interpolation. For the three-point interpolation coefficients the c_{-1}, c_0, c_1 quanties tabled are $\frac{-1}{2} p(1-p)$ [the minus sign being given in the column caption], $(1-p^2)$ and $\frac{1}{2} p(1+p)$. When p is a four-decimal-place number each of these three is an exact nine-decimal-place number, but since three-point interpolation will seldom yield accuracy to an additional seven, eight, or nine figures, though it will do so if third and higher order differences are negligibly small, the table of coefficients is published to five decimal places only. The coefficients are rounded off to the nearest fifth decimal place according to the rule given in the next paragraph. These five-decimal-place coefficients facilitate getting an answer correct usually to three, four, or five additional figures. As it seldom happens that three-point interpolation will yield more than five additional correct figures it would be, in general, inutile to retain more than five decimal places in the coefficients. When more than five additional correct figures are desired, four-, six-, or eight-point interpolation coefficients as given in Tables III, IV, and V may be used. A five-figure coefficient can be readily encompassed in a single immediate memory operation, but the writer has found that an operation calling for an eight-figure span frequently overtaxes clerks (and professors) and is conducive to inaccuracy and slower procedure.

Rule for rounding off the fifth decimal place in Lagrangian three-point interpolation coefficients:

Let $\tilde{c}$ = the correct (or non-rounded-off) value

Let c = the rounded-off value (i.e., the value tabled

Let e = the error, thus $\tilde{c} = c + e$

Let the function be quadric in t. Thus we have

Arguments	Tabled Values	Differences	Then tabled values may be written
a_{-1}	t_{-1}		t_{-1}
		Δ^I_{-1}	
a_0	t_0	Δ^{II}_{-1}	$t_{-1} + \Delta^I_{-1}$
		Δ^I_0 0	
a_1	t_1	Δ^{II}_{-1}	$t_{-1} + 2\Delta^I_{-1} + \Delta^{II}_{-1}$
		Δ^I_1	
a_2	t_2		$t_{-1} + 3\Delta^I_{-1} + 2\Delta^{II}_{-1}$

The interpolated value for a given p is

$$t = (\tilde{c}_{-1} - e_{-1})t_{-1} + (\tilde{c}_0 - e_0)(t_{-1} + \Delta^I_{-1}) + (\tilde{c}_1 - e_1)(t_{-1} + 2\Delta^I_{-1} + \Delta^{II}_{-1})$$

The error in this value, E, is

$$E = (e_{-1} + e_0 + e_1)t_{-1} + (e_0 + 2e_1)\Delta^I_{-1} + e_1 \Delta^{II}_{-1}$$

The rounding off must certainly be such that $(e_{-1} + e_0 + e_1) = 0$, so we may write $e_0 = -e_{-1} - e_1$, then

$$E = (-e_{-1} - e_1)\Delta^I_{-1} + e_1\Delta^{II}_{-1}$$

Clearly we should round off so that, in addition to $(e_{-1}+e_0+e_1)=0$, the quantity $(e_{-1} + e_1)$ shall be as small as possible. From [42] we see that

$$c_{-1} = \frac{-1}{2} p(1-p) = \frac{-p}{2} + \frac{p^2}{2} \text{ and } c_1 = \frac{1}{2} p(1+p) = \frac{p}{2} + \frac{p^2}{2}$$

Since $\frac{p}{2} > \frac{p^2}{2}$ and since $\frac{p}{2}$ is a figure of five decimal places the figures beyond the fifth decimal place of c_{-1} are complimentary to the figures beyond the fifth decimal place of c_1. Accordingly, if c_{-1} and c_1 are rounded to their nearest fifth decimal place, figures $(e_{-1} + e_1) = 0$, so that then $E = e_1\Delta^{II}_{-1}$, a small function of a second order difference only. Accordingly, the rounding-off rule that has been followed is *"Record c_{-1} and c_1 to their nearest five-decimal-place values and then adjust c_0 so that $e_{-1}+e_0+e_1 = 0$."*

As e_1 never exceeds .0000 05 the maximum rounding-off error is .0000 05Δ^{II}_{-1}. For example, for argument .7500 of Table I we find

Δ_{-1}^{II} = .0000 0007 so the maximum error due to rounding off of a three-point interpolated x (deviate in a unit normal distribution) is in the thirteenth decimal place which is far beyond the limits of this eight-place table and beyond the limits of utility of three-point interpolation had we a basic table to thirteen or more places. No purpose would here be served by exact (nine-place) three-point interpolation coefficients. The writer believes that the five-place values of Table II will serve well over 99 per cent of all situations in which three-point interpolation will serve.

Bearing upon Table III. In connection with Table II it was noted that no practical advantage results in using true, or nine-place, three-point interpolation coefficients for four-place p arguments, in that coefficients rounded off to five figures will yield as great accuracy as is permitted by the data to which three-point interpolation is applicable.

It will now be shown that no practical advantage results in using true, or ten-place, four-point interpolation coefficients for three decimal place p arguments, because four-point coefficients, rounded off to seven places, will yield as great accuracy as is permitted by the data to which four-point interpolation is applicable.

Assuming $\Delta^{iv} \cdot= 0$, for if it does not do so four-point interpolation is insufficient, and employing [43], we have t_p, the consequent for argument p:

$$t_p = \frac{1}{6} p(1-p^2)(2-p) \left[\frac{-1}{1+p} t_{-1} + \frac{3}{p}(t_{-1} + \Delta^I) + \frac{3}{1-p}(t_{-1} + 2\Delta^I + \Delta^{II}) \right.$$

$$\left. - \frac{1}{2-p}(t_{-1} + 3\Delta^I + 3\Delta^{II} + \Delta^{III}) \right] \quad \text{Equivalent to formula [43]} \quad [43a]$$

If the coefficients c_{-1}, c_0, c_1, and c_2 have errors e_{-1}, e_0, e_1, and e_2 respectively, the error, e_p, in t_p is

$$e_p = (e_{-1} + e_0 + e_1 + e_2)t_{-1} + (e_0 + 2e_1 + 3e_2)\Delta^I + (e_1 + 3e_2)\Delta^{II}$$

$$+ e_2 \Delta^{III} \quad \text{Equivalent to formula [52]} \quad [52a]$$

If we round off so as to make the coefficients of t_{-1}, Δ^I, and Δ^{II} zero, the final error in t_p is the very small quantity $e_2\Delta^{III}$. Let a true coefficient equal a rounded-off coefficient plus an error, thus $\tilde{c} = c + e$. It will be observed that

$$\tilde{c}_1 = [p + \frac{p^2}{2}] \quad - (\frac{p^3}{2})$$

$$3\tilde{c}_2 = [\quad -\frac{p}{2}] \quad + (\frac{p^3}{2})$$

$$\tilde{c}_0 = [1 - \frac{p}{2} - p^2] + (\frac{p^3}{2})$$

$$3\tilde{c}_{-1} = [-p + \frac{3p^2}{2}] - (\frac{p^3}{2})$$

When p is a magnitude of three decimal places the [ ] terms do not extend beyond the seventh decimal place. As $\frac{p^3}{2}$ is the compliment of $-\frac{p^3}{2}$, we observe that $\tilde{c}_1$ and $3\tilde{c}_2$ are complimentary for their portions that extend beyond the seventh decimal place. Similarly for $\tilde{c}_0$ and $3\tilde{c}_{-1}$. *Accordingly, if we round off $\tilde{c}_{-1}$ and $\tilde{c}_2$ to their nearest seventh decimal place values, the errors of this being e_{-1} and e_2, and round off $\tilde{c}_0$ and $\tilde{c}_1$ so that their errors are $-3e_{-1}$ and $-3e_2$ respectively, we will have seven place values for c_{-1}, c_0, c_1, and c_2.* We will also have made the coefficients of t_{-1}, Δ^{I}, and Δ^{II} in [52b] each equal to zero. The maximum error due to rounding off is the negligibly small quantity

$$\text{Max. } e_p = \text{Max. } e_2 \Delta^{III} = |.00000\ 005\ \Delta^{III}|$$

The procedure just described was followed in getting the coefficients entered in Table III.

Bearing upon all the Interpolation Tables. The six- and eight-point interpolation coefficients of Tables IV and V are similarly the total p functions which are the coefficients of the t's in formulas [45] and [47]. These are exact values as published to ten and eleven decimal places for p arguments of three, two, and one decimal places respectively. For the situations to which they apply they will commonly yield from seven to ten additional correct figures. Unless the user is gifted with an unusually long immediate memory span it is inadvisable to attempt to recall a ten or eleven figure coefficient through a single fixation. The coefficients of Tables II and III can ordinarily be encompassed in a single memory act.

In all the tables of interpolation coefficients the signs of the successive coefficients are indicated at the tops of the columns.

In all of the tables the sum of the coefficients = 1.00000 00000 0.

This is important. If the user employs these tables to a less number of decimal places than published, he should, for each row employed, assure himself that the sum of the coefficients as he uses them = 1.

In Tables II, III, and IV, for values of $p < .5$, the argument is at the left and the coefficients are in order from left to right.

In connection with Table II, having arguments a_{-1}, a_0, a_1, a_2, and tabled values t_{-1}, t_0, t_1, t_2, and a value of $p < .5$, the three tabled values t_{-1}, t_0, t_1 are operated upon. If, going from a_0 to a_1, the value of $p > .5$, then reverse the direction of interpolation so that p, going from a_1 to a_0, is less than .5 and the tabled values operated upon become t_2, t_1, t_0. Thus, for three-point interpolation a value of $p > .5$ is never called for.

In connection with Tables III and IV, for values of $p > .5$, enter at the right and read the coefficients in order from right to left.

We note that for interpolation using any odd number of points, p never need exceed .5. It can then be shown that the maximum error always occurs when $p = .5$, as also can be shown to be the case for all interpolation using an even number of points.

The error due to interpolation. The error inherent in an interpolation using a designated number of points is very approximately given by obtaining the difference between this value and that given using one more point. Thus, for linear interpolation the error $e^{ii} = t^{iii} - t^{ii}$. The maximum e^{ii} is designated E^{ii}, and similarly for the maximum error using a different number of points. We can express the maximum error of two-point interpolation as a function of a second-order difference, Δ_0^{II}; the maximum error of three-point interpolation as a function of a third order difference, Δ_{-1}^{III}; etc. These functions are given in the accompanying Table C.

TABLE C
ERRORS AND MAXIMUM ERRORS INHERENT
IN PARABOLIC INTERPOLATION OF DIFFERENT DEGREES

$$e^i = t^{ii} - t^i; \qquad \text{Maximum } e^i = E^i = \frac{1}{2}\Delta_0^{I} \quad \ldots \ldots [49]$$

$$e^{ii} = t^{iii} - t^{ii}; \qquad \text{Maximum } e^{ii} = E^{ii} = \frac{1}{8}\Delta_0^{II} \quad \ldots \ldots [50]$$

$$e^{iii} = t^{iv} - t^{iii}; \qquad \text{Maximum } e^{iii} = E^{iii} = \frac{1}{16}\Delta_{-1}^{III} \quad \ldots \ldots [51]$$

$$e^{iv} = t^{v} - t^{iv}; \qquad \text{Maximum } e^{iv} = E^{iv} = \frac{3}{128}\Delta_{-1}^{IV} \quad \ldots \ldots [52]$$

$$e^v = t^{vi} - t^v; \qquad \text{Maximum } e^v = E^v = \frac{3}{256}\Delta^V_{-2} \qquad [53]$$

$$e^{vi} = t^{vii} - t^{vi}; \qquad \text{Maximum } e^{vi} = E^{vi} = \frac{5}{1028}\Delta^{VI}_{-2} \qquad [54]$$

$$e^{vii} = t^{viii} - t^{vii}; \qquad \text{Maximum } e^{vii} = E^{vii} = \frac{5}{2048}\Delta^{VII}_{-3} \qquad [55]$$

$$e^{viii} = t^{ix} - t^{viii}; \qquad \text{Maximum } e^{viii} = E^{viii} = \frac{35}{32768}\Delta^{VIII}_{-3} \qquad [56]$$

These interpolation coefficient tables permit of greater accuracy than can be well demonstrated by use upon Table I. We will therefore illustrate their use in connection with a fifteen place table of logarithms. Let us find log π. To fifteen decimal places π = 3.14159 26535 89793 ...., and its logarithm is .49714 98726 94134. We will interpolate in a table having three figure arguments and tabled entries with rounding-off errors in the fifteenth decimal place.

Number	Log	Δ^I	Δ^{II}
3.10	.49136 16938 34273		
		.00139 86951 92565	
3.11	.49276 03890 26838		− 44902 00960
		.00139 42049 91605	
3.12	.49415 45940 18443		− 44614 63600
		.00138 97435 28005	
3.13	.49554 43375 46448		− 44330 01238
		.00138 53105 26767	
3.14	.49692 96480 73215		− 44048 10381
		.00138 09057 16386	
3.15	.49831 05537 89601		− 43768 87583
		.00137 65288 28803	
3.16	.49968 70826 18404		− 43492 29456
		.00137 21795 99347	
3.17	.50105 92622 17751		− 43218 32665
		.00136 78577 66682	
3.18	.50242 71199 84433		− 42946 93934
		.00136 35630 72748	
3.19	.50379 06830 57181		

Δ^{III}	Δ^{IV}	Δ^V	Δ^{VI}	Δ^{VII}
− .00000 00287 37360				
	− 2 74998			
− .00000 00284 62362		− 3493		
	− 2 71505		− 47	
− .00000 00281 90857		− 3446		+ 11
	− 2 68059		− 58	
− .00000 00279 22798		− 3388		− 5
	− 2 64671		− 53	
− .00000 00276 58127		− 3335		+ 6
	− 2 61336		− 59	
− .00000 00273 96791		− 3276		
	− 2 50860 −			
− .00000 00271 38731				

Due to the effect of the rounding-off error, seventh order differences are seen to be untrustworthy. Seven-point interpolation,—equivalent to using sixth order differences—is the highest order demanded by the data, but as we have tabled coefficient values for six-point, Table IV, and eight-point, Table V, and none for seven-point, we shall use the eight-point table and we may anticipate an error in the fifteenth place which is a slight augmentation of the rounding-off fifteenth place error in the basic table.

Employing eight-point interpolation, Table V, we obtain the seven values of the table herewith, the first six of which may be used in further six-point interpolation. The seventh value is unnecessary, but has been computed in order to show the error in fifth order differences attributable to the cumulative effect of the rounding-off error. The order of magnitude of δ^V, given by the approximate relationship [36], is .00000 00000 00000 3. We observe that the rounding-off error, a fifteenth decimal place error, is more potent than that given by [54].

No.	Log	δ^I	δ^{II}	δ^{III}	δ^{IV}	δ^V
3.139	.49679 13157 00042					
		.00013 83323 73173				
3.140	.49692 96480 73215		− 440 47883			
		.00013 82883 25290		− 28043		
3.141	.49706 79363 98505		− 440 19840		− 28	
		.00013 82443 05450		− 28015		− 4
3.142	.49720 61807 03955		− 439 91825		− 24	
		.00013 82003 13625		− 27991		+ 7
3.143	.49734 43810 17580		− 439 63834		− 31	
		.00013 81563 49791		− 27960		
3.144	.49748 25373 67371		− 439 35874			
		.00013 81124 13917				
3.145	.49762 06497 81288					

Using six-point interpolation, Table IV, we obtain the four following values which may be used in further four-point interpolation. We must expect an augmented error in d^{III} in the fifteenth place.

No.	Log	d^I	d^{II}	d^{III}
3.14158	.49714 81234 55457			
		13824 05653		
3.14159	.49714 95058 61110		− 4401	
		13824 01252		− 1
3.14160	.49715 08882 62362		− 4400	
		13823 96852		
3.14161	.49715 22706 59214			

Using four-point interpolation, Table III, we obtain the two following values which may be used in two-point interpolation.

	No.		Log
a_0	= 3.14159 265	t_0	=.49714 98721 97870
a_1	= 3.14159 266	t_1	=.49714 98735 80273

We must expect an error in the fifteenth place. The approximate second order difference, given by [32] in which i = 10000 00 and $|\Delta^{II}|$ =.00000 4, is .00000 00000 00000 004, so that linear interpolation is ample, but the error given by it will be consequent to the rounding-off error, i.e., a fifteenth (or possibly a fourteenth) place error and not that given by [50].

By linear interpolation we obtain t = .49714 98726 94124, with an expected error in the fifteenth or possibly the fourteenth place. Comparison with the true values reveals the error to be .00000 00000 00010.

In the interpolations just made we have used exact Lagrangian multipliers, so no rounding-off error consequent to rounding off the multipliers has been introduced. The three-point interpolation coefficients of Table II are exact for one and two decimal places of p only. *This table is not recommended for use when a consequent correct to more than five figures beyond that given by no interpolation is desired.* For example, an approximate value of log π without interpolation is log 3.14, which differs from log 3.15 in the third decimal place. If the problem is such that one is content with a logarithm correct to seven decimal places, which is five places additional to the number given by no interpolation at all, we can consider linear interpolation and, if that does not suffice, then three-point interpolation. We note that, since E^{ii} = .00000 05, linear interpolation does not guarantee the five additional decimal place accuracy. Using the coefficients of Table II, we obtain log 3.141593 = .49714 99213 00. Linear interpolation between this value and the tabled value of log 3.14 gives log 3.14159 26536 = .49714 98734 which, being in error by 8 in the tenth decimal place, has given an answer correct to 6 decimal places additional to that given by no interpolation at all. Also this is three decimal places additional to that given by linear interpolation (which yields log 3.14159 26536 = .49714 96). Table II is published to enable speedy interpolation rather than that of highest accuracy. The use of Table III is quite speedy and, in general, yields a pretty high order of accuracy.

SECTION VI: BEARING UPON TABLE VI
FOUR-PLACE χ^2 FUNCTIONS

Let us call an observed frequency in a uniquely defined class or cell, f; the theoretical frequency in this cell, $\tilde{f}$; the difference between them, d, the cell divergence; and $d^2/\tilde{f}$, the cell square-contingency. The sum of all such for all cells is χ^2 $(= \Sigma d^2/\tilde{f})$, the square-contingency, and this divided by the number of degrees of freedom in the system, n, is χ^2/n, the mean square-contingency. [The reader should note that this definition of mean square-contingency differs from that of certain earlier practice where χ^2/n', n' being the number of cells in the table, has been defined as the mean square-contingency.] If the theoretical cell frequencies are consequent to the right or true hypothesis, χ^2/n is a chance deviation from 0, differing in form of distribution for each separate number of degrees of freedom. P, the value tabled herewith, is the probability that, if the observed χ^2/n is a measure in this chance distribution of χ^2/n values, a greater value would arise as a matter of chance. Under a true hypothesis the mean value of χ^2/n is 1.00 and the median value, for which $P = .5000$, is slightly less than 1.00. Thus, if the hypothesis is correct, and the experiment repeated many times, resulting P values less than .5000 will occur as often as values greater than .5000. The value .5000 should be taken as the one which establishes the hypothesis with maximum likelihood. Very large values of P should be looked upon askance, just as should very small values.

But whatever the value, P is a final, interpretative, or terminal statistic. As such, it does not enter into further computation and does not call for accuracy to the number of decimal places desirable in intermediate statistics, that is, those arising in the logical process between the raw data and the crucial terminal statistic. A two-decimal-place value of P may well represent as great refinement as will lead to any difference in conclusions. The present table to four decimal places, enabling interpolation yielding two- or three-place accuracy, may thus be deemed generally adequate for all interpretative needs. Accuracy to an additional figure or two is available by use of the method of Section VIII giving P from a variance ratio.

For many purposes a knowledge of the value of χ^2/n suffices. For one degree of freedom $\sqrt{\chi^2/1}$ is distributed as half a unit normal distribution, and for n large Fisher and Yates[*] make the

[*] R. A. Fisher and F. Yates, STATISTICAL TABLES, 1938, page 27.

TABLE VI   25

useful observation that then $\sqrt{2\chi^2} - \sqrt{2n-1}$ "may be used as a normal deviate with unit variance." A still closer approximation and one serviceable for three or more degrees of freedom is given by Wilson and Hilferty*.

If $\chi_n^2(p)$ is the χ^2 with n degrees of freedom that as a matter of chance is exceeded p proportion of the time and if x_p is the x in a unit normal distribution that is exceeded p proportion of the time, Wilson and Hilferty give the following approximate relationship:

$$\frac{\chi_n^2(p)}{n} = \left(1 - \frac{2}{9n} + x_p \sqrt{\frac{2}{9n}} \right)^3 \quad \cdots \cdots [57]$$

The following table shows how nearly correct are the P values, using these approximations, if the number of degrees of freedom is 30, the highest number here tabled. Either approximation is so close that it is judged that additional values for higher numbers of degrees of freedom are not needed.

FOR 30 DEGREES OF FREEDOM THE VALUE OF P

χ^2/n	GIVEN BY TABLE II IS	RESULTING FROM THE FISHER APPROXIMATION IS	RESULTING FROM THE WILSON-HILFERTY APPROXIMATION IS
.25	1.0000−	.9999	1.0000−
.36	.9995	.9988	.9992
.49	.9913	.9960	.9912
.64	.9357	.9311	.9357
.81	.7583	.7611	.7587
1.00	.4657	.4742	.4657
1.21	.1984	.2006	.1981
1.44	.0589	.0533	.0562
1.69	.0118	.0085	.0105
1.96	.0013	.0008	.0013
2.25	.0001	.0000+	.0001

The form of Table VI is designed to facilitate interpolation for any desired value of χ^2 and any number of degrees of freedom. The accuracy of interpolation has been facilitated by (a) using $\chi/\sqrt{n}$ as the argument, where n is the number of degrees of freedom and χ is simply the $\sqrt{\chi^2}$. Using this argument, two- or three-place accuracy in P can very generally be obtained for values of χ and of n other than those tabled, by simple linear interpolation between four points as follows:

* Edwin B. Wilson and Margaret M. Hilferty, "The Distribution of Chi-square," PROCEEDINGS OF THE NATIONAL ACADEMY OF SCIENCES, Vol. 17, No. 12, December, 1931, pages 684-688.

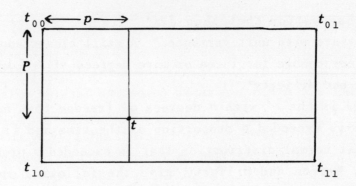

Given the four tabled entries t_{00}, t_{01}, t_{10}, t_{11}, let a value t be desired which corresponds to arguments p fraction of the distance from t_{00} to t_{01} and P fraction of the distance from t_{00} to t_{10}, as illustrated. Then, letting $q = 1 - p$ and $Q = 1 - P$, by two-way linear interpolation

$$t = qQt_{00} + qPt_{10} + Qpt_{01} + pPt_{11} \quad \ldots \quad \ldots \quad [58]$$

Quadric interpolation in $\chi / \sqrt{n}$, using Table II, followed, if need be, by linear interpolation in n, is also simple.

Where quadric interpolation is not employed it is frequently simpler to use $\chi^2 / \sqrt{n}$ as the argument instead of the argument $\chi / \sqrt{n}$ having equally spaced intervals.

SECTION VII: BEARING UPON TABLE VII

EIGHT-PLACE SQUARE ROOTS, CUBE ROOTS, AND NATURAL LOGARITHMS

The square and cube roots of all three digit numbers are given in this table and reference to the maximum interpolation errors shows that accuracy to 1 in the eighth significant figure is available by either linear or quadric interpolation.

Use of the table with simple machine computation enables the obtaining of still more accurate square and cube roots. Let N be the number whose square root, r, is sought. Let r' be the first approximation to this value as obtained from Table VII. Let $R' = N/r'$. Then an improved approximation is

$$r'' = \frac{R' + r'}{2} \quad \ldots \quad \ldots \quad \ldots \quad \ldots \quad [59]$$

This is substantially correct to twice as many digits as the number of digits of r' and r'' that agree. If not sufficiently accurate, the process may be repeated, using r'' as the approximate root.

A rapid method for obtaining a more precise cube root than the

TABLE VII   27

eight figure precision of linear or quadric interpolation is as follows: Let F be the number whose cube root, f, is desired. Let f', as found from Table VII, be the first approximation to this answer. Let $\phi' = F/f'^2$. A rapid computing machine method for obtaining ϕ' is to square f', transfer from the product dial to the keyboard and, by build-up multiplication, multiply by such a number, namely ϕ', as to yield F. Then an improved approximation to f is f''.

$$f'' = \frac{2}{3}f' + \frac{1}{3}\phi' = f' + \frac{\phi' - f'}{3} \quad \ldots \ldots [60]$$

If f' and f'' agree to j figures, then f' will in general be correct to $(2j + 1)$ figures. A rapid machine computation of $(\phi - f')/3$ is accomplished as follows: From Table VII determine f', a quantity slightly less than f. Set f' into the keyboard and square. Transfer f'^2 from the product dial to the keyboard. Clear product and rotation counter dials. Multiply f'^2 by f'. Clear the rotation counter dial. Now by build-up multiplication continue to multiply f'^2 by such a number, namely $(\phi' - f')$, as is necessary to produce F in the product dial. One-third of this quantity, showing in the rotation counter dial, added to f' is f''.

The natural logarithms of numbers from 1.00 to 10.00 are given. To obtain the natural logarithms of three or more decimal place numbers between these limits, linear, quadric, and higher interpolation procedures are available. To obtain the natural logarithms of numbers less than 1.00 or greater than 10.00, it is necessary first to take out 10 to such a power as to leave a number between 1.00 and 10.00. Then add the natural logarithm of this number to the natural logarithm of the power of 10 removed, as given in Table D herewith.

TABLE D. NATURAL LOGARITHMS OF POWERS OF 10

$$\ln 10^{-j} \qquad\qquad = -j(2.30258\ 50930)$$

.		.	
.		.	
.		.	
$\ln 10^{-3}$	$= \ln$	.001	$= -6.90775\ 52790$
$\ln 10^{-2}$	$= \ln$	.01	$= -4.60517\ 01860$
$\ln 10^{-1}$	$= \ln$	.1	$= -2.30258\ 50930$
$\ln 10^{0}$	$= \ln$	1	$= .00000\ 00000$
$\ln 10^{1}$	$= \ln$	10	$= 2.30258\ 50930$
$\ln 10^{2}$	$= \ln$	100	$= 4.60517\ 01860$
$\ln 10^{3}$	$= \ln$	1000	$= 6.90775\ 52790$
.			
.			
.			

$$\ln 10^{j} \qquad\qquad = j(2.30258\ 50930)$$

As an example we will find ln .0102. We write

$$.0102 = 1.02 \times 10^{-2}$$

$ln\ 10^{-2}$	$=$	$-$ **4.6051 7019**
$ln\ 1.02$	$=$	.0198 0263
$ln\ \ .0102$	$=$	$-$ **4.5853 6756**

SECTION VIII: BEARING UPON TABLE VIII AND UPON VARIANCE RATIO PROBABILITIES VIA A NORMALIZING TRANSFORMATION*

The problem of approximating a non-normal unimodal distribution, such as is the variance ratio distribution, by a function of the normal distribution has been attacked by many mathematicians with varying degrees of success. The degree of success is considered synonymous with the simplicity of the transformation, for any desired agreement can be attained if enough terms are kept in the transforming function.

The advantage of such a transformation lies not only in voiding the need for a wide variety of tables of percentage points, but also in simplifying thought in that a single probability attaches to the observed datum rather than some interval on the probability scale, as at present when percentage-point tables are used.

Notation:

$$F_{ij} = \frac{V_i}{V_j} = \text{a variance ratio in which } V_i, \text{ the numerator}$$

variance, is the sum of squared deviations from a mean divided by i, the number of degrees of freedom, and V_j is the similar denominator function based upon j degrees of freedom. These variances are, of course, independent. We shall designate the cube roots of F_{ij}, V_i, and V_j by f_{ij}, v_i, and v_j. P, or, if necessary for clarity, P_{ij}, is the probability that an F_{ij} as great as that observed would, under the hypothesis, arise as a matter of chance.

We may observe that the limit in one direction, of the F distribution, is $F_{i\infty}$ and that this is a χ^2/i distribution and that the limit in another direction is F_{1j}, which is the distribution of t^2 (Student's t squared). Thus a completely adequate normalizing distribution of F would avoid the need of probability tables of χ^2 and of t, as well as of percentage-point tables for other variance ratios.

Let us note that, in a scatter diagram with axes V_i and V_j,

* The writer gratefully acknowledges assistance received from Dr. Kenneth J. Arnold in the research connected with this section.

TABLE VIII   29

the variance ratio F_{ij} is shown by the straight line $F_{ij} V_j - V_i = 0$, and that the proportionate number of cases, or volume to the right of this line is P,—the probability sought. If we can perform a transformation that normalizes the marginal totals, i.e., the V_i and the V_j distributions (which, as noted, are χ_i^2/i and χ_j^2/j distributions), and that transforms the straight line F_{ij} into a straight line, the medians of the V_i and the V_j distributions will be transformed into medians of the new variables and the volume to the right of the new straight line will equal that to the right of $F_{ij} V_j - V_i = 0$ in the V_i, V_j surface. In terms of the new variables the bivariate distribution is of this type:

$$z = k \exp \frac{-1}{2} \left[\frac{(X - M_1)^2}{\sigma_1^2} + \frac{(Y - M_2)^2}{\sigma_2^2} \right]$$

and the transformed straight line is, say, $AX + BY + C = 0$. If we let $x = \dfrac{X - M_1}{\sigma_1}$ and $y = \dfrac{Y - M_2}{\sigma_2}$, these two equations become

$$z = k \exp \frac{-1}{2} \left[x^2 + y^2 \right]$$

and $ax + by + c = 0$. The volume sought is that to the right of the line $ax + by + c = 0$. The bivariate distribution is normal and of unit variance in both dimensions, so any array, parallel or not parallel to an axis, is a normal distribution of unit variance. The perpendicular distance from $ax + by + c = 0$ to the origin ($x = 0$, $y = 0$) equals $c / \sqrt{a^2 + b^2}$ and every array perpendicular to $ax + by + c = 0$ has the same proportion of cases to the right of this line, which is accordingly the proportion of the total volume to the right of this line. This proportion is q of Table I.

The χ^2 distribution is positively skewed, while its logarithm is negatively skewed. The intermediate distribution, that of $(\chi^2 / i)^{\frac{1}{3}}$, was found by Wilson and Hilferty to have, to a close approximation, a mean of $(1 - \dfrac{2}{9i})$, a variance of $2/9i$, negligible skewness except perhaps in the case of $i = 1$, and negligible diversion from mesokurtosis except perhaps in case $i < 4$. The following figures are taken from Wilson and Hilferty*:

* Edwin B. Wilson and Margaret M. Hilferty, "The Distribution of Chi-square," PROC. OF THE NAT. ACAD. OF SCI., Vol. 17, No. 12, 1931, pages 684-688.

TABLE E

THE $(\chi^2/i)^{\frac{1}{3}}$ DISTRIBUTION

i	Mean	W and H approximation $1-2/9i$	σ^2	W and H approximation $2/9i$	Pearson's β_1	β_2
1	.8024	.7778	.1870	.2222	.174	2.68
2	.8930	.8889	.1053	.1111	.028	2.73
3	.9272	.9259	.0723	.0741	.0086	2.80
10	.9778	.9778	.0222	.0222	.0014	2.97
30	.9926	.9926	.00741	.00741	.0000+	3.14

If we take the cube root of F_{ij}, namely $f_{ij} = v_i/v_j$, clearly f_{ij} is a straight line in the v_i, v_j plane and v_i and v_j are very nearly normally distributed except for small values of i or j, so the problem is solved except for these small values. A numerical check shows excellent agreement between the true values of P and those given utilizing the Wilson and Hilferty transformation if i and j are both greater than 3. In fact, there is very excellent agreement in the upper half of the curve, where $F_{ij} > 1$ and $P > .5$, no matter the value of i, if $j > 3$. This is the region of almost universal concern, but a slight further modification is here given, equation [63] which yields a very nearly correct value of P for all values of i and j and for values of F_{ij} both greater and smaller than 1.

Approximately normalizing with unit variance the V_i distribution by means of the Wilson-Hilferty transformation, we have

$$x_i = \frac{v_i - 1 + \frac{2}{9i}}{\sqrt{\frac{2}{9i}}} = 3 \, v_i \sqrt{\frac{i}{2}} \; - 3 \sqrt{\frac{i}{2}} + \frac{1}{3} \sqrt{\frac{2}{i}}$$

$$V_i^{\frac{1}{3}} = v_i = \frac{\sqrt{2} \, x_i}{3 \, \sqrt{i}} + 1 - \frac{2}{9i}$$

and similarly

$$v_j = \frac{\sqrt{2} \, x_j}{3 \, \sqrt{j}} + 1 - \frac{2}{9j}$$

$$f_{ij} \left(\frac{\sqrt{2} \, x_j}{3 \, \sqrt{j}} + 1 - \frac{2}{9j} \right) - \left(\frac{\sqrt{2} \, x_i}{3 \, \sqrt{i}} + 1 - \frac{2}{9i} \right) = 0$$

This is the equation of the f_{ij} line in the x_i, x_j plane. The

TABLE VIII   31

distance of this line from the origin, 0,0, is

$$d \;=\; \frac{-\,\theta_i \,+\, \theta_j \, f_{ij}}{\sqrt{\dfrac{1}{i} \,+\, \dfrac{1}{j} \, f_{ij}^{2}}} \qquad \text{The } d\text{-transformation approximately normalizing } F_{ij} \qquad [61]$$

which we will refer to as the d-transformation. In this

$$\theta_i \;=\; \frac{3}{\sqrt{2}} \,-\, \frac{\sqrt{2}}{3i}, \text{ and similarly for } \theta_j$$

$$\theta_i \;=\; 2.12132\ 03 \,-\, \frac{.47140\ 452}{i} \qquad \ldots \ldots \ldots \ldots \quad [62]$$

Table VIII herein gives values of θ_i for various values of i.

Though treating d as a deviate in a unit normal distribution yields good values of P for most situations, it was found by trial and retrial methods that a better value, in fact, excellent values even when i or j is 1 or 2, is P from x where

$$x \;=\; d(1 \,+\, \frac{.0800}{j^{3}} \, d^{4}) \qquad \text{The } xd \text{ normalizing transformation} \qquad [63]$$

When following this xd procedure we must so write F_{ij} that it is greater than 1. If interested in P from an $F_{ij} < 1$, we calculate P from F_{ji} and use the relationship $P_{ij} = 1 - P_{ji}$. The modifying factor $[1 + (.0800\ d^{4})/j^{3}]$ is not a moment or least-squares derivation but one meeting the requirement of simplicity and giving a good fit in the neighborhood of what seem, for experimental purposes, to be the crucial percentage points, namely, .99, .95, .90, .50, .10, .05, and .01.

In Table F herewith are comparative results for various percentage levels and various combinations of degrees of freedom. The first entry in each cell is P derived from d. A second entry which is P derived from x is given in case j is small, where there may be a significant difference from P derived from d.

TABLE F

P VALUES FOR VARIOUS F'S AND VARIOUS COMBINATIONS OF DEGREES OF FREEDOM

Correct values of P for various unrecorded F_{ij}'s

		.500	.250	.100	.050	.010	.0010
F_{11}	P from d	.500	.261	.131	.093	.061	.0518
	P from x	.500	.258	.104	.049	.012	.0055
F_{12}	P from d	.501	.251	.105	.058	.019	.0079
	P from x	.501	.251	.100	.047	.007	.0006
F_{21}	P from d	.499	.259	.132	.095	.062	.0520
	P from x	.499	.256	.105	.052	.013	.0056
$F_{1,10}$	P from d	.506	.249	.097	.048	.010	.0011
	P from x	.506	.249	.097	.048	.010	.0011
$F_{10,1}$	P from d	.494	.256	.132	.094	.062	.0520
	P from x	.494	.252	.105	.052	.103	.0057
$F_{1\infty}$	P from d	.507	.249	.096	.047	.010	.0012
	P from x	.507	.249	.096	.047	.010	.0012
$F_{\infty 1}$	P from d	.493	.255	.132	.094	.062	.0523
	P from x	.493	.251	.104	.052	.013	.0058
F_{22}	P from d	.500	.251	.106	.059	.021	.0087
	P from x	.500	.250	.101	.049	.008	.0008
$F_{2,10}$	P from d	.503	.250	.099	.049	.010	.0011
	P from x	.503	.250	.099	.049	.010	.0011
$F_{10,2}$	P from d	.497	.248	.106	.060	.021	.0088
	P from x	.497	.247	.100	.050	.009	.0009
$F_{2\infty}$	P from d	.500	.246	.095	.047	.009	.0010
	P from x	.500	.246	.095	.047	.009	.0010
$F_{\infty 2}$	P from d	.500	.249	.107	.061	.022	.0099
	P from x	.500	.248	.102	.051	.009	.0013

TABLE VIII   33

Herewith is a numerical illustration of the computational steps. Let $F_{24} = 10.00$. We desire the probability, P, that a variance ratio as large as this would arise as a matter of chance.

$$\sqrt[3]{F_{24}} = f_{24} = f \qquad\qquad = 2.1544$$

$$f^2 \qquad\qquad = 4.6414$$

$$\text{Denominator} = \sqrt{\frac{1}{i} + \frac{1}{j} f^2} = \sqrt{.5 + .25\ f^2} \qquad = 1.2886$$

$$\text{Numerator} = -\theta_i + \theta_j\ f = -1.885618 + 2.0034969\ f \qquad = 2.4307$$

$$d = \text{Numerator/Denominator} \qquad\qquad = 1.8863$$

$$1 + \frac{.08}{j^3} d^4 = 1 + .00125\ d^4 \qquad\qquad = 1.0158$$

$$x = d\ (1 + \frac{.08}{j^3} d^4) \qquad\qquad = 1.9161$$

$$P \text{ from table of normal distribution} \qquad = .0277$$

$$\begin{array}{l} P \text{ found using Pearson's Tables of the Incomplete} \\ \text{Beta-Function} \end{array} \qquad = .02780$$

In case $j = \infty$ the computation is still simpler, for then $V_j = 1$ and $1/j = 0$. In this case $F_{i\infty} = V_i$. The test is now that of χ_i^2, a χ^2 with i degrees of freedom, for $iV_i = \chi_i^2$. For illustration, let $V_4 = 2.00$.

$$\sqrt[3]{F_{4\infty}} = f \qquad\qquad = 1.2599$$

$$\text{Denominator} = \sqrt{\frac{1}{i}} \qquad\qquad = .5000$$

$$\text{Numerator} = -\theta_1 + \theta_\infty f = -2.0034969 + 2.1213220\ f \qquad = .6692$$

$$x = d = \text{Numerator/Denominator} \qquad = 1.3384$$

$$P \text{ from table of normal distribution} \qquad = .0904$$

This illustrates the use of this method in lieu of the χ^2 distribution.

An illustration of its use in lieu of "Student's" t has already been provided in Table F. The t^2 distribution is that of F_{1j}. Let $t_{1,10} = 1.812$, which is the value given at the .100 level in Fisher's Table of the t-distribution. We have

$$F_{1,10} = t^2_{1,10} = 3.2833$$

Proceeding as before we obtain $P = .097$, as recorded in Table F.

Fisher's z is related to the variance ratio by the equation $e^{2z} = F$ and all the purposes of percentage points in z are identically served by the same percentage points in F.

An incomplete beta-function is also related to F. The x of Pearson's* Table of the Incomplete Beta-Function and of Thompson's† Table of Percentage Points of the Incomplete Beta-Function is related to F thus

$$x_{ij} = \frac{1}{1 + \dfrac{i}{j} F_{ij}}$$

Also i = Thompson's ν_1 = Pearson's $2q$
Also j = Thompson's ν_2 = Pearson's $2p$

substituting V_i/V_j for F_{ij}

$$x_{ij} = \frac{j\ V_j}{i\ V_i + j\ V_j}$$

The quantity iV_i is a variance prior to division by the number of degrees of freedom. Thus, if $(iV_i + jV_j)$ is a total variance which can be split into independent parts iV_i and jV_j, then x_{ij} of Pearson or Thompson is the ratio of a part variance to a total variance, whereas F_{ij} is the ratio of one part variance to the other part variance. We thus see that P from F, or from the d or the xd transformations, serves the same field as P for the incomplete beta-function. However, since Pearson required of his tables an accuracy of five significant figures, and Thompson's tables are to five decimal places in x, we can expect an accuracy in P of the same general order, that is, to the fourth or fifth decimal place. Both degrees of accuracy mentioned are under the proviso, which is always subject to question, that the original measures which have led to the denominator variance of the variance ratio (usually errors of measurement) are, in the population of such measures, normally distributed.

* Karl Pearson, Editor, TABLES OF THE INCOMPLETE-BETA FUNCTION, 1934.

† Catherine M. Thompson, "Table of Percentage Points of the χ^2 Distribution," BIOMETRIKA, Vol. 32, 1941, pages 187-191.

, "Table of Percentage Points of the Incomplete Beta-Function," BIOMETRIKA, Vol. 32, 1941, pages 151-153.

TABLES

Table I   Supplementary Values

.9999   **.0001**

p	x	z	q	
.9999	3.7190 1649	.0003 9584 8	**.0001**	[1 in 10 thousand]
.9999 5	3.8905 919	.0002 0607 072	.0000 5	
.9999 9	4.2648 908	.0000 4478 7331	.0000 1	[1 in 100 thousand]
.9999 95	4.4171 734	.0000 2312 3795	.0000 05	
.9999 99	4.7534 243	.0000 0494 8328 2	.0000 01	[1 in a million]
.9999 995	4.8916 385	.0000 0254 0881 0	.0000 005	
.9999 999	5.1993 376	.0000 0053 7954 87	.0000 001	[1 in 10 million]
.9999 9995	5.3267 24	.0000 0027 5153 1	.0000 0005	
.9999 9999	5.6120 01	.0000 0005 7803 52	.0000 0001	[1 in 100 million]
.9999 9999 5	5.7307 3	.0000 0002 9479 8	.0000 0000 5	
.9999 9999 9	5.9978 1	.0000 0000 6156 53	**.0000 0000 1**	[1 in a U. S. billion]

$P=.9975 \begin{cases} E^{-ii}=.0000,0048 \quad E^{ii}=.000058 \quad\quad .0000,017 \quad\quad .0000,027 \quad\quad .0000,038 \quad\quad .0000,0000+ \\ E^{-iii}=.0000,0000,9 \quad E^{iii}=.0000,024 \quad .0000,0003 \quad\quad .0000,0008 \quad\quad .0000,0011 \end{cases}$

$p=.9985 \begin{cases} E^{-ii}=.0000,0084 \quad E^{ii}=.00015 \quad\quad\quad .0000,027 \quad\quad .0000,060 \quad\quad .0000,085 \quad\quad .0000,0000+ \\ E^{-iii}=.0000,00033 \quad E^{iii}=.000011 \quad\quad .0000,0008 \quad\quad .0000,0029 \quad\quad .0000,0042 \end{cases}$

Table I

Normal Distribution, Simple Correlation, and Probability Functions

p = the larger proportion in a dichotomized unit
normal distribution;
or a proportion > .5;
or a correlation coefficient > .5.

$q = 1 - p$.

x = the distance from the mean to the point of
dichotomy in the unit normal distribution.

z = the ordinate at the point of dichotomy in the
unit normal distribution.

ϵ^{ii} and ϵ^{iii} are the maximal two- and three-point interpolation errors, as given by formulas [50] and [51] when p, or q, are the arguments. ϵ^{ii} is positive in connection with x and negative in connection with z.

ϵ^{-ii} and ϵ^{-iii} are the maximal inverse two- and three-point interpolation errors in p, or q, when x is the argument, as given by formulas [6] and [9].

TABLE I

.5000   .5000

p	x	z	√pq	√1−p²	√1−q²	q
.5000	.0000 0000	.3989 4228	.5000 0000	.8660 2540	.8660 2540	**.5000**
.5001	02 5066	4227	.4999 9999	.8659 6766	0 8313	.4999
.5002	05 0133	4223	9996	9 0990	1 4084	.4998
.5003	07 5199	4217	9991	8 5213	1 9854	.4997
.5004	10 0265	4208	9984	7 9434	2 5622	.4996
.5005	12 5331	4197	9975	7 3654	3 1389	.4995
.5006	15 0398	4183	9964	6 7872	3 7154	.4994
.5007	17 5464	4167	9951	6 2088	4 2917	.4993
.5008	20 0530	4148	9936	5 6303	4 8679	.4992
.5009	22 5597	4127	9919	5 0516	5 4440	.4991
.5010	.0025 0663	.3989 4103	.4999 9900	.8654 4728	.8666 0198	**.4990**
.5011	27 5729	4076	9879	3 8939	6 5956	.4989
.5012	30 0796	4048	9856	3 3147	7 1712	.4988
.5013	32 5862	4016	9831	2 7355	7 7466	.4987
.5014	35 0929	3982	9804	2 1560	8 3219	.4986
.5015	37 5995	3946	9775	1 5764	8 8970	.4985
.5016	40 1062	3907	9744	0 9967	.8669 4720	.4984
.5017	42 6128	3866	9711	.8650 4168	.8670 0468	.4983
.5018	45 1195	3822	9676	.8649 8368	0 6214	.4982
.5019	47 6261	3776	9639	9 2566	1 1959	.4981
.5020	.0050 1328	.3989 3727	.4999 9600	.8648 6762	.8671 7703	**.4980**
.5021	52 6394	3675	9559	8 0957	2 3445	.4979
.5022	55 1461	3621	9516	7 5150	2 9185	.4978
.5023	57 6528	3565	9471	6 9342	3 4924	.4977
.5024	60 1594	3506	9424	6 3532	4 0662	.4976
.5025	62 6661	3445	9375	5 7721	4 6398	.4975
.5026	65 1728	3381	9324	5 1908	5 2132	.4974
.5027	67 6795	3314	9271	4 6094	5 7865	.4973
.5028	70 1862	3245	9216	4 0278	6 3596	.4972
.5029	72 6929	3174	9159	3 4460	6 9326	.4971
.5030	.0075 1996	.3989 3100	.4999 9100	.8642 8641	.8677 5054	**.4970**
.5031	77 7063	3024	9039	2 2820	8 0781	.4969
.5032	80 2130	2945	8976	1 6998	8 6506	.4968
.5033	82 7197	2863	8911	1 1175	9 2229	.4967
.5034	85 2264	2779	8844	.8640 5349	.8679 7952	.4966
.5035	87 7331	2693	8775	.8639 9523	.8680 3672	.4965
.5036	90 2398	2604	8704	9 3694	0 9391	.4964
.5037	92 7466	2512	8631	8 7864	1 5109	.4963
.5038	95 2533	2418	8556	8 2033	2 0825	.4962
.5039	.0097 7601	2322	8479	7 6200	2 6539	.4961
.5040	.0100 2668	.3989 2223	.4999 8400	.8637 0365	.8683 2252	**.4960**
.5041	02 7736	2121	8319	6 4529	3 7963	.4959
.5042	05 2803	2017	8236	5 8692	4 3673	.4958
.5043	07 7871	1911	8151	5 2852	4 9382	.4957
.5044	10 2939	1802	8064	4 7012	5 5089	.4956
.5045	12 8007	1690	7975	4 1169	6 0794	.4955
.5046	15 3075	1576	7884	3 5325	6 6498	.4954
.5047	17 8143	1459	7791	2 9480	7 2200	.4953
.5048	20 3211	1340	7696	2 3633	7 7901	.4952
.5049	22 8279	1219	7599	1 7784	8 3600	.4951
.5050	.0125 3347	.3989 1095	.4999 7500	.8631 1934	.8688 9297	**.4950**

E⁻ⁱⁱ= Eⁱⁱ=.0000,0000+ .0000,0000+ .0000,0000+ .0000,0000+ .0000,0000+
.0000,0000,0+

TABLE I

p	x	z	$\sqrt{pq}$	$\sqrt{1-p^2}$	$\sqrt{1-q^2}$	q
.5050	.0125 3347	.3989 1095	.4999 7500	.8631 1934	.8688 9297	.4950
.5051	27 8415	0968	7399	0 6083	.8689 4994	.4949
.5052	30 3484	0839	7296	.8630 0229	.8690 0688	.4948
.5053	32 8552	0707	7191	.8629 4375	0 6381	.4947
.5054	35 3621	0573	7084	8 8518	1 2073	.4946
.5055	37 8689	0437	6975	8 2660	1 7763	.4945
.5056	40 3758	0298	6864	7 6801	2 3451	.4944
.5057	42 8827	0156	6751	7 0940	2 9138	.4943
.5058	45 3896	.3989 0012	6636	6 5078	3 4824	.4942
.5059	47 8965	.3988 9865	6519	5 9213	4 0508	.4941
.5060	.0150 4034	.3988 9716	.4999 6400	.8625 3348	.8694 6190	.4940
.5061	52 9103	9564	6279	4 7481	5 1871	.4939
.5062	55 4172	9410	6156	4 1612	5 7551	.4938
.5063	57 9241	9254	6031	3 5741	6 3228	.4937
.5064	60 4311	9094	5904	2 9870	6 8905	.4936
.5065	62 9380	8933	5775	2 3996	7 4580	.4935
.5066	65 4450	8768	5644	1 8121	8 0253	.4934
.5067	67 9520	8602	5511	1 2244	8 5925	.4933
.5068	70 4590	8433	5376	0 6366	9 1595	.4932
.5069	72 9660	8261	5239	.8620 0487	.8699 7264	.4931
.5070	.0175 4730	.3988 8087	.4999 5100	.8619 4605	.8700 2931	.4930
.5071	77 9800	7910	4959	8 8723	0 8597	.4929
.5072	80 4870	7731	4816	8 2838	1 4261	.4928
.5073	82 9941	7549	4671	7 6952	1 9924	.4927
.5074	85 5011	7365	4524	7 1065	2 5585	.4926
.5075	88 0082	7178	4375	6 5176	3 1244	.4925
.5076	90 5153	6989	4224	5 9285	3 6903	.4924
.5077	93 0224	6797	4071	5 3393	4 2559	.4923
.5078	95 5295	6603	3916	4 7499	4 8214	.4922
.5079	.0198 0366	6406	3759	4 1604	5 3868	.4921
.5080	.0200 5437	.3988 6207	.4999 3600	.8613 5707	.8705 9520	.4920
.5081	03 0508	6005	3439	2 9808	6 5170	.4919
.5082	05 5580	5800	3276	2 3908	7 0819	.4918
.5083	08 0652	5594	3111	1 8007	7 6467	.4917
.5084	10 5723	5384	2943	1 2104	8 2113	.4916
.5085	13 0795	5173	2774	0 6199	8 7757	.4915
.5086	15 5867	4958	2603	.8610 0293	9 3400	.4914
.5087	18 0939	4741	2430	.8609 4385	.8709 9042	.4913
.5088	20 6012	4522	2255	8 8475	.8710 4682	.4912
.5089	23 1084	4300	2078	8 2564	1 0320	.4911
.5090	.0225 6157	.3988 4076	.4999 1899	.8607 6652	.8711 5957	.4910
.5091	28 1230	3849	1718	7 0738	2 1593	.4909
.5092	30 6302	3620	1535	6 4822	2 7227	.4908
.5093	33 1375	3388	1350	5 8905	3 2859	.4907
.5094	35 6449	3153	1163	5 2986	3 8490	.4906
.5095	38 1522	2916	0974	4 7066	4 4119	.4905
.5096	40 6595	2677	0783	4 1144	4 9747	.4904
.5097	43 1669	2435	0590	3 5220	5 5373	.4903
.5098	45 6743	2191	0395	2 9295	6 0998	.4902
.5099	48 1817	1944	.4999 0198	2 3368	6 6621	.4901
.5100	.0250 6891	.3988 1694	.4998 9999	.8601 7440	.8717 2243	.4900

$E^{-ii} =$ $E^{ii} = .0000,0000+$ $.0000,0000+$ $.0000,0000+$ $.0000,0000+$ $.0000,0000+$
$.0000,0000,0+$

TABLE I

.5100   .4900

p	x	z	√pq	√1−p²	√1−q²	q
.5100	.0250 6891	.3988 1694	.4998 9999	.8601 7440	.8717 2243	**.4900**
.5101	53 1965	1442	9798	1 1510	7 7864	.4899
.5102	55 7039	1188	9595	.8600 5579	8 3482	.4898
.5103	58 2114	0931	9390	.8599 9646	8 9100	.4897
.5104	60 7189	0671	9183	9 3711	.8719 4715	.4896
.5105	63 2264	0409	8974	8 7775	.8720 0330	.4895
.5106	65 7339	.3988 0145	8763	8 1838	0 5942	.4894
.5107	68 2414	.3987 9878	8550	7 5898	1 1554	.4893
.5108	70 7489	9608	8335	6 9958	1 7163	.4892
.5109	73 2565	9336	8118	6 4015	2 2772	.4891
.5110	.0275 7641	.3987 9062	.4998 7899	.8595 8071	.8722 8378	**.4890**
.5111	78 2716	8785	7677	5 2126	3 3984	.4889
.5112	80 7793	8505	7454	4 6179	3 9587	.4888
.5113	83 2869	8223	7229	4 0230	4 5190	.4887
.5114	85 7945	7939	7002	3 4280	5 0790	.4886
.5115	88 3022	7652	6773	2 8328	5 6389	.4885
.5116	90 8099	7362	6542	2 2374	6 1987	.4884
.5117	93 3176	7070	6309	1 6419	6 7583	.4883
.5118	95 8253	6776	6074	1 0463	7 3178	.4882
.5119	.0298 3330	6479	5837	.8590 4505	7 8771	.4881
.5120	.0300 8408	.3987 6179	.4998 5598	.8589 8545	.8728 4363	**.4880**
.5121	03 3485	5877	5357	9 2583	8 9953	.4879
.5122	05 8563	5572	5114	8 6621	.8729 5542	.4878
.5123	08 3641	5265	4869	8 0656	.8730 1129	.4877
.5124	10 8720	4956	4622	7 4690	0 6715	.4876
.5125	13 3798	4643	4373	6 8722	1 2299	.4875
.5126	15 8877	4329	4121	6 2753	1 7881	.4874
.5127	18 3956	4012	3868	5 6782	2 3462	.4873
.5128	20 9035	3692	3613	5 0810	2 9042	.4872
.5129	23 4114	3370	3356	4 4836	3 4620	.4871
.5130	.0325 9194	.3987 3045	.4998 3097	.8583 8861	.8734 0197	**.4870**
.5131	28 4273	2718	2836	3 2884	4 5772	.4869
.5132	30 9353	2388	2573	2 6905	5 1346	.4868
.5133	33 4433	2056	2308	2 0925	5 6918	.4867
.5134	35 9514	1721	2041	1 4943	6 2489	.4866
.5135	38 4594	1384	1772	0 8959	6 8058	.4865
.5136	40 9675	1044	1501	.8580 2974	7 3625	.4864
.5137	43 4756	0702	1227	.8579 6988	7 9191	.4863
.5138	45 9837	0358	0952	9 1000	8 4756	.4862
.5139	48 4919	.3987 0010	0675	8 5010	9 0319	.4861
.5140	.0351 0000	.3986 9661	.4998 0396	.8577 9018	.8739 5881	**.4860**
.5141	53 5082	9308	.4998 0115	7 3025	.8740 1441	.4859
.5142	56 0164	8954	.4997 9832	6 7031	0 7000	.4858
.5143	58 5246	8596	9547	6 1035	1 2557	.4857
.5144	61 0329	8236	9260	5 5037	1 8113	.4856
.5145	63 5412	7874	8971	4 9038	2 3667	.4855
.5146	66 0495	7509	8679	4 3037	2 9219	.4854
.5147	68 5578	7142	8386	3 7035	3 4771	.4853
.5148	71 0661	6772	8091	3 1031	4 0320	.4852
.5149	73 5745	6400	7794	2 5025	4 5868	.4851
.5150	.0376 0829	.3986 6025	.4997 7495	.8571 9018	.8745 1415	**.4850**

E⁻ⁱⁱ=   Eⁱⁱ=.0000,0000+   .0000,0000+   .0000,0000+   .0000,0000+   .0000,0000+
.0000,0000+

TABLE I

Page 41

.5150   .4850

p	x	z	$\sqrt{pq}$	$\sqrt{1-p^2}$	$\sqrt{1-q^2}$	q
.5150	.0376 0829	.3986 6025	.4997 7495	.8571 9018	.8745 1415	**.4850**
.5151	78 5913	5648	7194	1 3009	5 6960	.4849
.5152	81 0997	5268	6891	0 6999	6 2504	.4848
.5153	83 6082	4886	6586	.8570 0987	6 8046	.4847
.5154	86 1167	4501	6278	.8569 4973	7 3587	.4846
.5155	88 6252	4113	5969	8 8958	7 9126	.4845
.5156	91 1337	3724	5658	8 2941	8 4664	.4844
.5157	93 6423	3331	5345	7 6923	9 0200	.4843
.5158	96 1509	2936	5030	7 0903	.8749 5735	.4842
.5159	.0398 6595	2539	4713	6 4881	.8750 1268	.4841
.5160	.0401 1681	.3986 2139	.4997 4393	.8565 8858	.8750 6800	**.4840**
.5161	03 6768	1736	4072	5 2834	1 2330	.4839
.5162	06 1854	1332	3749	4 6807	1 7859	.4838
.5163	08 6942	0924	3424	4 0779	2 3386	.4837
.5164	11 2029	0514	3097	3 4750	2 8912	.4836
.5165	13 7117	.3986 0102	2768	2 8719	3 4436	.4835
.5166	16 2204	.3985 9687	2436	2 2686	3 9959	.4834
.5167	18 7293	9269	2103	1 6652	4 5480	.4833
.5168	21 2381	8849	1768	1 0616	5 1000	.4832
.5169	23 7470	8427	1431	.8560 4579	5 6518	.4831
.5170	.0426 2559	.3985 8002	.4997 1092	.8559 8540	.8756 2035	**.4830**
.5171	28 7648	7574	0750	9 2499	6 7550	.4829
.5172	31 2737	7144	0407	8 6457	7 3064	.4828
.5173	33 7827	6712	.4997 0062	8 0413	7 8577	.4827
.5174	36 2917	6277	.4996 9715	7 4368	8 4088	.4826
.5175	38 8007	5839	9366	6 8321	8 9597	.4825
.5176	41 3098	5399	9014	6 2272	.8759 5105	.4824
.5177	43 8189	4957	8661	5 6222	.8760 0611	.4823
.5178	46 3280	4511	8306	5 0170	0 6116	.4822
.5179	48 8371	4064	7949	4 4117	1 1620	.4821
.5180	.0451 3463	.3985 3614	.4996 7589	.8553 8062	.8761 7122	**.4820**
.5181	53 8555	3161	7228	3 2005	2 2622	.4819
.5182	56 3647	2706	6865	2 5947	2 8121	.4818
.5183	58 8740	2248	6500	1 9887	3 3619	.4817
.5184	61 3832	1788	6133	1 3826	3 9115	.4816
.5185	63 8926	1326	5763	0 7763	4 4609	.4815
.5186	66 4019	0861	5392	.8550 1698	5 0102	.4814
.5187	68 9113	.3985 0393	5019	.8549 5632	5 5594	.4813
.5188	71 4207	.3984 9923	4644	8 9564	6 1084	.4812
.5189	73 9301	9450	4266	8 3495	6 6572	.4811
.5190	.0476 4396	.3984 8975	.4996 3887	.8547 7424	.8767 2059	**.4810**
.5191	78 9490	8497	3506	7 1351	7 7545	.4809
.5192	81 4586	8017	3122	6 5277	8 3029	.4808
.5193	83 9681	7534	2737	5 9201	8 8512	.4807
.5194	86 4777	7049	2350	5 3124	9 3993	.4806
.5195	88 9873	6561	1961	4 7045	.8769 9473	.4805
.5196	91 4970	6071	1569	4 0964	.8770 4951	.4804
.5197	94 0066	5578	1176	3 4882	1 0428	.4803
.5198	96 5163	5083	0781	2 8798	1 5903	.4802
.5199	.0499 0261	4585	.4996 0383	2 2713	2 1377	.4801
.5200	.0501 5358	.3984 4085	.4995 9984	.8541 6626	.8772 6849	**.4800**

$E^{-ii} = $ $E^{ii} = .0000,0000+$ .0000,0000+   .0000,0000+   .0000,0000+   .0000,0000+
.0000,0000 0+

TABLE I

.5200   **.4800**

p	x	z	$\sqrt{pq}$	$\sqrt{1-p^2}$	$\sqrt{1-q^2}$	q
.5200	.0501 5358	.3984 4085	.4995 9984	.8541 6626	.8772 6849	.4800
.5201	04 0456	3582	9583	1 0537	3 2320	.4799
.5202	06 5555	3077	9179	.8540 4447	3 7789	.4798
.5203	09 0653	2569	8774	.8539 8355	4 3257	.4797
.5204	11 5752	2059	8367	9 2262	4 8723	.4796
.5205	14 0851	1546	7957	8 6167	5 4188	.4795
.5206	16 5951	1031	7546	8 0070	5 9651	.4794
.5207	19 1051	.3984 0513	7133	7 3972	6 5113	.4793
.5208	21 6151	.3983 9992	6717	6 7872	7 0574	.4792
.5209	24 1252	9470	6300	6 1771	7 6033	.4791
.5210	.0526 6353	.3983 8944	.4995 5881	.8535 5668	.8778 1490	.4790
.5211	29 1454	8416	5459	4 9563	8 6946	.4789
.5212	31 6555	7886	5036	4 3457	9 2401	.4788
.5213	34 1657	7353	4610	3 7349	.8779 7854	.4787
.5214	36 6760	6817	4183	3 1239	.8780 3305	.4786
.5215	39 1862	6280	3754	2 5128	0 8755	.4785
.5216	41 6965	5739	3322	1 9015	1 4204	.4784
.5217	44 2068	5196	2889	1 2901	1 9651	.4783
.5218	46 7172	4651	2453	0 6785	2 5097	.4782
.5219	49 2276	4103	2016	.8530 0668	3 0541	.4781
.5220	.0551 7380	.3983 3552	.4995 1577	.8529 4548	.8783 5984	.4780
.5221	54 2485	2999	1135	8 8428	4 1425	.4779
.5222	56 7590	2444	0692	8 2305	4 6864	.4778
.5223	59 2695	1886	.4995 0246	7 6181	5 2303	.4777
.5224	61 7801	1325	.4994 9799	7 0056	5 7740	.4776
.5225	64 2907	0762	9349	6 3928	6 3175	.4775
.5226	66 8013	.3983 0197	8898	5 7800	6 8609	.4774
.5227	69 3120	.3982 9629	8444	5 1669	7 4041	.4773
.5228	71 8227	9058	7989	4 5537	7 9472	.4772
.5229	74 3335	8485	7531	3 9403	8 4901	.4771
.5230	.0576 8443	.3982 7909	.4994 7072	.8523 3268	.8789 0329	.4770
.5231	79 3551	7331	6610	2 7131	.8789 5756	.4769
.5232	81 8659	6751	6147	2 0993	.8790 1181	.4768
.5233	84 3768	6168	5681	1 4853	0 6604	.4767
.5234	86 8878	5582	5214	0 8711	1 2026	.4766
.5235	89 3987	4994	4744	.8520 2567	1 7447	.4765
.5236	91 9097	4403	4273	.8519 6422	2 2866	.4764
.5237	94 4208	3810	3799	9 0276	2 8284	.4763
.5238	96 9318	3214	3324	8 4128	3 3700	.4762
.5239	.0599 4430	2616	2846	7 7978	3 9115	.4761
.5240	.0601 9541	.3982 2015	.4994 2367	.8517 1826	.8794 4528	.4760
.5241	04 4653	1412	1885	6 5673	4 9940	.4759
.5242	06 9765	0806	1402	5 9519	5 5350	.4758
.5243	09 4878	.3982 0198	0916	5 3362	6 0759	.4757
.5244	11 9991	.3981 9587	.4994 0429	4 7204	6 6166	.4756
.5245	14 5105	8974	.4993 9939	4 1045	7 1572	.4755
.5246	17 0218	8358	9447	3 4884	7 6977	.4754
.5247	19 5333	7740	8954	2 8721	8 2379	.4753
.5248	22 0447	7119	8458	2 2556	8 7781	.4752
.5249	24 5562	6496	7961	1 6390	9 3181	.4751
.5250	.0627 0678	.3981 5870	.4993 7461	.8511 0223	.8799 8580	.4750

$E^{-ii}=$ $E^{ii}=$.0000,0000+   .0000,0000+   .0000,0000+   .0000,0000+   .0000,0000+
.0000,0000,0+

TABLE I

p	x	z	√pq	√1−p²	√1−q²	q
.5250	.0627 0678	.3981 5870	.4993 7461	.8511 0223	.8799 8580	**.4750**
.5251	29 5794	5242	6959	.8510 4053	.8800 3977	.4749
.5252	32 0910	4611	6456	.8509 7882	0 9372	.4748
.5253	34 6026	3978	5950	9 1710	1 4766	.4747
.5254	37 1143	3342	5442	8 5536	2 0159	.4746
.5255	39 6261	2704	4933	7 9360	2 5550	.4745
.5256	42 1379	2063	4421	7 3183	3 0940	.4744
.5257	44 6497	1419	3907	6 7004	3 6328	.4743
.5258	47 1615	0773	3392	6 0823	4 1715	.4742
.5259	49 6735	.3981 0125	2874	5 4641	4 7100	.4741
.5260	.0652 1854	.3980 9474	.4993 2354	.8504 8457	.8805 2484	**.4740**
.5261	54 6974	8821	1833	4 2271	5 7867	.4739
.5262	57 2094	8165	1309	3 6084	6 3248	.4738
.5263	59 7215	7506	0783	2 9895	6 8627	.4737
.5264	62 2336	6845	.4993 0255	2 3705	7 4005	.4736
.5265	64 7457	6182	.4992 9726	1 7513	7 9382	.4735
.5266	67 2579	5516	9194	1 1319	8 4757	.4734
.5267	69 7702	4847	8660	.8500 5124	9 0131	.4733
.5268	72 2824	4176	8124	.8499 8927	.8809 5503	.4732
.5269	74 7948	3503	7587	9 2729	.8810 0873	.4731
.5270	.0677 3071	.3980 2827	.4992 7047	.8498 6528	.8810 6243	**.4730**
.5271	79 8195	2148	6505	8 0327	1 1610	.4729
.5272	82 3320	1467	5961	7 4123	1 6977	.4728
.5273	84 8445	0783	5415	6 7918	2 2342	.4727
.5274	87 3570	.3980 0097	4868	6 1711	2 7705	.4726
.5275	89 8696	.3979 9409	4318	5 5503	3 3067	.4725
.5276	92 3822	8717	3766	4 9293	3 8427	.4724
.5277	94 8949	8024	3212	4 3082	4 3787	.4723
.5278	97 4076	7328	2656	3 6868	4 9144	.4722
.5279	.0699 9203	6629	2098	3 0653	5 4500	.4721
.5280	.0702 4331	.3979 5928	.4992 1538	.8492 4437	.8815 9855	**.4720**
.5281	04 9460	5224	0977	1 8219	6 5208	.4719
.5282	07 4589	4518	.4992 0413	1 1999	7 0560	.4718
.5283	09 9718	3809	.4991 9847	.8490 5778	7 5910	.4717
.5284	12 4848	3098	9279	.8489 9555	8 1259	.4716
.5285	14 9978	2384	8709	9 3330	8 6606	.4715
.5286	17 5109	1668	8137	8 7104	9 1952	.4714
.5287	20 0240	0949	7563	8 0876	.8819 7296	.4713
.5288	22 5371	.3979 0228	6987	7 4646	.8820 2639	.4712
.5289	25 0503	.3978 9504	6409	6 8415	0 7981	.4711
.5290	.0727 5636	.3978 8778	.4991 5829	.8486 2182	.8821 3321	**.4710**
.5291	30 0769	8049	5247	5 5948	1 8660	.4709
.5292	32 5902	7318	4663	4 9712	2 3997	.4708
.5293	35 1036	6584	4077	4 3474	2 9332	.4707
.5294	37 6170	5847	3489	3 7235	3 4667	.4706
.5295	40 1305	5109	2899	3 0994	3 9999	.4705
.5296	42 6440	4367	2307	2 4751	4 5331	.4704
.5297	45 1576	3623	1713	1 8507	5 0661	.4703
.5298	47 6712	2877	1117	1 2261	5 5989	.4702
.5299	50 1849	2128	.4991 0519	.8480 6013	6 1316	.4701
.5300	.0752 6986	.3978 1377	.4990 9919	.8479 9764	.8826 6641	**.4700**

E⁻ⁱᴸ= Eⁱᴸ=.0000,0000+ .0000,0000+ .0000,0000+ .0000,0000+ .0000,0000+
.0000,0000,0+

TABLE I

.5300   **.4700**

p	x	z	√pq	√1−p²	√1−q²	q
.5300	.0752 6986	.3978 1377	.4990 9919	.8479 9764	.8826 6641	**.4700**
.5301	55 2124	.3978 0623	9317	9 3513	7 1966	.4699
.5302	57 7262	.3977 9866	8713	8 7261	7 7288	.4698
.5303	60 2401	9107	8107	8 1007	8 2609	.4697
.5304	62 6540	8346	7498	7 4751	8 7929	.4696
.5305	65 2679	7582	6888	6 8494	9 3247	.4695
.5306	67 7819	6815	6276	6 2235	.8829 8564	.4694
.5307	70 2960	6046	5662	5 5974	.8830 3879	.4693
.5308	72 8101	5275	5046	4 9712	0 9193	.4692
.5309	75 3242	4500	4428	4 3448	1 4506	.4691
.5310	.0777 8384	.3977 3724	.4990 3807	.8473 7182	.8831 9817	**.4690**
.5311	80 3527	2945	3185	3 0915	2 5126	.4689
.5312	82 8670	2163	2561	2 4646	3 0434	.4688
.5313	85 3813	1379	1935	1 8375	3 5741	.4687
.5314	87 8957	.3977 0592	1307	1 2103	4 1046	.4686
.5315	90 4101	.3976 9803	0676	.8470 5829	4 6350	.4685
.5316	92 9246	9012	.4990 0044	.8469 9554	5 1652	.4684
.5317	95 4392	8217	.4989 9410	9 3277	5 6953	.4683
.5318	.0797 9538	7421	8774	8 6998	6 2252	.4682
.5319	.0800 4684	6621	8135	8 0717	6 7550	.4681
.5320	.0802 9831	.3976 5820	.4989 7495	.8467 4435	.8837 2847	**.4680**
.5321	05 4979	5016	6853	6 8152	7 8142	.4679
.5322	08 0127	4209	6208	6 1866	8 3435	.4678
.5323	10 5275	3399	5562	5 5579	8 8727	.4677
.5324	13 0424	2588	4914	4 9291	9 4018	.4676
.5325	15 5574	1773	4263	4 3000	.8839 9307	.4675
.5226	18 0724	0957	3611	3 6708	.8840 4595	.4674
.5327	20 5874	.3976 0137	2956	3 0415	0 9881	.4673
.5328	23 1025	.3975 9315	2300	2 4119	1 5166	.4672
.5329	25 6177	8491	1642	1 7823	2 0450	.4671
.5330	.0828 1329	.3975 7664	.4989 0981	.8461 1524	.8842 5732	**.4670**
.5331	30 6482	6835	.4989 0319	.8460 5224	3 1012	.4669
.5332	33 1635	6003	.4988 9654	.8459 8922	3 6291	.4668
.5333	35 6789	5168	8988	9 2618	4 1569	.4667
.5334	38 1943	4332	8319	8 6313	4 6845	.4666
.5335	40 7098	3492	7649	8 0007	5 2120	.4665
.5336	43 2253	2650	6976	7 3698	5 7393	.4664
.5337	45 7409	1806	6302	6 7388	6 2665	.4663
.5338	48 2565	0959	5625	6 1076	6 7935	.4662
.5339	50 7722	.3975 0109	4947	5 4763	7 3204	.4661
.5340	.0853 2879	.3974 9257	.4988 4266	.8454 8448	.8847 8472	**.4660**
.5341	55 8037	8403	3583	4 2131	8 3738	.4659
.5342	58 3196	7545	2899	3 5813	8 9003	.4658
.5343	60 8355	6686	2212	2 9492	9 4266	.4657
.5344	63 3515	5824	1524	2 3171	.8849 9528	.4656
.5345	65 8675	4959	0833	1 6847	.8850 4788	.4655
.5346	68 3835	4092	.4988 0140	1 0522	1 0047	.4654
.5347	70 8997	3222	.4987 9446	.8450 4196	1 5304	.4653
.5348	73 4159	2350	8749	.8449 7867	2 0560	.4652
.5349	75 9321	1476	8050	9 1537	2 5815	.4651
.5350	.0878 4484	.3974 0598	.4987 7350	.8448 5206	.8853 1068	**.4650**

E^{-iL} = E^{iL} .0000,0000+ ·   .0000,0000+   .0000,0000+   .0000,0000+   .0000,0000+
.0000,0000,0+

TABLE I

.5350 .4650

p	x	z	$\sqrt{pq}$	$\sqrt{1-p^2}$	$\sqrt{1-q^2}$	q
.5350	.0878 4484	.3974 0598	.4987 7350	.8448 5206	.8853 1068	**.4650**
.5351	80 9647	.3973 9719	6647	7 8873	3 6320	.4649
.5352	83 4811	8836	5942	7 2538	4 1570	.4648
.5353	85 9976	7952	5235	6 6201	4 6819	.4647
.5354	88 5141	7064	4527	5 9863	5 2066	.4646
.5355	91 0307	6175	3816	5 3523	5 7312	.4645
.5356	93 5473	5282	3103	4 7181	6 2556	.4644
.5357	96 0640	4388	2388	4 0838	6 7799	.4643
.5358	.0898 5807	3490	1671	3 4493	7 3041	.4642
.5359	.0901 0975	2590	0952	2 8146	7 8281	.4641
.5360	.0903 6144	.3973 1688	.4987 0232	.8442 1798	.8858 3520	**.4640**
.5361	06 1313	.3973 0783	.4986 9509	1 5448	8 8757	.4639
.5362	08 6483	.3972 9876	8784	0 9097	9 3993	.4638
.5363	11 1653	8966	8057	.8440 2743	.8859 9227	.4637
.5364	13 6824	8054	7328	.8439 6389	.8860 4460	.4636
.5365	16 1995	7139	6597	9 0032	0 9692	.4635
.5366	18 7167	6221	5864	8 3674	1 4922	.4634
.5367	21 2340	5301	5129	7 7314	2 0151	.4633
.5368	23 7513	4379	4392	7 0952	2 5378	.4632
.5369	26 2687	3454	3653	6 4589	3 0604	.4631
.5370	.0928 7861	.3972 2526	.4986 2912	.8435 8224	.8863 5828	**.4630**
.5371	31 3036	1596	2169	5 1858	4 1051	.4629
.5372	33 8211	.3972 0663	1424	4 5490	4 6272	.4628
.5373	36 3387	.3971 9728	.4986 0677	3 9120	5 1492	.4627
.5374	38 8564	8791	.4985 9928	3 2748	5 6711	.4626
.5375	41 3741	7851	9177	2 6375	6 1928	.4625
.5376	43 8919	6908	8424	2 0000	6 7144	.4624
.5377	46 4098	5963	7668	1 3623	7 2358	.4623
.5378	48 9277	5015	6911	0 7245	7 7571	.4622
.5379	51 4457	4065	6152	.8430 0865	8 2782	.4621
.5380	.0953 9637	.3971 3112	.4985 5391	.8429 4484	.8868 7992	**.4620**
.5381	56 4818	2157	4628	8 8101	9 3201	.4619
.5382	58 9999	1199	3862	8 1716	.8869 8408	.4618
.5383	61 5181	.3971 0239	3095	7 5329	.8870 3614	.4617
.5384	64 0364	.3970 9276	2326	6 8941	0 8818	.4616
.5385	66 5548	8311	1555	6 2551	1 4021	.4615
.5386	69 0731	7343	0781	5 6159	1 9222	.4614
.5387	71 5916	6373	.4985 0006	4 9766	2 4422	.4613
.5388	74 1101	5400	.4984 9229	4 3371	2 9621	.4612
.5389	76 6287	4425	8449	3 6975	3 4818	.4611
.5390	.0979 1473	.3970 3447	.4984 7668	.8423 0576	.8874 0014	**.4610**
.5391	81 6660	2466	6885	2 4176	4 5208	.4609
.5392	84 1848	1483	6099	1 7775	5 0401	.4608
.5393	86 7036	.3970 0498	5312	1 1372	5 5592	.4607
.5394	89 2225	.3969 9510	4522	.8420 4967	6 0782	.4606
.5395	91 7415	8520	3731	.8419 8560	6 5970	.4605
.5396	94 2605	7527	2937	9 2152	7 1157	.4604
.5397	96 7796	6531	2142	8 5742	7 6343	.4603
.5398	.0999 2987	5533	1344	7 9330	8 1527	.4602
.5399	.1001 8179	4532	.4984 0545	7 2917	8 6710	.4601
.5400	.1004 3372	.3969 3529	.4983 9743	.8416 6502	.8879 1892	**.4600**

$E^{-ii} =$ $E^{ii} = .0000,0000+$ $.0000,0000+$ $.0000,0000+$ $.0000,0000+$ $.0000,0000+$
.0000,0000,1

TABLE I
.5400 .4600

p	x	z	$\sqrt{pq}$	$\sqrt{1-p^2}$	$\sqrt{1-q^2}$	q
.5400	.1004 3372	.3969 3529	.4983 9743	.8416 6502	.8879 1892	**.4600**
.5401	06 8565	2524	8940	6 0085	.8879 7071	.4599
.5402	09 3759	1516	8134	5 3667	.8880 2250	.4598
.5403	11 8954	.3969 0505	7326	4 7247	0 7427	.4597
.5404	14 4149	.3968 9492	6517	4 0825	1 2603	.4596
.5405	16 9345	8476	5705	3 4401	1 7777	.4595
.5406	19 4542	7458	4891	2 7976	2 2950	.4594
.5407	21 9739	6437	4076	2 1550	2 8121	.4593
.5408	24 4937	5414	3258	1 5121	3 3291	.4592
.5409	27 0135	4388	2438	0 8691	3 8460	.4591
.5410	.1029 5334	.3968 3360	.4983 1616	.8410 2259	.8884 3627	**.4590**
.5411	32 0534	2329	.4983 0793	.8409 5826	4 8792	.4589
.5412	34 5735	1296	.4982 9967	8 9391	5 3957	.4588
.5413	37 0936	.3968 0260	9139	8 2954	5 9119	.4587
.5414	39 6138	.3967 9222	8309	7 6515	6 4281	.4586
.5415	42 1340	8181	7477	7 0075	6 9441	.4585
.5416	44 6543	7137	6634	6 3633	7 4599	.4584
.5417	47 1747	6092	5808	5 7189	7 9756	.4583
.5418	49 6951	5043	4970	5 0744	8 4912	.4582
.5419	52 2156	3992	4130	4 4297	9 0066	.4581
.5420	.1054 7362	.3967 2939	.4982 3288	.8403 7849	.8889 5219	**.4580**
.5421	57 2569	1883	2444	3 1398	.8890 0371	.4579
.5422	59 7776	.3967 0824	1598	2 4946	0 5521	.4578
.5423	62 2984	.3966 9763	.4982 0750	1 8493	1 0669	.4577
.5424	64 8192	8700	.4981 9900	1 2037	1 5816	.4576
.5425	67 3401	7634	9048	.8400 5580	2 0962	.4575
.5426	69 8611	6565	8193	.8399 9121	2 6106	.4574
.5427	72 3821	5494	7337	9 2661	3 1249	.4573
.5428	74 9033	4420	6479	8 6199	3 6391	.4572
.5429	77 4244	3344	5619	7 9735	4 1531	.4571
.5430	.1079 9457	.3966 2265	.4981 4757	.8397 3270	.8894 6669	**.4570**
.5431	82 4670	1184	3893	6 6802	5 1807	.4569
.5432	84 9884	.3966 0100	3026	6 0333	5 6942	.4568
.5433	87 5099	.3965 9014	2158	5 3863	6 2077	.4567
.5434	90 0314	7925	1288	4 7391	6 7210	.4566
.5435	92 5530	6834	.4981 0416	4 0917	7 2341	.4565
.5436	95 0747	5740	.4980 9541	3 4441	7 7471	.4564
.5437	.1097 5964	4644	8665	2 7964	8 2600	.4563
.5438	.1100 1182	3545	7787	2 1485	8 7727	.4562
.5439	02 6401	2444	6906	1 5004	9 2853	.4561
.5440	.1105 1620	.3965 1340	.4980 6024	.8390 8522	.8899 7978	**.4560**
.5441	07 6841	.3965 0233	5139	.8390 2038	.8900 3101	.4559
.5442	10 2061	.3964 9124	4253	.8389 5552	0 8222	.4558
.5443	12 7283	8013	3364	8 9064	1 3342	.4557
.5444	15 2505	6899	2474	8 2575	1 8461	.4556
.5445	17 7728	5782	1581	7 6084	2 3578	.4555
.5446	20 2952	4663	.4980 0687	6 9592	2 8694	.4554
.5447	22 8176	3542	.4979 9790	6 3097	3 3809	.4553
.5448	25 3402	2418	8892	5 6601	3 8922	.4552
.5449	27 8627	1291	7991	5 0104	4 4033	.4551
.5450	.1130 3854	.3964 0162	.4979 7088	.8384 3604	.8904 9144	**.4550**

$E^{-ii} =$ $E^{ii} = $ .0000,0000+   .0000,0000+   .0000,0000+   .0000,0000+   .0000,0000+
.0000,0000,1

TABLE I

p	x	z	$\sqrt{pq}$	$\sqrt{1-p^2}$	$\sqrt{1-q^2}$	q
.5450	.1130 3854	.3964 0162	.4979 7088	.8384 3604	.8904 9144	**.4550**
.5451	32 9081	.3963 9030	6184	3 7103	5 4253	.4549
.5452	35 4309	7896	5277	3 0601	5 9360	.4548
.5453	37 9538	6760	4368	2 4096	6 4466	.4547
.5454	40 4768	5620	3457	1 7590	6 9571	.4546
.5455	42 9998	4479	2545	1 1082	7 4674	.4545
.5456	45 5229	3334	1630	.8380 4573	7 9775	.4544
.5457	48 0460	2188	.4979 0713	.8379 8061	8 4876	.4543
.5458	50 5693	.3963 1038	.4978 9794	9 1548	8 9975	.4542
.5459	53 0926	.3962 9886	8873	8 5034	.8909 5072	.4541
.5460	.1155 6160	.3962 8732	.4978 7950	.8377 8518	.8910 0168	**.4540**
.5461	58 1394	7575	7025	7 1999	0 5263	.4539
.5462	60 6630	6416	6098	6 5480	1 0356	.4538
.5463	63 1866	5254	5169	5 8958	1 5448	.4537
.5464	65 7102	4089	4238	5 2435	2 0539	.4536
.5465	68 2340	2922	3305	4 5910	2 5628	.4535
.5466	70 7578	1753	2370	3 9384	3 0715	.4534
.5467	73 2817	.3962 0581	1433	3 2856	3 5801	.4533
.5468	75 8057	.3961 9406	.4978 0494	2 6326	4 0886	.4532
.5469	78 3298	8229	.4977 9553	1 9794	4 5970	.4531
.5470	.1180 8539	.3961 7050	.4977 8610	.8371 3261	.8915 1052	**.4530**
.5471	83 3781	5868	7665	0 6726	5 6132	.4529
.5472	85 9024	4683	6717	.8370 0189	6 1211	.4528
.5473	88 4267	3496	5768	.8369 3650	6 6289	.4527
.5474	90 9512	2306	4817	8 7110	7 1365	.4526
.5475	93 4757	.3961 1114	3864	8 0568	7 6440	.4525
.5476	96 0002	.3960 9919	2908	7 4025	8 1514	.4524
.5477	.1198 5249	8722	1951	6 7479	8 6586	.4523
.5478	.1201 0496	7522	0992	6 0932	9 1657	.4522
.5479	03 5745	6320	.4977 0030	5 4384	.8919 6726	.4521
.5480	.1206 0993	.3960 5115	.4976 9067	.8364 7833	.8920 1794	**.4520**
.5481	08 6243	3908	8101	4 1281	0 6860	.4519
.5482	11 1493	2698	7134	3 4727	1 1925	.4518
.5483	13 6745	1485	6164	2 8172	1 6989	.4517
.5484	16 1997	.3960 0270	5193	2 1614	2 2051	.4516
.5485	18 7249	.3959 9053	4219	1 5055	2 7112	.4515
.5486	21 2503	7833	3243	0 8495	3 2171	.4514
.5487	23 7757	6610	2266	.8360 1932	3 7229	.4513
.5488	26 3012	5385	1286	.8359 5368	4 2286	.4512
.5489	28 8268	4158	.4976 0304	8 8802	4 7341	.4511
.5490	.1231 3525	.3959 2928	.4975 9321	.8358 2235	.8925 2395	**.4510**
.5491	33 8782	1695	8335	7 5666	5 7447	.4509
.5492	36 4040	.3959 0460	7347	6 9095	6 2498	.4508
.5493	38 9299	.3958 9222	6357	6 2522	6 7548	.4507
.5494	41 4559	7982	5366	5 5948	7 2596	.4506
.5495	43 9820	6739	4372	4 9372	7 7643	.4505
.5496	46 5081	5494	3376	4 2794	8 2688	.4504
.5497	49 0343	4246	2378	3 6214	8 7732	.4503
.5498	51 5606	2996	1378	2 9633	9 2775	.4502
.5499	54 0870	1743	.4975 0376	2 3050	.8929 7816	.4501
.5500	.1256 6135	.3958 0488	.4974 9372	.8351 6465	.8930 2855	**.4500**

$E^{-ii}=$ $E^{ii}=$.0000,0000+   .0000,0000+   .0000,0000+   .0000,0000+   .0000,0000+
.0000,0000+

TABLE I

.5500　　　　　　　　　　　　　　　　　　　　　　　　　　　　**.4500**

p	x	z	$\sqrt{pq}$	$\sqrt{1-p^2}$	$\sqrt{1-q^2}$	q
.5500	.1256 6135	.3958 0488	.4974 9372	.8351 6465	.8930 2855	**.4500**
.5501	59 1400	.3957 9230	8366	0 9879	0 7894	.4499
.5502	61 6666	7970	7358	.8350 3291	1 2931	.4498
.5503	64 1933	6707	6348	.8349 6701	1 7966	.4497
.5504	66 7201	5441	5335	9 0110	2 3000	.4496
.5505	69 2470	4173	4321	8 3516	2 8033	.4495
.5506	71 7739	2903	3305	7 6921	3 3064	.4494
.5507	74 3009	1630	2287	7 0325	3 8094	.4493
.5508	76 8280	.3957 0354	1267	6 3726	4 3123	.4492
.5509	79 3552	.3956 9076	.4974 0244	5 7126	4 8150	.4491
.5510	.1281 8825	.3956 7795	.4973 9220	.8345 0524	.8935 3176	**.4490**
.5511	84 4098	6512	8194	4 3921	5 8200	.4489
.5512	86 9373	5227	7165	3 7315	6 3223	.4488
.5513	89 4648	3938	6135	3 0708	6 8244	.4487
.5514	91 9924	2648	5102	2 4100	7 3264	.4486
.5515	94 5200	1354	4068	1 7489	7 8283	.4485
.5516	97 0478	.3956 0059	3031	1 0877	8 3300	.4484
.5517	.1299 5757	.3955 8760	1993	.8340 4263	8 8316	.4483
.5518	.1302 1036	7459	.4973 0952	.8339 7647	9 3331	.4482
.5519	04 6316	6156	.4972 9910	9 1030	.8939 8344	.4481
.5520	.1307 1597	.3955 4850	.4972 8865	.8338 4411	.8940 3356	**.4480**
.5521	09 6879	3542	7818	7 7790	0 8366	.4479
.5522	12 2161	2231	6769	7 1168	1 3375	.4478
.5523	14 7445	.3955 0917	5719	6 4543	1 8382	.4477
.5524	17 2729	.3954 9601	4666	5 7917	2 3388	.4476
.5525	19 8014	8283	3611	5 1290	2 8393	.4475
.5526	22 3300	6962	2554	4 4660	3 3396	.4474
.5527	24 8587	5638	1495	3 8029	3 8398	.4473
.5528	27 3874	4312	.4972 0434	3 1396	4 3399	.4472
.5529	29 9163	2983	.4971 9371	2 4762	4 8398	.4471
.5530	.1332 4452	.3954 1652	.4971 8306	.8331 8125	.8945 3396	**.4470**
.5531	34 9743	.3954 0318	7239	1 1487	5 8392	.4469
.5532	37 5034	.3953 8982	6170	.8330 4847	6 3387	.4468
.5533	40 0326	7643	5099	.8329 8206	6 8380	.4467
.5534	42 5618	6302	4026	9 1563	7 3373	.4466
.5535	45 0912	4958	2951	8 4918	7 8363	.4465
.5536	47 6206	3612	1874	7 8271	8 3353	.4464
.5537	50 1502	2263	.4971 0795	7 1622	8 8341	.4463
.5538	52 6798	.3953 0912	.4970 9713	6 4972	9 3327	.4462
.5539	55 2095	.3952 9558	8630	5 8320	.8949 8312	.4461
.5540	.1357 7393	.3952 8201	.4970 7545	.8325 1667	.8950 3296	**.4460**
.5541	60 2692	6842	6457	4 5011	0 8278	.4459
.5542	62 7992	5481	5368	3 8354	1 3259	.4458
.5543	65 3292	4117	4276	3 1695	1 8239	.4457
.5544	67 8594	2750	3183	2 5035	2 3217	.4456
.5545	70 3896	1381	2087	1 8372	2 8194	.4455
.5546	72 9199	.3952 0009	.4970 0990	1 1708	3 3169	.4454
.5547	75 4503	.3951 8635	.4969 9890	.8320 5043	3 8143	.4453
.5548	77 9808	7258	8789	.8319 8375	4 3116	.4452
.5549	80 5114	5879	7685	9 1706	4 8087	.4451
.5550	.1383 0421	.3951 4497	.4969 6579	.8318 5035	.8955 3057	**.4450**

E^{-iL}　E^{iL}.0000,0000+　.0000,0000+　.0000,0000+　.0000,0000+　.0000,0000+
.0000,0001

TABLE I

p	x	z	$\sqrt{pq}$	$\sqrt{1-p^2}$	$\sqrt{1-q^2}$	q
.5550	.1383 0421	.3951 4497	.4969 6579	.8318 5035	.8955 3057	.4450
.5551	85 5728	3113	5472	7 8362	5 8025	.4449
.5552	88 1037	1726	4362	7 1687	6 2992	.4448
.5553	90 6346	.3951 0337	3250	6 5011	6 7958	.4447
.5554	93 1657	.3950 8945	2136	5 8333	7 2922	.4446
.5555	95 6968	7550	.4969 1020	5 1654	7 7885	.4445
.5556	.1398 2280	6154	.4968 9902	4 4972	8 2847	.4444
.5557	.1400 7593	4754	8782	3 8289	8 7807	.4443
.5558	03 2907	3352	7660	3 1604	9 2765	.4442
.5559	05 8221	1947	6536	2 4917	.8959 7723	.4441
.5560	.1408 3537	.3950 0540	.4968 5410	.8311 8229	.8960 2679	.4440
.5561	10 8854	.3949 9131	4282	1 1539	0 7633	.4439
.5562	13 4171	7719	3152	.8310 4847	1 2586	.4438
.5563	15 9489	6304	2020	.8309 8153	1 7538	.4437
.5564	18 4809	4887	.4968 0886	9 1458	2 2488	.4436
.5565	21 0129	3467	.4967 9749	8 4761	2 7437	.4435
.5566	23 5450	2045	8611	7 8062	3 2385	.4434
.5567	26 0772	.3949 0620	7471	7 1361	3 7331	.4433
.5568	28 6095	.3948 9193	6328	6 4659	4 2276	.4432
.5569	31 1419	7763	5184	5 7955	4 7219	.4431
.5570	.1433 6744	.3948 6330	.4967 4037	.8305 1249	.8965 2161	.4430
.5571	36 2069	4895	2889	4 4542	5 7102	.4429
.5572	38 7396	3458	1738	3 7832	6 2041	.4428
.5573	41 2723	2018	.4967 0586	3 1121	6 6979	.4427
.5574	43 8052	.3948 0575	.4966 9431	2 4408	7 1915	.4426
.5575	46 3381	.3947 9130	8275	1 7694	7 6850	.4425
.5576	48 8711	7683	7116	1 0978	8 1784	.4424
.5577	51 4043	6232	5955	.8300 4260	8 6716	.4423
.5578	53 9375	4780	4792	.8299 7540	9 1647	.4422
.5579	56 4708	3325	3628	9 0818	.8969 6577	.4421
.5580	.1459 0042	.3947 1867	.4966 2461	.8298 4095	.8970 1505	.4420
.5581	61 5377	.3947 0407	1292	7 8370	0 6432	.4419
.5582	64 0713	.3946 8944	.4966 0121	7 1643	1 1357	.4418
.5583	66 6050	7478	.4965 8948	6 4914	1 6281	.4417
.5584	69 1388	6011	7773	5 8184	2 1204	.4416
.5585	71 6726	4540	6596	5 1452	2 6125	.4415
.5586	74 2066	3067	5417	4 4718	3 1045	.4414
.5587	76 7407	1592	4235	3 7983	3 5963	.4413
.5588	79 2748	.3946 0114	3052	3 0245	4 0880	.4412
.5589	81 8091	.3945 8633	1867	2 3506	4 5796	.4411
.5590	.1484 3434	.3945 7150	.4965 0680	.8291 6765	.8975 0710	.4410
.5591	86 8778	5665	.4964 9490	1 0023	5 5623	.4409
.5592	89 4124	4176	8299	.8290 3279	6 0535	.4408
.5593	91 9470	2686	7106	.8289 6532	6 5445	.4407
.5594	94 4818	.3945 1192	5910	8 9785	7 0354	.4406
.5595	97 0166	.3944 9697	4713	8 3035	7 5261	.4405
.5596	.1499 5515	8198	3513	7 6284	8 0167	.4404
.5597	.1502 0865	6698	2312	6 9531	8 5072	.4403
.5598	04 6216	5194	.4964 1108	6 2776	8 9975	.4402
.5599	07 1568	3688	.4963 9902	5 6019	9 4877	.4401
.5600	.1509 6922	.3944 2180	.4963 8695	.8284 9261	.8979 9777	.4400

$E^{-iL}=$ $E^{iL}=$.0000,0000+ .0000,0000+ .0000,0000+ .0000,0000+ .0000,0000+
.0000,0000,1

TABLE I

.5600   **.4400**

p	x	z	$\sqrt{pq}$	$\sqrt{1-p^2}$	$\sqrt{1-q^2}$	q
.5600	.1509 6922	.3944 2180	.4963 8695	.8284 9261	.8979 9777	**.4400**
.5601	12 2276	.3944 0669	7485	4 2501	.8980 4676	.4399
.5602	14 7631	.3943 9155	6273	3 5739	0 9574	.4398
.5603	17 2987	7639	5059	2 8975	1 4470	.4397
.5604	19 8344	6121	3843	2 2210	1 9365	.4396
.5605	22 3702	4600	2625	1 5442	2 4259	.4395
.5606	24 9060	3076	1405	0 8673	2 9151	.4394
.5607	27 4420	1550	.4963 0183	.8280 1903	3 4042	.4393
.5608	29 9781	.3943 0021	.4962 8959	.8279 5130	3 8931	.4392
.5609	32 5143	.3942 8490	7733	8 8356	4 3819	.4391
.5610	.1535 0506	.3942 6956	.4962 6505	.8278 1580	.8984 8706	**.4390**
.5611	37 5870	5420	5275	7 4802	5 3591	.4389
.5612	40 1235	3881	4043	6 8023	5 8475	.4388
.5613	42 6600	2340	2808	6 1242	6 3358	.4387
.5614	45 1967	.3942 0796	1572	5 4458	6 8239	.4386
.5615	47 7335	.3941 9249	.4962 0334	4 7674	7 3119	.4385
.5616	50 2704	7700	.4961 9093	4 0887	7 7997	.4384
.5617	52 8074	6149	7851	3 4099	8 2874	.4383
.5618	55 3444	4595	6606	2 7309	8 7750	.4382
.5619	57 8816	3038	5360	2 0517	9 2624	.4381
.5620	.1560 4189	.3941 1479	.4961 4111	.8271 3723	.8989 7497	**.4380**
.5621	62 9563	.3940 9917	2860	0 6928	.8990 2369	.4379
.5622	65 4938	8353	1608	.8270 0131	0 7239	.4378
.5623	68 0314	6786	.4961 0353	.8269 3332	1 2108	.4377
.5624	70 5691	5217	.4960 9096	8 6531	1 6975	.4376
.5625	73 1068	3645	7837	7 9728	2 1841	.4375
.5626	75 6447	2071	6576	7 2924	2 6706	.4374
.5627	78 1827	.3940 0494	5313	6 6118	3 1569	.4373
.5628	80 7208	.3939 8914	4048	5 9310	3 6431	.4372
.5629	83 2590	7332	2781	5 2501	4 1291	.4371
.5630	.1585 7973	.3939 5748	.4960 1512	.8264 5690	.8994 6151	**.4370**
.5631	88 3357	4161	.4960 0241	3 8876	5 1008	.4369
.5632	90 8742	2571	.4959 8968	3 2062	5 5865	.4368
.5633	93 4128	.3939 0979	7692	2 5245	6 0720	.4367
.5634	95 9515	.3938 9384	6415	1 8427	6 5573	.4366
.5635	.1598 4903	7787	5136	1 1606	7 0426	.4365
.5636	.1601 0292	6187	3854	.8260 4784	7 5277	.4364
.5637	03 5682	4585	2571	.8259 7961	8 0126	.4363
.5638	06 1073	2980	.4959 1286	9 1135	8 4974	.4362
.5639	08 6466	.3938 1373	.4958 9998	8 4308	8 9821	.4361
.5640	.1611 1859	.3937 9763	.4958 8708	.8257 7479	.8999 4667	**.4360**
.5641	13 7253	8150	7417	7 0648	.8999 9511	.4359
.5642	16 2648	6535	6123	6 3815	.9000 4353	.4358
.5643	18 8045	4918	4827	5 6981	0 9195	.4357
.5644	21 3442	3298	3530	5 0145	1 4034	.4356
.5645	23 8841	1675	2230	4 3307	1 8873	.4355
.5646	26 4240	.3937 0050	.4958 0928	3 6467	2 3710	.4354
.5647	28 9641	.3936 8422	.4957 9624	2 9626	2 8546	.4353
.5648	31 5042	6792	8318	2 2782	3 3380	.4352
.5649	34 0445	5159	7010	1 5937	3 8214	.4351
.5650	.1636 5849	.3936 3524	.4957 5700	.8250 9090	.9004 3045	**.4350**

$E^{-iL}_{.0000,0000,1}$ E^{iL}.0000,0000+   .0000,0000+   .0000,0000+   .0000,0000+   .0000,0000+

TABLE I

p	x	z	$\sqrt{pq}$	$\sqrt{1-p^2}$	$\sqrt{1-q^2}$	q
.5650	.1636 5849	.3936 3524	.4957 5700	.8250 9090	.9004 3045	**.4350**
.5651	39 1253	1886	4388	.8250 2242	4 7876	.4349
.5652	41 6659	.3936 0246	3073	.8249 5391	5 2705	.4348
.5653	44 2066	.3935 8603	1757	8 8539	5 7532	.4347
.5654	46 7474	6957	.4957 0439	8 1685	6 2358	.4346
.5655	49 2883	5309	.4956 9118	7 4829	6 7183	.4345
.5656	51 8293	3659	7796	6 7972	7 2007	.4344
.5657	54 3704	2006	6472	6 1113	7 6829	.4343
.5658	56 9116	.3935 0350	5145	5 4252	8 1650	.4342
.5659	59 4530	.3934 8692	3816	4 7389	8 6469	.4341
.5660	.1661 9944	.3934 7031	.4956 2486	.8244 0524	.9009 1287	**.4340**
.5661	64 5359	5368	.4956 1153	3 3658	.9009 6104	.4339
.5662	67 0776	3702	.4955 9818	2 6789	.9010 0919	.4338
.5663	69 6193	2034	8482	1 9919	0 5733	.4337
.5664	72 1612	.3934 0363	7143	1 3048	1 0545	.4336
.5665	74 7032	.3933 8689	5802	.8240 6174	1 5357	.4335
.5666	77 2453	7013	4459	.8239 9299	2 0166	.4334
.5667	79 7875	5335	3114	9 2421	2 4975	.4333
.5668	82 3298	3654	1767	8 5542	2 9782	.4332
.5669	84 8722	1970	.4955 0418	7 8662	3 4588	.4331
.5670	.1687 4147	.3933 0284	.4954 9067	.8237 1779	.9013 9392	**.4330**
.5671	89 9573	.3932 8595	7713	6 4895	4 4195	.4329
.5672	92 5000	6904	6358	5 8009	4 8997	.4328
.5673	95 0429	5210	5001	5 1121	5 3797	.4327
.5674	.1697 5858	3514	3641	4 4231	5 8596	.4326
.5675	.1700 1289	1815	2280	3 7340	6 3393	.4325
.5676	02 6721	.3932 0114	.4954 0916	3 0446	6 8190	.4324
.5677	05 2154	.3931 8410	.4953 9551	2 3551	7 2984	.4323
.5678	07 7587	6703	8183	1 6654	7 7778	.4322
.5679	10 3022	4994	6814	0 9756	8 2570	.4321
.5680	.1712 8459	.3931 3283	.4953 5442	.8230 2855	.9018 7361	**.4320**
.5681	15 3896	.3931 1569	4068	.8229 5953	9 2150	.4319
.5682	17 9334	.3930 9852	2692	8 9049	.9019 6938	.4318
.5683	20 4774	8133	.4953 1314	8 2143	.9020 1724	.4317
.5684	23 0214	6411	.4952 9934	7 5236	0 6510	.4316
.5685	25 5656	4687	8552	6 8326	1 1294	.4315
.5686	28 1099	2960	7168	6 1415	1 6076	.4314
.5687	30 6543	.3930 1231	5782	5 4502	2 0857	.4313
.5688	33 1988	.3929 9499	4394	4 7587	2 5637	.4312
.5689	35 7434	7764	3004	4 0671	3 0416	.4311
.5690	.1738 2881	.3929 6027	.4952 1611	.8223 3752	.9023 5193	**.4310**
.5691	40 8330	4288	.4952 0217	2 6832	3 9968	.4309
.5692	43 3779	2546	.4951 8821	1 9910	4 4743	.4308
.5693	45 9230	.3929 0801	7422	1 2986	4 9516	.4307
.5694	48 4682	.3928 9054	6022	.8220 6061	5 4287	.4306
.5695	51 0135	7304	4619	.8219 9133	5 9058	.4305
.5696	53 5589	5552	3214	9 2204	6 3827	.4304
.5697	56 1044	3797	1808	8 5273	6 8594	.4303
.5698	58 6500	2039	.4951 0399	7 8340	7 3360	.4302
.5699	61 1958	.3928 0280	.4950 8988	7 1406	7 8125	.4301
.5700	.1763 7416	.3927 8517	.4950 7575	.8216 4469	.9028 2889	**.4300**

$E^{-ii}=$ .0000,0000,1 $E^{ii}=$.0000,0000+   .0000,0000+   .0000,0000+   .0000,0000+   .0000,0000+

TABLE I

p	x	z	$\sqrt{pq}$	$\sqrt{1-p^2}$	$\sqrt{1-q^2}$	q
.5700	.1763 7416	.3927 8517	.4950 7575	.8216 4469	.9028 2889	.4300
.5701	66 2876	6752	0 6160	5 7531	8 7651	.4299
.5702	68 8337	4984	0 4743	5 0591	9 2412	.4298
.5703	71 3799	3214	0 3324	4 3649	.9029 7171	.4297
.5704	73 9262	.3927 1442	0 1903	3 6706	.9030 1929	.4296
.5705	76 4727	.3926 9667	.4950 0480	2 9760	0 6686	.4295
.5706	79 0192	7889	.4949 9055	2 2813	1 1441	.4294
.5707	81 5659	6108	9 7627	1 5864	1 6195	.4293
.5708	84 1127	4326	9 6198	0 8913	2 0948	.4292
.5709	86 6596	2540	9 4766	.8210 1960	2 5699	.4291
.5710	.1789 2066	.3926 0752	.4949 3333	.8209 5006	.9033 0449	.4290
.5711	91 7537	.3925 8962	9 1897	8 8050	3 5197	.4289
.5712	94 3010	7169	9 0460	8 1092	3 9945	.4288
.5713	96 8483	5373	8 9020	7 4132	4 4690	.4287
.5714	.1799 3958	3575	8 7578	6 7170	4 9435	.4286
.5715	.1801 9434	.3925 1774	8 6134	6 0207	5 4178	.4285
.5716	04 4911	.3924 9971	8 4689	5 3241	5 8920	.4284
.5717	07 0390	8165	8 3241	4 6274	6 3660	.4283
.5718	09 5869	6357	8 1791	3 9305	6 8399	.4882
.5719	12 1350	4546	8 0339	3 2334	7 3137	.4281
.5720	.1814 6832	.3924 2733	.4947 8884	.8202 5362	.9037 7873	.4280
.5721	17 2315	.3924 0917	7 7428	1 8388	8 2608	.4279
.5722	19 7799	.3923 9098	7 5970	1 1411	8 7342	.4278
.5723	· 22 3284	7277	7 4510	.8200 4433	9 2074	.4277
.5724	24 8771	5454	7 3047	.8199 7454	.9039 6805	.4276
.5725	27 4259	3628	7 1583	9 0472	.9040 1535	.4275
.5726	29 9747	.3923 1799	7 0116	8 3489	0 6263	.4274
.5727	32 5238	.3922 9968	6 8648	7 6503	1 0990	.4273
.5728	35 0729	8134	6 7177	6 9516	1 5715	.4272
.5729	37 6221	6298	6 5704	6 2527	2 0440	.4271
.5730	.1840 1715	.3922 4459	.4946 4229	.8195 5537	.9042 5162	.4270
.5731	42 7210	2617	6 2753	4 8544	2 9884	.4269
.5732	45 2706	.3922 0773	6 1274	4 1550	3 4604	.4268
.5733	47 8203	.3921 8927	5 9793	3 4554	3 9323	.4267
.5734	50 3702	7078	5 8310	2 7556	4 4040	.4266
.5735	52 9202	5226	5 6825	2 0556	4 8756	.4265
.5736	55 4702	3372	5 5337	1 3554	5 3471	.4264
.5737	58 0205	.3921 1515	5 3848	.8190 6551	5 8184	.4263
.5738	60 5708	.3920 9656	5 2357	.8189 9546	6 2896	.4262
.5739	63 1212	7794	5 0863	9 2539	6 7607	.4261
.5740	.1865 6718	.3920 5929	.4944 9368	.8188 5530	.9047 2316	.4260
.5741	68 2225	4062	4 7871	7 8519	7 7024	.4259
.5742	70 7733	2193	4 6371	7 1507	8 1731	.4258
.5743	73 3243	.3920 0321	4 4869	6 4492	8 6436	.4257
.5744	75 8753	.3919 8446	4 3366	5 7476	9 1140	.4256
.5745	78 4265	6569	4 1860	5 0458	.9049 5842	.4255
.5746	80 9778	4689	4 0352	4 3438	.9050 0544	.4254
.5747	83 5292	2807	3 8842	3 6417	0 5243	.4253
.5748	86 0808	.3919 0922	3 7330	2 9393	0 9942	.4252
.5749	88 6325	.3918 9035	3 5816	2 2368	1 4639	.4251
.5750	.1891 1843	.3918 7145	.4943 4300	.8181 5341	.9051 9335	.4250

E^{-ii}= E^{iL}.0000,0000+　.0000,0000+　.0000,0000+　.0000,0000+　.0000,0000+
.0000,0000,1

TABLE I

.5750 .4250

p	x	z	$\sqrt{pq}$	$\sqrt{1-p^2}$	$\sqrt{1-q^2}$	q
.5750	.1891 1843	.3918 7145	.4943 4300	.8181 5341	.9051 9335	**.4250**
.5751	93 7362	5253	3 2782	0 8312	2 4029	.4249
.5752	96 2882	3358	3 1261	.8180 1281	2 8723	.4248
.5753	.1898 8404	.3918 1460	2 9739	.8179 4249	3 3414	.4247
.5754	.1901 3927	.3917 9560	2 8215	8 7214	3 8105	.4246
.5755	03 9451	7657	2 6688	8 0178	4 2794	.4245
.5756	06 4976	5752	2 5160	7 3140	4 7481	.4244
.5757	09 0503	3844	2 3629	6 6100	5 2168	.4243
.5758	11 6031	1934	2 2096	5 9058	5 6853	.4242
.5759	14 1560	.3917 0021	2 0562	5 2015	6 1537	.4241
.5760	.1916 7090	.3916 8106	.4941 9025	.8174 4969	.9056 6219	**.4240**
.5761	19 2622	6188	1 7486	3 7922	7 0900	.4239
.5762	21 8155	4267	1 5945	3 0873	7 5579	.4238
.5763	24 3689	2344	1 4402	2 3822	8 0258	.4237
.5764	26 9224	.3916 0418	1 2857	1 6769	8 4935	.4236
.5765	29 4761	.3915 8490	1 1309	0 9715	8 9610	.4235
.5766	32 0299	6559	0 9760	.8170 2658	9 4285	.4234
.5767	34 5838	4626	0 8209	.8169 5600	.9059 8957	.4233
.5768	37 1378	2690	0 6655	8 8540	.9060 3629	.4232
.5769	39 6920	.3915 0752	0 5100	8 1478	0 8299	.4231
.5770	.1942 2463	.3914 8811	.4940 3542	.8167 4415	.9061 2968	**.4230**
.5771	44 8007	6867	0 1983	6 7349	1 7636	.4229
.5772	47 3553	4921	.4940 0421	6 0282	2 2302	.4228
.5773	49 9099	2973	.4939 8857	5 3212	2 6967	.4227
.5774	52 4647	.3914 1022	9 7291	4 6141	3 1630	.4226
.5775	55 0197	.3913 9068	9 5723	3 9068	3 6292	.4225
.5776	57 5747	7111	9 4154	3 1994	4 0953	.4224
.5777	60 1299	5153	9 2581	2 4917	4 5613	.4223
.5778	62 6852	3191	9 1007	1 7839	5 0271	.4222
.5779	65 2406	.3913 1227	8 9431	1 0758	5 4928	.4221
.5780	.1967 7962	.3912 9261	.4938 7853	.8160 3676	.9065 9583	**.4220**
.5781	70 3519	7292	8 6272	.8159 6592	6 4237	.4219
.5782	72 9077	5320	8 4690	8 9507	6 8890	.4218
.5783	75 4637	3346	8 3105	8 2419	7 3541	.4217
.5784	78 0198	.3912 1369	8 1519	7 5330	7 8191	.4216
.5785	80 5760	.3911 9390	7 9930	6 8238	8 2840	.4215
.5786	83 1323	7408	7 8339	6 1145	8 7488	.4214
.5787	85 6888	5424	7 6747	5 4050	9 2134	.4213
.5788	88 2454	3437	7 5152	4 6953	.9069 6778	.4212
.5789	90 8021	.3911 1447	7 3555	3 9855	.9070 1422	.4211
.5790	.1993 3590	.3910 9455	.4937 1956	.8153 2754	.9070 6064	**.4210**
.5791	95 9160	7460	7 0354	2 5652	1 0704	.4209
.5792	.1998 4731	5463	6 8751	1 8548	1 5344	.4208
.5793	.2001 0304	3463	6 7146	1 1442	1 9982	.4207
.5794	03 5877	.3910 1461	6 5539	.8150 4334	2 4618	.4206
.5795	06 1452	.3909 9456	6 3929	.8149 7224	2 9254	.4205
.5796	08 7029	7449	6 2318	9 0112	3 3888	.4204
.5797	11 2607	5439	6 0704	8 2999	3 8520	.4203
.5798	13 8186	3426	5 9088	7 5884	4 3152	.4202
.5799	16 3766	.3909 1411	5 7471	6 8766	4 7782	.4201
.5800	.2018 9348	.3908 9394	.4935 5851	.8146 1647	.9075 2410	**.4200**

$E^{-ii}=$.0000,0000,1 $E^{ii}=$.0000,0000+ .0000,0000+ .0000,0000+ .0000,0000+

TABLE I

.5800 .4200

p	x	z	$\sqrt{pq}$	$\sqrt{1-p^2}$	$\sqrt{1-q^2}$	q
.5800	.2018 9348	.3908 9394	.4935 5851	.8146 1647	.9075 2410	.4200
.5801	21 4931	7373	5 4229	5 4527	5 7038	.4199
.5802	24 0515	5351	5 2605	4 7404	6 1664	.4198
.5803	26 6101	3325	5 0979	4 0279	6 6288	.4197
.5804	29 1688	.3908 1297	4 9351	3 3153	7 0912	.4196
.5805	31 7276	.3907 9267	4 7720	2 6025	7 5534	.4195
.5806	34 2866	7234	4 6088	1 8895	8 0154	.4194
.5807	36 8457	5198	4 4454	1 1763	8 4774	.4193
.5808	39 4049	3160	4 2817	.8140 4629	8 9391	.4192
.5809	41 9643	.3907 1119	4 1179	.8139 7493	9 4008	.4191
.5810	.2044 5238	.3906 9076	.4933 9538	.8139 0356	.9079 8623	.4190
.5811	47 0835	7030	3 7895	8 3216	.9080 3237	.4189
.5812	49 6432	4982	3 6250	7 6075	0 7850	.4188
.5813	52 2031	2931	3 4603	6 8932	1 2461	.4187
.5814	54 7632	.3906 0878	3 2955	6 1787	1 7071	.4186
.5815	57 3233	.3905 8822	3 1303	5 4640	2 1680	.4185
.5816	59 8836	6763	2 9650	4 7492	2 6287	.4184
.5817	62 4441	4702	2 7995	4 0341	3 0893	.4183
.5818	65 0047	2638	2 6338	3 3189	3 5497	.4182
.5819	67 5654	.3905 0572	2 4678	2 6035	4 0101	.4181
.5820	.2070 1262	.3904 8503	.4932 3017	.8131 8878	.9084 4703	.4180
.5821	72 6872	6432	2 1353	1 1721	4 9303	.4179
.5822	75 2483	4358	1 9688	.8130 4561	5 3903	.4178
.5823	77 8096	2281	1 8020	.8129 7399	5 8500	.4177
.5824	80 3710	.3904 0202	1 6350	9 0236	6 3097	.4176
.5825	82 9325	.3903 8120	1 4678	8 3070	6 7692	.4175
.5826	85 4942	6036	1 3004	7 5903	7 2286	.4174
.5827	88 0560	3949	1 1328	6 8734	7 6879	.4173
.5828	90 6179	.3903 1860	0 9650	6 1563	8 1470	.4172
.5829	93 1800	.3902 9768	0 7970	5 4390	8 6060	.4171
.5830	.2095 7422	.3902 7674	.4930 6288	.8124 7215	.9089 0649	.4170
.5831	.2098 3046	5577	0 4603	4 0039	9 5236	.4169
.5832	.2100 8671	3477	0 2917	3 2860	.9089 9822	.4168
.5833	03 4297	.3902 1375	.4930 1228	2 5680	.9090 4406	.4167
.5834	05 9925	.3901 9270	.4929 9538	1 8498	0 8990	.4166
.5835	08 5554	7163	9 7845	1 1314	1 3572	.4165
.5836	11 1184	5053	9 6150	.8120 4128	1 8152	.4164
.5837	13 6816	2941	9 4453	.8119 6940	2 2731	.4163
.5838	16 2449	.3901 0826	9 2754	8 9751	2 7309	.4162
.5839	18 8084	.3900 8708	9 1053	8 2559	3 1886	.4161
.5840	.2121 3720	.3900 6588	.4928 9350	.8117 5366	.9093 6461	.4160
.5841	23 9357	4465	8 7644	6 8170	4 1035	.4159
.5842	26 4996	2340	8 5937	6 0973	4 5608	.4158
.5843	29 0636	.3900 0212	8 4228	5 3774	5 0179	.4157
.5844	31 6278	.3899 8082	8 2516	4 6574	5 4749	.4156
.5845	34 1921	5949	8 0803	3 9371	5 9318	.4155
.5846	36 7565	3814	7 9087	3 2166	6 3885	.4154
.5847	39 3211	.3899 1676	7 7369	2 4960	6 8451	.4153
.5848	41 8858	.3898 9535	7 5649	1 7751	7 3016	.4152
.5849	44 4507	7392	7 3927	1 0541	7 7579	.4151
.5850	.2147 0157	.3898 5246	.4927 2203	.8110 3329	.9098 2141	.4150

$E^{-ii}=$ .0000,0000,1 $E^{ii}=$.0000,0000+   .0000,0000+   .0000,0000+   .0000,0000+   .0000,0000+

TABLE I

.5850   **.4150**

p	x	z	$\sqrt{pq}$	$\sqrt{1-p^2}$	$\sqrt{1-q^2}$	q
.5850	.2147 0157	.3898 5246	.4927 2203	.8110 3329	.9098 2141	**.4150**
.5851	49 5808	3098	7 0477	.8109 6115	8 6702	.4149
.5852	52 1461	.3898 0947	6 8749	8 8899	9 1261	.4148
.5853	54 7115	.3897 8794	6 7018	8 1682	.9099 5819	.4147
.5854	57 2771	6638	6 5286	7 4462	.9100 0376	.4146
.5855	59 8428	4479	6 3551	6 7241	0 4931	.4145
.5856	62 4087	2318	6 1815	6 0017	0 9485	.4144
.5857	64 9747	.3897 0154	6 0076	5 2792	1 4038	.4143
.5858	67 5408	.3896 7988	5 8335	4 5565	1 8589	.4142
.5859	70 1071	5819	5 6592	3 8336	2 3139	.4141
.5860	.2172 6735	.3896 3648	.4925 4847	.8103 1105	.9102 7688	**.4140**
.5861	75 2401	.3896 1474	5 3100	2 3872	3 2235	.4139
.5862	77 8068	.3895 9297	5 1351	1 6638	3 6782	.4138
.5863	80 3736	7118	4 9600	0 9401	4 1326	.4137
.5864	82 9406	4937	4 7847	.8100 2163	4 5870	.4136
.5865	85 5078	2752	4 6091	.8099 4923	5 0412	.4135
.5866	88 0751	.3895 0565	4 4334	8 7681	5 4953	.4134
.5867	90 6425	.3894 8376	4 2574	8 0437	5 9492	.4133
.5868	93 2101	6184	4 0812	7 3191	6 4030	.4132
.5869	95 7778	3990	3 9049	6 5943	6 8567	.4131
.5870	.2198 3456	.3894 1793	.4923 7283	.8095 8693	.9107 3103	**.4130**
.5871	.2200 9136	.3893 9593	3 5515	5 1442	7 7637	.4129
.5872	03 4818	7391	3 3745	4 4188	8 2169	.4128
.5873	06 0501	5186	3 1972	3 6933	8 6701	.4127
.5874	08 6185	2979	3 0198	2 9676	9 1231	.4126
.5875	11 1871	.3893 0769	2 8422	2 2417	.9109 5760	.4125
.5876	13 7559	.3892 8556	2 6643	1 5156	.9110 0288	.4124
.5877	16 3248	6341	2 4863	0 7893	0 4814	.4123
.5878	18 8938	4124	2 3080	.8090 0628	0 9339	.4122
.5879	21 4629	.3892 1904	2 1295	.8089 3361	1 3862	.4121
.5880	.2224 0323	.3891 9681	.4921 9508	.8088 6093	.9111 8385	**.4120**
.5881	26 6017	7455	1 7719	7 8822	2 2905	.4119
.5882	29 1714	5228	1 5928	7 1550	2 7425	.4118
.5883	31 7411	2997	1 4135	6 4276	3 1943	.4117
.5884	34 3110	.3891 0764	1 2340	5 7000	3 6460	.4116
.5885	36 8811	.3890 8528	1 0543	4 9722	4 0976	.4115
.5886	39 4513	6290	0 8743	4 2442	4 5490	.4114
.5887	42 0216	4050	0 6942	3 5160	5 0003	.4113
.5888	44 5921	.3890 1806	0 5138	2 7876	5 4515	.4112
.5889	47 1628	.3889 9560	0 3332	2 0591	5 9025	.4111
.5890	.2249 7336	.3889 7312	.4920 1524	.8081 3303	.9116 3534	**.4110**
.5891	52 3045	5061	.4919 9714	.8080 6014	6 8042	.4109
.5892	54 8756	2807	9 7902	.8079 8723	7 2548	.4108
.5893	57 4469	.3889 0551	9 6088	9 1430	7 7054	.4107
.5894	60 0183	.3888 8292	9 4272	8 4135	8 1557	.4106
.5895	62 5898	6031	9 2454	7 6838	8 6060	.4105
.5896	65 1615	3767	9 0633	6 9539	9 0561	.4104
.5897	67 7333	.3888 1501	8 8811	6 2238	9 5061	.4103
.5898	70 3053	.3887 9232	8 6986	5 4935	.9119 9559	.4102
.5899	72 8775	6960	8 5159	4 7631	.9120 4056	.4101
.5900	.2275 4498	.3887 4686	.4918 3331	.8074 0324	.9120 8552	**.4100**

$E^{-ii} =$ $E^{il} =$ .0000,0000+   .0000,0000+   .0000,0000+   .0000,0000+   .0000,0000+
.0000,0000,1

Table I

p	x	z	$\sqrt{pq}$	$\sqrt{1-p^2}$	$\sqrt{1-q^2}$	q
.5900	.2275 4498	.3887 4686	.4918 3331	.8074 0324	.9120 8552	**.4100**
.5901	78 0222	2409	8 1500	3 3016	1 3047	.4099
.5902	80 5948	.3887 0130	7 9667	2 5706	1 7540	.4098
.5903	83 1675	.3886 7848	7 7831	1 8394	2 2032	.4097
.5904	85 7404	5564	7 5994	1 1080	2 6522	.4096
.5905	88 3135	3277	7 4155	.8070 3764	3 1012	.4095
.5906	90 8867	.3886 0987	7 2313	.8069 6446	3 5500	.4094
.5907	93 4600	.3885 8695	7 0470	8 9126	3 9986	.4093
.5908	96 0335	6400	6 8624	8 1805	4 4472	.4092
.5909	.2298 6072	4103	6 6776	7 4481	4 8956	.4091
.5910	.2301 1810	.3885 1803	.4916 4927	.8066 7156	.9125 3438	**.4090**
.5911	03 7550	.3884 9500	6 3075	5 9828	5 7920	.4089
.5912	06 3291	7195	6 1220	5 2499	6 2400	.4088
.5913	08 9033	4888	5 9364	4 5168	6 6878	.4087
.5914	11 4778	2578	5 7506	3 7835	7 1356	.4086
.5915	14 0523	.3884 0265	5 5646	3 0500	7 5832	.4085
.5916	16 6271	.3883 7949	5 3783	2 3163	8 0307	.4084
.5917	19 2019	5632	5 1919	1 5824	8 4780	.4083
.5918	21 7770	3311	5 0052	0 8483	8 9252	.4082
.5919	24 3522	.3883 0988	4 8183	.8060 1141	9 3723	.4081
.5920	.2326 9275	.3882 8662	.4914 6312	.8059 3796	.9129 8193	**.4080**
.5921	29 5030	6334	4 4439	8 6450	.9130 2661	.4079
.5922	32 0786	4003	4 2564	7 9102	0 7128	.4078
.5923	34 6544	.3882 1670	4 0687	7 1751	1 1593	.4077
.5924	37 2304	.3881 9334	3 8807	6 4399	1 6058	.4076
.5925	39 8065	6996	3 6926	5 7045	2 0521	.4075
.5926	42 3828	4654	3 5042	4 9689	2 4982	.4074
.5927	44 9592	.3881 2311	3 3157	4 2331	2 9443	.4073
.5928	47 5358	.3880 9965	3 1269	3 4971	3 3902	.4072
.5929	50 1125	7616	2 9379	2 7610	3 8359	.4071
.5930	.2352 6894	.3880 5264	.4912 7487	.8052 0246	.9134 2816	**.4070**
.5931	55 2665	2910	2 5593	1 2880	4 7271	.4069
.5932	57 8437	.3880 0554	2 3697	.8050 5513	5 1725	.4068
.5933	60 4210	.3879 8195	2 1799	.8049 8143	5 6177	.4067
.5934	62 9985	5833	1 9898	9 0772	6 0628	.4066
.5935	65 5762	3469	1 7996	8 3399	6 5078	.4065
.5936	68 1540	.3879 1102	1 6091	7 6024	6 9527	.4064
.5937	70 7320	.3878 8732	1 4184	6 8647	7 3974	.4063
.5938	73 3102	6360	1 2275	6 1268	7 8420	.4062
.5939	75 8885	3986	1 0364	5 3887	8 2864	.4061
.5940	.2378 4670	.3878 1609	.4910 8451	.8044 6504	.9138 7308	**.4060**
.5941	81 0456	.3877 9229	0 6536	3 9119	9 1750	.4059
.5942	83 6244	6846	0 4619	3 1733	.9139 6190	.4058
.5943	86 2033	4462	0 2700	2 4344	.9140 0630	.4057
.5944	88 7824	.3877 2074	.4910 0778	1 6953	0 5068	.4056
.5945	91 3616	.3876 9684	.4909 8854	0 9561	0 9504	.4055
.5946	93 9411	7291	9 6929	.8040 2167	1 3940	.4054
.5947	96 5206	4896	9 5001	8039 4770	1 8374	.4053
.5948	.2399 1004	2498	9 3071	8 7372	2 2807	.4052
.5949	.2401 6803	.3876 0098	9 1139	7 9972	2 7238	.4051
.5950	.2404 2603	.3875 7695	.4908 9205	.8037 2570	.9143 1668	**.4050**

$E^{-ii}=$ $E^{ii}=$ .0000,0000+ .0000,0000+ .0000,0000+ .0000,0000+ .0000,0000+
.000,0000,1

TABLE I

p	x	z	$\sqrt{pq}$	$\sqrt{1-p^2}$	$\sqrt{1-q^2}$	q
.5950	.2404 2603	.3875 7695	.4908 9205	.8037 2570	.9143 1668	.4050
.5951	06 8405	5289	8 7268	6 5166	3 6097	.4049
.5952	09 4209	2881	8 5330	5 7760	4 0525	.4048
.5953	12 0014	.3875 0471	8 3389	5 0352	4 4951	.4047
.5954	14 5821	.3874 8057	8 1447	4 2942	4 9376	.4046
.5955	17 1630	5641	7 9502	3 5531	5 3800	.4045
.5956	19 7440	3223	7 7555	2 8117	5 8222	.4044
.5957	22 3252	.3874 0802	7 5606	2 0702	6 2643	.4043
.5958	24 9065	.3873 8378	7 3655	1 3284	6 7063	.4042
.5959	27 4880	5952	7 1702	.8030 5865	7 1481	.4041
.5960	.2430 0697	.3873 3523	.4906 9746	.8029 8443	.9147 5898	.4040
.5961	32 6515	.3873 1092	6 7789	9 1020	8 0314	.4039
.5962	35 2335	.3872 8658	6 5829	8 3595	8 4729	.4038
.5963	37 8156	6221	6 3868	7 6168	8 9142	.4037
.5964	40 3979	3782	6 1904	6 8739	9 3554	.4036
.5965	42 9804	.3872 1341	5 9938	6 1308	.9149 7964	.4035
.5966	45 5631	.3871 8896	5 7970	5 3875	.9150 2374	.4034
.5967	48 1459	6450	5 6000	4 6440	0 6782	.4033
.5968	50 7288	4000	5 4027	3 9003	1 1188	.4032
.5969	53 3119	.3871 1548	5 2053	3 1564	1 5594	.4031
.5970	.2455 8952	.3870 9093	.4905 0076	.8022 4124	.9151 9998	.4030
.5971	58 4787	6636	4 8098	1 6681	2 4401	.4029
.5972	61 0623	4177	4 6117	0 9236	2 8802	.4028
.5973	63 6461	.3870 1714	4 4134	.8020 1790	3 3202	.4027
.5974	66 2300	.3869 9249	4 2149	.8019 4341	3 7601	.4026
.5975	68 8141	6782	4 0162	8 6891	4 1999	.4025
.5976	71 3984	4312	3 8173	7 9439	4 6395	.4024
.5977	73 9829	.3869 1839	3 6182	7 1985	5 0790	.4023
.5978	76 5675	.3868 9364	3 4188	6 4528	5 5183	.4022
.5979	79 1522	6886	3 2192	5 7070	5 9576	.4021
.5980	.2481 7372	.3868 4405	.4903 0195	.8014 9610	.9156 3967	.4020
.5981	84 3223	.3868 1922	2 8195	4 2148	6 8356	.4019
.5982	86 9076	.3867 9437	2 6193	3 4684	7 2745	.4018
.5983	89 4930	6948	2 4189	2 7218	7 7132	.4017
.5984	92 0786	4458	2 2183	1 9750	8 1518	.4016
.5985	94 6644	.3867 1964	2 0174	1 2281	8 5902	.4015
.5986	97 2503	.3866 9468	1 8164	.8010 4809	9 0286	.4014
.5987	.2499 8364	6970	1 6151	.8009 7335	9 4667	.4013
.5988	.2502 4227	4469	1 4137	8 9860	.9159 9048	.4012
.5989	05 0091	.3866 1965	1 2120	8 2382	.9160 3427	.4011
.5990	.2507 5957	.3865 9459	.4901 0101	.8007 4902	.9160 7805	.4010
.5991	10 1825	6950	0 8080	6 7421	1 2182	.4009
.5992	12 7694	4438	0 6057	5 9938	1 6557	.4008
.5993	15 3565	.3865 1924	0 4031	5 2452	2 0932	.4007
.5994	17 9438	.3864 9408	.4900 2004	4 4965	2 5304	.4006
.5995	20 5313	6888	.4899 9974	3 7476	2 9676	.4005
.5996	23 1189	4367	9 7943	2 9984	3 4046	.4004
.5997	25 7066	.3864 1842	9 5909	2 2491	3 8415	.4003
.5998	28 2946	.3863 9315	9 3873	1 4996	4 2783	.4002
.5999	30 8827	6786	9 1835	0 7499	4 7149	.4001
.6000	.2533 4710	.3863 4253	.4898 9795	.8000 0000	.9165 1514	.4000

$E^{-ii} =$
.0000,0000,1 $E^{ii} =$ .0000,0000+   .0000,0000+   .0000,0000+   .0000,0000+   .0000,0000+

TABLE I

.6000 .4000

p	x	z	$\sqrt{pq}$	$\sqrt{1-p^2}$	$\sqrt{1-q^2}$	q
.6000	.2533 4710	.3863 4253	.4898 9795	.8000 0000	.9165 1514	.4000
.6001	36 0595	3 1719	8 7753	.7999 2499	5 5878	.3999
.6002	38 6481	2 9181	8 5708	8 4996	6 0240	.3998
.6003	41 2369	2 6641	8 3662	7 7491	6 4601	.3997
.6004	43 8259	2 4099	8 1613	6 9984	6 8961	.3996
.6005	46 4150	2 1554	7 9562	6 2476	7 3319	.3995
.6006	49 0043	1 9006	7 7509	5 4965	7 7677	.3994
.6007	51 5938	1 6456	7 5454	4 7452	8 2033	.3993
.6008	54 1835	1 3903	7 3397	3 9937	8 6387	.3992
.6009	56 7733	1 1347	7 1338	3 2421	9 0741	.3991
.6010	.2559 3633	.3860 8789	.4896 9276	.7992 4902	.9169 5093	.3990
.6011	61 9535	0 6229	6 7212	1 7382	.9169 9443	.3989
.6012	64 5438	0 3665	6 5147	0 9859	.9170 3793	.3988
.6013	67 1343	.3860 1099	6 3079	.7990 2335	0 8141	.3987
.6014	69 7250	.3859 8531	6 1009	.7989 4808	1 2488	.3986
.6015	72 3159	9 5960	5 8937	8 7280	1 6833	.3985
.6016	74 9069	9 3386	5 6863	7 9750	2 1177	.3984
.6017	77 4981	9 0810	5 4786	7 2217	2 5520	.3983
.6018	80 0895	8 8231	5 2708	6 4683	2 9862	.3982
.6019	82 6811	8 5650	5 0627	5 7147	3 4202	.3981
.6020	.2585 2728	.3858 3066	.4894 8544	.7984 9609	.9173 8542	.3980
.6021	87 8647	8 0479	4 6460	4 2068	4 2879	.3979
.6022	90 4568	7 7890	4 4373	3 4526	4 7216	.3978
.6023	93 0490	7 5299	4 2283	2 6982	5 1551	.3977
.6024	95 6414	7 2704	4 0192	1 9436	5 5885	.3976
.6025	.2598 2340	7 0107	3 8099	1 1888	6 0217	.3975
.6026	.2600 8268	6 7508	3 6003	.7980 4338	6 4549	.3974
.6027	03 4197	6 4906	3 3905	.7979 6786	6 8879	.3973
.6028	06 0128	6 2301	3 1806	8 9232	7 3207	.3972
.6029	08 6061	5 9694	2 9704	8 1676	7 7535	.3971
.6030	.2611 1996	.3855 7084	.4892 7600	.7977 4119	.9178 1861	.3970
.6031	13 7932	5 4471	2 5493	6 6559	8 6186	.3969
.6032	16 3871	5 1856	2 3385	5 8997	9 0509	.3968
.6033	18 9811	4 9238	2 1275	5 1433	9 4832	.3967
.6034	21 5752	4 6618	1 9162	4 3867	.9179 9153	.3966
.6035	24 1696	4 3995	1 7047	3 6300	.9180 3472	.3965
.6036	26 7641	4 1370	1 4930	2 8730	0 7791	.3964
.6037	29 3588	3 8742	1 2811	2 1158	1 2108	.3963
.6038	31 9537	3 6111	1 0690	1 3585	1 6423	.3962
.6039	34 5488	3 3478	0 8567	.7970 6009	2 0738	.3961
.6040	.2637 1440	.3853 0842	.4890 6441	.7969 8432	.9182 5051	.3960
.6041	39 7394	2 8204	0 4314	9 0852	2 9363	.3959
.6042	42 3350	2 5563	0 2184	8 3271	3 3674	.3958
.6043	44 9308	2 2919	.4890 0052	7 5687	3 7983	.3957
.6044	47 5267	2 0273	.4889 7918	6 8102	4 2291	.3956
.6045	50 1228	1 7624	9 5782	6 0514	4 6598	.3955
.6046	52 7191	1 4972	9 3644	5 2925	5 0903	.3954
.6047	55 3156	1 2318	9 1503	4 5333	5 5207	.3953
.6048	57 9122	0 9662	8 9361	3 7740	5 9510	.3952
.6049	60 5091	0 7003	8 7216	3 0144	6 3812	.3951
.6050	.2663 1061	.3850 4341	.4888 5069	.7962 2547	.9186 8112	.3950

E^{-ii}= E^{ii}=.0000,0000+ .0000,0000+ .0000,0000+ .0000,0000+ .0000,0000+
.0000,0000,1

TABLE I

p	x	z	$\sqrt{pq}$	$\sqrt{1-p^2}$	$\sqrt{1-q^2}$	q
.6050	.2663 1061	.3850 4341	.4888 5069	.7962 2547	.9186 8112	**.3950**
.6051	65 7033	.3850 1676	8 2920	1 4948	7 2411	.3949
.6052	68 3007	.3849 9009	8 0769	.7960 7346	7 6709	.3948
.6053	70 8982	9 6340	7 8616	.7959 9743	8 1005	.3947
.6054	73 4960	9 3668	7 6461	9 2138	8 5300	.3946
.6055	76 0939	9 0993	7 4303	8 4531	8 9594	.3945
.6056	78 6920	8 8315	7 2143	7 6921	9 3887	.3944
.6057	81 2903	8 5635	6 9982	6 9310	.9189 8178	.3943
.6058	83 8887	8 2953	6 7818	6 1697	.9190 2468	.3942
.6059	86 4874	8 0268	6 5652	5 4082	0 6757	.3941
.6060	.2689 0862	.3847 7580	.4886 3483	.7954 6464	.9191 1044	**.3940**
.6061	91 6852	7 4889	6 1313	3 8845	1 5330	.3939
.6062	94 2844	7 2196	5 9140	3 1224	1 9615	.3938
.6063	96 8838	6 9501	5 6966	2 3601	2 3898	.3937
.6064	.2699 4833	6 6803	5 4789	1 5976	2 8181	.3936
.6065	.2702 0831	6 4102	5 2610	0 8349	3 2462	.3935
.6066	04 6830	6 1398	5 0429	.7950 0719	3 6741	.3934
.6067	07 2831	5 8693	4 8246	.7949 3088	4 1020	.3933
.6068	09 8834	5 5984	4 6060	8 5455	4 5297	.3932
.6069	12 4838	5 3273	4 3873	7 7820	4 9573	.3931
.6070	.2715 0845	.3845 0559	.4884 1683	.7947 0183	.9195 3847	**.3930**
.6071	17 6853	4 7843	3 9491	6 2544	5 8120	.3929
.6072	20 2864	4 5124	3 7297	5 4903	6 2392	.3928
.6073	22 8876	4 2402	3 5101	4 7260	6 6663	.3927
.6074	25 4890	3 9678	3 2903	3 9615	7 0932	.3926
.6075	28 0905	3 6951	3 0702	3 1968	7 5200	.3925
.6076	30 6923	3 4222	2 8500	2 4319	7 9467	.3924
.6077	33 2943	3 1490	2 6295	1 6668	8 3733	.3923
.6078	35 8964	2 8755	2 4088	0 9015	8 7997	.3922
.6079	38 4987	2 6018	2 1879	.7940 1360	9 2260	.3921
.6080	.2741 1012	.3842 3278	.4881 9668	.7939 3703	.9199 6522	**.3920**
.6081	43 7039	2 0536	1 7455	8 6043	.9200 0782	.3919
.6082	46 3068	1 7791	1 5239	7 8382	0 5041	.3918
.6083	48 9098	1 5043	1 3022	7 0719	0 9299	.3917
.6084	51 5130	1 2293	1 0802	6 3054	1 3556	.3916
.6085	54 1165	0 9540	0 8580	5 5387	1 7811	.3915
.6086	56 7201	0 6785	0 6356	4 7718	2 2065	.3914
.6087	59 3239	0 4027	0 4130	4 0047	2 6317	.3913
.6088	61 9279	.3840 1266	.4880 1902	3 2374	3 0569	.3912
.6089	64 5320	.3839 8503	.4879 9671	2 4699	3 4819	.3911
.6090	.2767 1364	.3839 5737	.4879 7438	.7931 7022	.9203 9068	**.3910**
.6091	69 7409	9 2968	9 5204	0 9343	4 3315	.3909
.6092	72 3457	9 0197	9 2967	.7930 1662	4 7562	.3908
.6093	74 9506	8 7424	9 0728	.7929 3979	5 1807	.3907
.6094	77 5557	8 4647	8 8486	8 6294	5 6050	.3906
.6095	80 1610	8 1869	8 6243	7 8607	6 0293	.3905
.6096	82 7665	7 9087	8 3997	7 0918	6 4534	.3904
.6097	85 3721	7 6303	8 1750	6 3227	6 8774	.3903
.6098	87 9780	7 3516	7 9500	5 5534	7 3012	.3902
.6099	90 5841	7 0727	7 7248	4 7838	7 7250	.3901
.6100	.2793 1903	.3836 7935	.4877 4994	.7924 0141	.9208 1486	**.3900**

$E^{-ii} = .000,000,1$ $E^{ii} = .0000,0000+$ .0000,0000+   .0000,0000+   .0000,0000+   .0000,0000+

TABLE I

p	x	z	$\sqrt{pq}$	$\sqrt{1-p^2}$	$\sqrt{1-q^2}$	q
.6100	.2793 1903	.3836 7935	.4877 4994	.7924 0141	.9208 1486	**.3900**
.6101	95 7967	6 5141	7 2737	3 2442	8 5720	.3899
.6102	.2798 4034	6 2344	7 0479	2 4741	8 9954	.3898
.6103	.2801 0102	5 9544	6 8218	1 7038	9 4186	.3897
.6104	03 6172	5 6742	6 5955	0 9333	.9209 8417	.3896
.6105	06 2244	5 3937	6 3690	.7920 1626	.9210 2647	.3895
.6106	08 8318	5 1129	6 1423	.7919 3916	0 6875	.3894
.6107	11 4394	4 8319	5 9154	8 6205	1 1102	.3893
.6108	14 0472	4 5506	5 6883	7 8492	1 5328	.3892
.6109	16 6551	4 2691	5 4609	7 0777	1 9552	.3891
.6110	.2819 2633	.3833 9873	.4875 2333	.7916 3060	.9212 3775	**.3890**
.6111	21 8716	3 7052	5 0055	5 5340	2 7997	.3889
.6112	24 4802	3 4229	4 7775	4 7619	3 2218	.3888
.6113	27 0889	3 1403	4 5493	3 9896	3 6437	.3887
.6114	29 6978	2 8575	4 3209	3 2170	4 0656	.3886
.6115	32 3069	2 5744	4 0922	2 4443	4 4872	.3885
.6116	34 9162	2 2910	3 8634	1 6714	4 9088	.3884
.6117	37 5257	2 0074	3 6343	0 8982	5 3302	.3883
.6118	40 1354	1 7235	3 4050	.7910 1249	5 7515	.3882
.6119	42 7453	1 4394	3 1755	.7909 3514	6 1727	.3881
.6120	.2845 3554	.3831 1550	.4872 9457	.7908 5776	.9216 5937	**.3880**
.6121	47 9657	0 8703	2 7158	7 8037	7 0146	.3879
.6122	50 5761	0 5854	2 4856	7 0295	7 4354	.3878
.6123	53 1868	0 3002	2 2552	6 2552	7 8561	.3877
.6124	55 7977	.3830 0148	2 0246	5 4806	8 2766	.3876
.6125	58 4087	.3829 7291	1 7938	4 7059	8 6970	.3875
.6126	61 0200	9 4431	1 5628	3 9309	9 1173	.3874
.6127	63 6314	9 1568	1 3315	3 1558	9 5375	.3873
.6128	66 2430	8 8704	1 1001	2 3804	.9219 9575	.3872
.6129	68 8549	8 5836	0 8684	1 6048	.9220 3774	.3871
.6130	.2871 4669	.3828 2966	.4870 6365	.7900 8291	.9220 7971	**.3870**
.6131	74 0791	8 0093	0 4044	.7900 0531	1 2168	.3869
.6132	76 6916	7 7218	.4870 1721	.7899 2769	1 6363	.3868
.6133	79 3042	7 4340	.4869 9395	8 5006	2 0557	.3867
.6134	81 9170	7 1459	9 7068	7 7240	2 4749	.3866
.6135	84 5300	6 8576	9 4738	6 9472	2 8941	.3865
.6136	87 1433	6 5690	9 2406	6 1702	3 3131	.3864
.6137	89 7567	6 2802	9 0072	5 3930	3 7319	.3863
.6138	92 3703	5 9910	8 7736	4 6156	4 1507	.3862
.6139	94 9841	5 7017	8 5397	3 8380	4 5693	.3861
.6140	.2897 5981	.3825 4121	.4868 3057	.7893 0602	.9224 9878	**.3860**
.6141	.2900 2123	5 1222	8 0714	2 2822	5 4062	.3859
.6142	02 8267	4 8320	7 8369	1 5040	5 8244	.3858
.6143	05 4413	4 5416	7 6022	.7890 7256	6 2425	.3857
.6144	08 0561	4 2509	7 3673	.7889 9470	6 6605	.3856
.6145	10 6711	3 9600	7 1321	9 1682	7 0784	.3855
.6146	13 2863	3 6688	6 8968	8 3892	7 4961	.3854
.6147	15 9016	3 3773	6 6612	7 6100	7 9137	.3853
.6148	18 5172	3 0856	6 4254	6 8305	8 3312	.3852
.6149	21 1330	2 7936	6 1894	6 0509	8 7485	.3851
.6150	.2923 7490	.3822 5014	.4865 9531	.7885 2711	.9229 1657	**.3850**

$E^{-ii} =$ $E^{ii} = .0000,0000+$ $.0000,0000+$ $.0000,0000+$ $.0000,0000+$ $.0000,0000+$
.0000,0000,1

TABLE I

p	x	z	$\sqrt{pq}$	$\sqrt{1-p^2}$	$\sqrt{1-q^2}$	q
.6150	.2923 7490	.3822 5014	.4865 9531	.7885 2711	.9229 1657	**.3850**
.6151	26 3652	2 2089	5 7167	4 4910	9 5828	.3849
.6152	28 9816	1 9161	5 4800	3 7108	.9229 9998	.3848
.6153	31 5982	1 6231	5 2432	2 9304	.9230 4166	.3847
.6154	34 2149	1 3298	5 0061	2 1497	0 8333	.3846
.6155	36 8319	1 0362	4 7688	1 3689	1 2499	.3845
.6156	39 4491	0 7424	4 5312	.7880 5878	1 6664	.3844
.6157	42 0665	0 4483	4 2935	.7879 8065	2 0827	.3843
.6158	44 6841	.3820 1540	4 0555	9 0251	2 4989	.3842
.6159	47 3019	.3819 8594	3 8173	8 2434	2 9150	.3841
.6160	.2949 9199	.3819 5645	.4863 5789	.7877 4615	.9233 3309	**.3840**
.6161	52 5381	9 2694	3 3403	6 6794	3 7467	.3839
.6162	55 1565	8 9740	3 1015	5 8972	4 1624	.3838
.6163	57 7751	8 6784	2 8624	5 1147	4 5780	.3837
.6164	60 3939	8 3825	2 6232	4 3320	4 9934	.3836
.6165	63 0129	8 0863	2 3837	3 5491	5 4088	.3835
.6166	65 6321	7 7899	2 1440	2 7660	5 8239	.3834
.6167	68 2515	7 4932	1 9041	1 9827	6 2390	.3833
.6168	70 8711	7 1962	1 6639	1 1991	6 6539	.3832
.6169	73 4910	6 8990	1 4236	.7870 4154	7 0687	.3831
.6170	.2976 1110	.3816 6015	.4861 1830	.7869 6315	.9237 4834	**.3830**
.6171	78 7312	6 3038	0 9422	8 8474	7 8980	.3829
.6172	81 3517	6 0058	0 7012	8 0630	8 3124	.3828
.6173	83 9723	5 7075	0 4600	7 2785	8 7267	.3827
.6174	86 5932	5 4090	.4860 2185	6 4938	9 1409	.3826
.6175	89 2142	5 1102	.4859 9769	5 7088	9 5549	.3825
.6176	91 8355	4 8111	9 7350	4 9236	.9239 9688	.3824
.6177	94 4570	4 5118	9 4929	4 1383	.9240 3826	.3823
.6178	97 0786	4 2123	9 2506	3 3527	0 7963	.3822
.6179	.2999 7005	3 9124	9 0080	2 5669	1 2098	.3821
.6180	.3002 3226	.3813 6123	.4858 7653	.7861 7810	.9241 6232	**.3820**
.6181	04 9449	3 3120	8 5223	0 9948	2 0365	.3819
.6182	07 5674	3 0113	8 2791	.7860 2084	2 4497	.3818
.6183	10 1901	2 7104	8 0357	.7859 4218	2 8627	.3817
.6184	12 8130	2 4093	7 7921	8 6350	3 2756	.3816
.6185	15 4361	2 1079	7 5482	7 8480	3 6884	.3815
.6186	18 0594	1 8062	7 3042	7 0608	4 1010	.3814
.6187	20 6830	1 5043	7 0599	6 2734	4 5136	.3813
.6188	23 3067	1 2021	6 8154	5 4857	4 9260	.3812
.6189	25 9306	0 8996	6 5707	4 6979	5 3382	.3811
.6190	.3028 5548	.3810 5969	.4856 3258	.7853 9099	.9245 7504	**.3810**
.6191	31 1792	.3810 2939	6 0806	3 1216	6 1624	.3809
.6192	33 8037	.3809 9906	5 8353	2 3332	6 5743	.3808
.6193	36 4285	9 6871	5 5897	1 5445	6 9860	.3807
.6194	39 0535	9 3834	5 3439	.7850 7556	7 3977	.3806
.6195	41 6787	9 0793	5 0978	.7849 9666	7 8092	.3805
.6196	44 3041	8 7750	4 8516	9 1773	8 2206	.3804
.6197	46 9298	8 4705	4 6051	8 3878	8 6318	.3803
.6198	49 5556	8 1656	4 3585	7 5981	9 0430	.3802
.6199	52 1816	7 8606	4 1116	6 8082	9 4540	.3801
.6200	.3054 8079	.3807 5552	.4853 8644	.7846 0181	.9249 8649	**.3800**

$E^{-ii} =$ .0000,0000,1 $E^{ii} =$ .0000,0000+   .0000,0000+   .0000,0000+   .0000,0000+   .0000,0000+

TABLE I

.6200   **.3800**

p	x	z	$\sqrt{pq}$	$\sqrt{1-p^2}$	$\sqrt{1-q^2}$	q
.6200	.3054 8079	.3807 5552	.4853 8644	.7846 0181	.9249 8649	**.3800**
.6201	57 4344	7 2496	3 6171	5 2278	.9250 2756	.3799
.6202	60 0610	6 9437	3 3696	4 4373	0 6862	.3798
.6203	62 6879	6 6376	3 1218	3 6465	1 0967	.3797
.6204	65 3150	6 3312	2 8738	2 8556	1 5071	.3796
.6205	67 9423	6 0245	2 6256	2 0645	1 9174	.3795
.6206	70 5698	5 7176	2 3771	1 2731	2 3275	.3794
.6207	73 1975	5 4104	2 1285	.7840 4816	2 7375	.3793
.6208	75 8255	5 1029	1 8796	.7839 7898	3 1474	.3792
.6209	78 4536	4 7952	1 6306	8 9978	3 5571	.3791
.6210	.3081 0820	.3804 4873	.4851 3812	.7838 1056	.9253 9667	**.3790**
.6211	83 7106	4 1790	1 1317	7 3133	4 3762	.3789
.6212	86 3394	3 8705	0 8820	6 5207	4 7856	.3788
.6213	88 9684	3 5618	0 6320	5 7279	5 1948	.3787
.6214	91 5976	3 2527	0 3818	4 9348	5 6039	.3786
.6215	94 2271	2 9434	.4850 1314	4 1416	6 0129	.3785
.6216	96 8567	2 6339	.4849 8808	3 3482	6 4218	.3784
.6217	.3099 4866	2 3241	9 6300	2 5546	6 8305	.3783
.6218	.3102 1167	2 0140	9 3789	1 7607	7 2391	.3782
.6219	04 7470	1 7036	9 1277	0 9667	7 6476	.3781
.6220	.3107 3775	.3801 3930	.4848 8762	.7830 1724	.9258 0560	**.3780**
.6221	10 0082	1 0822	8 6244	.7829 3779	8 4642	.3779
.6222	12 6392	0 7710	8 3725	8 5833	8 8723	.3778
.6223	15 2703	0 4596	8 1204	7 7884	9 2803	.3777
.6224	17 9017	.3800 1480	7 8680	6 9933	.9259 6881	.3776
.6225	20 5333	.3799 8361	7 6154	6 1980	.9260 0958	.3775
.6226	23 1651	9 5239	7 3626	5 4025	0 5034	.3774
.6227	25 7971	9 2114	7 1096	4 6068	0 9109	.3773
.6228	28 4293	8 8987	6 8563	3 8108	1 3183	.3772
.6229	31 0617	8 5857	6 6028	3 0147	1 7255	.3771
.6230	.3133 6944	.3798 2725	.4846 3491	.7822 2184	.9262 1326	**.3770**
.6231	36 3273	7 9590	6 0952	1 4218	2 5396	.3769
.6232	38 9604	7 6452	5 8411	.7820 6250	2 9464	.3768
.6233	41 5937	7 3312	5 5868	.7819 8281	3 3531	.3767
.6234	44 2272	7 0169	5 3322	9 0309	3 7597	.3766
.6235	46 8610	6 7024	5 0774	8 2335	4 1662	.3765
.6236	49 4949	6 3875	4 8224	7 4359	4 5725	.3764
.6237	52 1291	6 0725	4 5672	6 6381	4 9787	.3763
.6238	54 7635	5 7571	4 3117	5 8401	5 3848	.3762
.6239	57 3982	5 4415	4 0560	5 0418	5 7908	.3761
.6240	.3160 0330	.3795 1256	.4843 8002	.7814 2434	.9266 1966	**.3760**
.6241	62 6681	4 8095	3 5441	3 4448	6 6023	.3759
.6242	65 3034	4 4931	3 2877	2 6459	7 0079	.3758
.6243	67 9389	4 1764	3 0312	1 8468	7 4134	.3757
.6244	70 5746	3 8595	2 7744	1 0476	7 8187	.3756
.6245	73 2105	3 5423	2 5174	.7810 2481	8 2239	.3755
.6246	75 8467	3 2249	2 2602	.7809 4484	8 6290	.3754
.6247	78 4831	2 9072	2 0028	8 6485	9 0340	.3753
.6248	81 1197	2 5892	1 7451	7 8484	9 4388	.3752
.6249	83 7565	2 2709	1 4873	7 0480	.9269 8435	.3751
.6250	.3186 3936	.3791 9524	.4841 2292	.7806 2475	.9270 2481	**.3750**

$\varepsilon^{-ii}=$ $\varepsilon^{ii}=$.0000,0000+ .0000,0000+   .0000,0000+   .0000,0000+   .0000,0000+
.0000,0000,1

TABLE I

.6250   **.3750**

p	x	z	$\sqrt{pq}$	$\sqrt{1-p^2}$	$\sqrt{1-q^2}$	q
.6250	.3186 3936	.3791 9524	.4841 2292	.7806 2475	.9270 2481	**.3750**
.6251	89 0309	1 6337	0 9709	5 4468	0 6526	.3749
.6252	91 6684	1 3146	0 7123	4 6458	1 0569	.3748
.6253	94 3061	0 9953	0 4536	3 8446	1 4611	.3747
.6254	96 9441	0 6758	.4840 1946	3 0433	1 8652	.3746
.6255	.3199 5822	0 3559	.4839 9354	2 2417	2 2691	.3745
.6256	.3202 2206	.3790 0358	9 6760	1 4399	2 6730	.3744
.6257	04 8592	.3789 7155	9 4164	.7800 6379	3 0767	.3743
.6258	07 4981	9 3949	9 1565	.7799 8356	3 4803	.3742
.6259	10 1371	9 0740	8 8965	9 0332	3 8837	.3741
.6260	.3212 7764	.3788 7528	.4838 6362	.7798 2306	.9274 2870	**.3740**
.6261	15 4159	8 4314	8 3757	7 4277	4 6902	.3739
.6262	18 0556	8 1098	8 1149	6 6247	5 0933	.3738
.6263	20 6956	7 7878	7 8540	5 8214	5 4963	.3737
.6264	23 3357	7 4656	7 5928	5 0179	5 8991	.3736
.6265	25 9761	7 1432	7 3314	4 2142	6 3018	.3735
.6266	28 6168	6 8204	7 0698	3 4103	6 7044	.3734
.6267	31 2576	6 4974	6 8079	2 6062	7 1068	.3733
.6268	33 8987	6 1742	6 5459	1 8018	7 5091	.3732
.6269	36 5400	5 8506	6 2836	0 9973	7 9113	.3731
.6270	.3239 1815	.3785 5269	.4836 0211	.7790 1926	.9278 3134	**.3730**
.6271	41 8233	5 2028	5 7584	.7789 3876	8 7154	.3729
.6272	44 4653	4 8785	5 4954	8 5824	9 1172	.3728
.6273	47 1075	4 5539	5 2323	7 7770	9 5189	.3727
.6274	49 7499	4 2291	4 9689	6 9714	.9279 9205	.3726
.6275	52 3926	3 9040	4 7053	6 1656	.9280 3219	.3725
.6276	55 0355	3 5786	4 4414	5 3596	0 7232	.3724
.6277	57 6786	3 2530	4 1774	4 5534	1 1244	.3723
.6278	60 3219	2 9271	3 9131	3 7469	1 5255	.3722
.6279	62 9655	2 6009	3 6486	2 9403	1 9265	.3721
.6280	.3265 6093	.3782 2745	.4833 3839	.7782 1334	.9282 3273	**.3720**
.6281	68 2533	1 9478	3 1190	1 3263	2 7280	.3719
.6282	70 8976	1 6208	2 8538	.7780 5190	3 1286	.3718
.6283	73 5420	1 2936	2 5884	.7779 7115	3 5290	.3717
.6284	76 1867	0 9661	2 3228	8 9038	3 9293	.3716
.6285	78 8317	0 6384	2 0570	8 0958	4 3295	.3715
.6286	81 4768	.3780 3103	1 7910	7 2877	4 7296	.3714
.6287	84 1222	.3779 9821	1 5247	6 4793	5 1296	.3713
.6288	86 7679	9 6535	1 2582	5 6708	5 5294	.3712
.6289	89 4137	9 3247	0 9915	4 8620	5 9291	.3711
.6290	.3292 0598	.3778 9956	.4830 7246	.7774 0530	.9286 3287	**.3710**
.6291	94 7061	8 6663	0 4574	3 2438	6 7281	.3709
.6292	97 3527	8 3367	.4830 1901	2 4344	7 1274	.3708
.6293	.3299 9995	8 0068	.4829 9225	1 6247	7 5266	.3707
.6294	.3302 6465	7 6767	9 6546	0 8149	7 9257	.3706
.6295	05 2938	7 3463	9 3866	.7770 0048	8 3247	.3705
.6296	07 9412	7 0156	9 1183	.7769 1946	8 7235	.3704
.6297	10 5889	6 6847	8 8499	8 3841	9 1222	.3703
.6298	13 2369	6 3535	8 5812	7 5734	9 5208	.3702
.6299	15 8851	6 0221	8 3122	6 7625	.9289 9192	.3701
.6300	.3318 5335	.3775 6903	.4828 0431	.7765 9513	.9290 3175	**.3700**

$E^{-ii}=$ $E^{ii}=$.0000,0000+   .0000,0000+   .0000,0000+   .0000,0000+   .0000,0000+
.0000,00001

TABLE I

.6300   **.3700**

p	x	z	$\sqrt{pq}$	$\sqrt{1-p^2}$	$\sqrt{1-q^2}$	q
.6300	.3318 5335	.3775 6903	.4828 0431	.7765 9513	.9290 3175	**.3700**
.6301	21 1821	5 3584	7 7737	5 1400	0 7157	.3699
.6302	23 8310	5 0261	7 5041	4 3284	1 1138	.3698
.6303	26 4801	4 6936	7 2343	3 5167	1 5118	.3697
.6304	29 1295	4 3608	6 9643	2 7047	1 9096	.3696
.6305	31 7790	4 0278	6 6940	1 8925	2 3073	.3695
.6306	34 4288	3 6945	6 4235	1 0801	2 7049	.3694
.6307	37 0789	3 3609	6 1528	.7760 2675	3 1023	.3693
.6308	39 7292	3 0270	5 8819	.7759 4546	3 4997	.3692
.6309	42 3797	2 6929	5 6107	8 6416	3 8969	.3691
.6310	.3345 0304	.3772 3586	.4825 3394	.7757 8283	.9294 2939	**.3690**
.6311	47 6814	2 0239	5 0678	7 0148	4 6909	.3689
.6312	50 3326	1 6890	4 7960	6 2011	5 0877	.3688
.6313	52 9840	1 3539	4 5239	5 3872	5 4844	.3687
.6314	55 6357	1 0184	4 2517	4 5731	5 8810	.3686
.6315	58 2876	0 6827	3 9792	3 7588	6 2775	.3685
.6316	60 9398	0 3468	3 7065	2 9442	6 6738	.3684
.6317	63 5922	.3770 0105	3 4335	2 1294	7 0700	.3683
.6318	66 2448	.3769 6741	3 1604	1 3145	7 4661	.3682
.6319	68 8977	9 3373	2 8870	.7750 4993	7 8621	.3681
.6320	.3371 5508	.3769 0003	.4822 6134	.7749 6839	.9298 2579	**.3680**
.6321	74 2041	8 6630	2 3396	8 8682	8 6536	.3679
.6322	76 8577	8 3254	2 0655	8 0524	9 0492	.3678
.6323	79 5115	7 9876	1 7913	7 2363	9 4447	.3677
.6324	82 1656	7 6495	1 5168	6 4201	.9299 8400	.3676
.6325	84 8199	7 3112	1 2421	5 6036	.9300 2352	.3675
.6326	87 4744	6 9726	0 9671	4 7869	0 6303	.3674
.6327	90 1292	6 6337	0 6920	3 9700	1 0253	.3673
.6328	92 7842	6 2945	0 4166	3 1528	1 4201	.3672
.6329	95 4394	5 9551	.4820 1410	2 3355	1 8148	.3671
.6330	.3398 0949	.3765 6155	.4819 8651	.7741 5179	.9302 2094	**.3670**
.6331	.3400 7506	5 2755	9 5891	.7740 7002	2 6039	.3669
.6332	03 4066	4 9353	9 3128	.7739 8822	2 9982	.3668
.6333	06 0628	4 5948	9 0363	9 0640	3 3924	.3667
.6334	08 7192	4 2541	8 7596	8 2455	3 7865	.3666
.6335	11 3759	3 9131	8 4826	7 4269	4 1805	.3665
.6336	14 0329	3 5718	8 2055	6 6080	4 5744	.3664
.6337	16 6901	3 2303	7 9281	5 7890	4 9681	.3663
.6338	19 3475	2 8885	7 6505	4 9697	5 3617	.3662
.6339	22 0051	2 5464	7 3726	4 1502	5 7552	.3661
.6340	.3424 6630	.3762 2041	.4817 0946	.7733 3305	.9306 1485	**.3660**
.6341	27 3211	1 8615	6 8163	2 5105	6 5417	.3659
.6342	29 9795	1 5186	6 5378	1 6904	6 9348	.3658
.6343	32 6381	1 1755	6 2590	0 8700	7 3278	.3657
.6344	35 2970	0 8321	5 9801	.7730 0494	7 7207	.3656
.6345	37 9561	0 4884	5 7009	.7729 2286	8 1134	.3655
.6346	40 6155	.3760 1445	5 4215	8 4076	8 5060	.3654
.6347	43 2751	.3759 8003	5 1418	7 5864	8 8985	.3653
.6348	45 9349	9 4558	4 8620	6 7649	9 2908	.3652
.6349	48 5950	9 1111	4 5819	5 9432	.9309 6831	.3651
.6350	.3451 2553	.3758 7661	.4814 3016	.7725 1214	.9310 0752	**.3650**

$E^{-ii} =$
.0000,0000,1
   $E^{ii} =$ .0000,0000+   .0000,0000+   .0000,0000+   .0000,0000+   .0000,0000+

TABLE I

.6350   .3650

p	x	z	$\sqrt{pq}$	$\sqrt{1-p^2}$	$\sqrt{1-q^2}$	q
.6350	.3451 2553	.3758 7661	.4814 3016	.7725 1214	.9310 0752	**.3650**
.6351	53 9159	8 4209	4 0211	4 2993	0 4672	.3649
.6352	56 5767	8 0753	3 7403	3 4769	0 8590	.3648
.6353	59 2377	7 7295	3 4594	2 6544	1 2508	.3647
.6354	61 8990	7 3835	3 1782	1 8316	1 6424	.3646
.6355	64 5606	7 0372	2 8967	1 0087	2 0339	.3645
.6356	67 2224	6 6906	2 6151	.7720 1855	2 4252	.3644
.6357	69 8844	6 3437	2 3332	.7719 3621	2 8165	.3643
.6358	72 5467	5 9966	2 0511	8 5385	3 2076	.3642
.6359	75 2092	5 6492	1 7688	7 7146	3 5986	.3641
.6360	.3477 8720	.3755 3016	.4811 4863	.7716 8906	.9313 9895	**.3640**
.6361	80 5350	4 9536	1 2035	6 0663	4 3802	.3639
.6362	83 1983	4 6055	0 9205	5 2418	4 7709	.3638
.6363	85 8618	4 2570	0 6373	4 4171	5 1614	.3637
.6364	88 5256	3 9083	0 3538	3 5922	5 5517	.3636
.6365	91 1896	3 5593	.4810 0702	2 7670	5 9420	.3635
.6366	93 8539	3 2100	.4809 7863	1 9416	6 3321	.3634
.6367	96 5184	2 8605	9 5022	1 1161	6 7221	.3633
.6368	.3499 1831	2 5107	9 2178	.7710 2903	7 1120	.3632
.6369	.3501 8481	2 1607	8 9332	.7709 4642	7 5018	.3631
.6370	.3504 5134	.3751 8104	.4808 6485	.7708 6380	.9317 8914	**.3630**
.6371	07 1789	1 4598	8 3634	7 8116	8 2809	.3629
.6372	09 8447	1 1089	8 0782	6 9849	8 6703	.3628
.6373	12 5107	0 7578	7 7927	6 1580	9 0596	.3627
.6374	15 1769	0 4064	7 5070	5 3309	9 4487	.3626
.6375	17 8434	.3750 0548	7 2211	4 5036	.9319 8377	.3625
.6376	20 5102	.3749 7029	6 9350	3 6760	.9320 2266	.3624
.6377	23 1772	9 3507	6 6486	2 8482	0 6154	.3623
.6378	25 8444	8 9982	6 3620	2 0203	1 0040	.3622
.6379	28 5119	8 6455	6 0752	1 1921	1 3925	.3621
.6380	.3531 1797	.3748 2925	.4805 7882	.7700 3636	.9321 7809	**.3620**
.6381	33 8477	7 9393	5 5009	.7699 5350	2 1692	.3619
.6382	36 5160	7 5858	5 2134	8 7061	2 5574	.3618
.6383	39 1845	7 2320	4 9257	7 8770	2 9454	.3617
.6384	41 8532	6 8779	4 6378	7 0477	3 3333	.3616
.6385	44 5222	6 5236	4 3496	6 2182	3 7211	.3615
.6386	47 1915	6 1690	4 0612	5 3885	4 1088	.3614
.6387	49 8610	5 8142	3 7726	4 5585	4 4963	.3613
.6388	52 5308	5 4590	3 4837	3 7284	4 8837	.3612
.6389	55 2008	5 1037	3 1947	2 8980	5 2710	.3611
.6390	.3557 8711	.3744 7480	.4802 9054	.7692 0673	.9325 6582	**.3610**
.6391	60 5416	4 3921	2 6158	1 2365	6 0452	.3609
.6392	63 2124	4 0359	2 3261	.7690 4055	6 4321	.3608
.6393	65 8835	3 6794	2 0361	.7689 5742	6 8189	.3607
.6394	68 5548	3 3227	1 7459	8 7427	7 2056	.3606
.6395	71 2263	2 9657	1 4555	7 9110	7 5921	.3605
.6396	73 8981	2 6085	1 1649	7 0790	7 9786	.3604
.6397	76 5702	2 2509	0 8740	6 2469	8 3649	.3603
.6398	79 2425	1 8932	0 5829	5 4145	8 7510	.3602
.6399	81 9151	1 5351	0 2916	4 5819	9 1371	.3601
.6400	.3584 5879	.3741 1768	.4800 0000	.7683 7491	.9329 5230	**.3600**

$E^{-ii}=$ $E^{ii}=$.0000,0000+   .0000,0000+   .0000,0000+   .0000,0000+   .0000,0000+
.0000,0000,1

TABLE I

p	x	z	√pq	√1−p²	√1−q²	q
.6400	.3584 5879	.3741 1768	.4800 0000	.7683 7491	.9329 5230	.3600
.6401	87 2610	0 8182	.4799 7082	2 9160	.9329 9088	.3599
.6402	89 9343	0 4593	9 4162	2 0828	.9330 2945	.3598
.6403	92 6079	.3740 1002	9 1240	1 2493	0 6801	.3597
.6404	95 2818	.3739 7408	8 8315	.7680 4156	1 0655	.3596
.6405	.3597 9559	9 3811	8 5388	.7679 5817	1 4509	.3595
.6406	.3600 6302	9 0212	8 2459	8 7476	1 8360	.3594
.6407	03 3049	8 6610	7 9528	7 9132	2 2211	.3593
.6408	05 9798	8 3005	7 6594	7 0786	2 6061	.3592
.6409	08 6549	7 9398	7 3658	6 2438	2 9909	.3591
.6410	.3611 3303	.3737 5788	.4797 0720	.7675 4088	.9333 3756	.3590
.6411	14 0060	7 2175	6 7780	4 5735	3 7602	.3589
.6412	16 6819	6 8560	6 4837	3 7381	4 1446	.3588
.6413	19 3581	6 4942	6 1892	2 9024	4 5290	.3587
.6414	22 0345	6 1321	5 8945	2 0665	4 9132	.3586
.6415	24 7112	5 7698	5 5995	1 2303	5 2973	.3585
.6416	27 3882	5 4072	5 3044	.7670 3940	5 6812	.3584
.6417	30 0654	5 0443	5 0090	.7669 5574	6 0651	.3583
.6418	32 7429	4 6812	4 7133	8 7206	6 4488	.3582
.6419	35 4206	4 3178	4 4175	7 8836	6 8324	.3581
.6420	.3638 0986	.3733 9541	.4794 1214	.7667 0464	.9337 2159	.3580
.6421	40 7769	3 5902	3 8251	6 2089	7 5992	.3579
.6422	43 4554	3 2259	3 5286	5 3712	7 9824	.3578
.6423	46 1341	2 8615	3 2318	4 5333	8 3655	.3577
.6424	48 8132	2 4967	2 9348	3 6952	8 7485	.3576
.6425	51 4925	2 1317	2 6376	2 8568	9 1314	.3575
.6426	54 1721	1 7664	2 3401	2 0183	9 5141	.3574
.6427	56 8519	1 4009	2 0425	1 1795	.9339 8967	.3573
.6428	59 5319	1 0351	1 7446	.7660 3405	.9340 2792	.3572
.6429	62 2123	0 6690	1 4464	.7659 5012	0 6616	.3571
.6430	.3664 8929	.3730 3026	.4791 1481	.7658 6618	.9341 0438	.3570
.6431	67 5738	.3729 9360	0 8495	7 8221	1 4260	.3569
.6432	70 2549	9 5691	0 5507	6 9822	1 8080	.3568
.6433	72 9363	9 2019	.4790 2517	6 1420	2 1898	.3567
.6434	75 6180	8 8345	.4789 9524	5 3017	2 5716	.3566
.6435	78 2999	8 4668	9 6529	4 4611	2 9532	.3565
.6436	80 9821	8 0988	9 3532	3 6203	3 3347	.3564
.6437	83 6646	7 7306	9 0532	2 7793	3 7161	.3563
.6438	86 3473	7 3621	8 7531	1 9381	4 0974	.3562
.6439	89 0303	6 9933	8 4527	1 0966	4 4785	.3561
.6440	.3691 7136	.3726 6243	.4788 1520	.7650 2549	.9344 8595	.3560
.6441	94 3971	6 2550	7 8512	.7649 4130	5 2404	.3559
.6442	97 0809	5 8854	7 5501	8 5708	5 6212	.3558
.6443	.3699 7650	5 5156	7 2488	7 7285	6 0019	.3557
.6444	.3702 4493	5 1455	6 9473	6 8859	6 3824	.3556
.6445	05 1339	4 7751	6 6455	6 0431	6 7628	.3555
.6446	07 8188	4 4044	6 3435	5 2001	7 1431	.3554
.6447	10 5039	4 0335	6 0413	4 3568	7 5233	.3553
.6448	13 1893	3 6623	5 7388	3 5133	7 9033	.3552
.6449	15 8750	3 2909	5 4361	2 6696	8 2832	.3551
.6450	.3718 5609	.3722 9192	.4785 1332	.7641 8257	.9348 6630	.3550

$E^{-ii}=$ $E^{ii}=$.0000,0000+ .0000,0000+ .0000,0000+ .0000,0000+ .0000,0000+
.0000,0000,1

TABLE I

.6450   **.3550**

p	x	z	√pq	√(1−p²)	√(1−q²)	q
.6450	.3718 5609	.3722 9192	.4785 1332	.7641 8257	.9348 6630	.3550
.6451	21 2471	2 5472	4 8301	0 9815	9 0427	.3549
.6452	23 9336	2 1749	4 5267	.7640 1372	9 4222	.3548
.6453	26 6203	1 8024	4 2231	.7639 2926	.9349 8017	.3547
.6454	29 3073	1 4296	3 9193	8 4477	.9350 1810	.3546
.6455	31 9946	1 0565	3 6153	7 6027	0 5601	.3545
.6456	34 6821	0 6832	3 3110	6 7574	0 9392	.3544
.6457	37 3699	.3720 3096	3 0065	5 9119	1 3181	.3543
.6458	40 0580	.3719 9357	2 7017	5 0662	1 6970	.3542
.6459	42 7464	9 5616	2 3968	4 2203	2 0757	.3541
.6460	.3745 4350	.3719 1872	.4782 0916	.7633 3741	.9352 4542	**.3540**
.6461	48 1239	8 8125	1 7862	2 5277	2 8327	.3539
.6462	50 8131	8 4376	1 4805	1 6811	3 2110	.3538
.6463	53 5025	8 0623	1 1746	.7630 8342	3 5892	.3537
.6464	56 1922	7 6869	0 8685	.7629 9872	3 9673	.3536
.6465	58 8822	7 3111	0 5622	9 1399	4 3452	.3535
.6466	61 5725	6 9351	.4780 2556	8 2923	4 7231	.3534
.6467	64 2630	6 5588	.4779 9488	7 4446	5 1008	.3533
.6468	66 9538	6 1822	9 6418	6 5966	5 4784	.3532
.6469	69 6449	5 8054	9 3346	5 7484	5 8559	.3531
.6470	.3772 3362	.3715 4283	.4779 0271	.7624 9000	.9356 2332	**.3530**
.6471	75 0278	5 0509	8 7194	4 0514	6 6104	.3529
.6472	77 7197	4 6733	8 4115	3 2025	6 9875	.3528
.6473	80 4119	4 2954	8 1033	2 3534	7 3645	.3527
.6474	83 1043	3 9172	7 7949	1 5041	7 7414	.3526
.6475	85 7970	3 5388	7 4863	.7620 6545	8 1181	.3525
.6476	88 4900	3 1600	7 1774	.7619 8047	8 4948	.3524
.6477	91 1833	2 7811	6 8683	8 9547	8 8712	.3523
.6478	93 8768	2 4018	6 5590	8 1045	9 2476	.3522
.6479	96 5706	2 0223	6 2495	7 2540	.9359 6239	.3521
.6480	.3799 2647	.3711 6425	.4775 9397	.7616 4034	.9360 0000	**.3520**
.6481	.3801 9591	1 2624	5 6297	5 5524	0 3760	.3519
.6482	04 6537	0 8821	5 3195	4 7013	0 7519	.3518
.6483	07 3486	0 5015	5 0090	3 8499	1 1277	.3517
.6484	10 0438	.3710 1206	4 6983	2 9984	1 5033	.3516
.6485	12 7393	.3709 7395	4 3874	2 1465	1 8788	.3515
.6486	15 4350	9 3581	4 0762	1 2945	2 2542	.3514
.6487	18 1311	8 9764	3 7649	.7610 4422	2 6295	.3513
.6488	20 8274	8 5945	3 4533	.7609 5897	3 0046	.3512
.6489	23 5239	8 2122	3 1414	8 7370	3 3797	.3511
.6490	.3826 2208	.3707 8298	.4772 8293	.7607 8841	.9363 7546	**.3510**
.6491	28 9179	7 4470	2 5171	7 0309	4 1294	.3509
.6492	31 6153	7 0640	2 2045	6 1775	4 5040	.3508
.6493	34 3130	6 6807	1 8918	5 3239	4 8786	.3507
.6494	37 0110	6 2971	1 5788	4 4700	5 2530	.3506
.6495	39 7093	5 9133	1 2656	3 6159	5 6273	.3505
.6496	42 4078	5 5292	0 9521	2 7616	6 0015	.3504
.6497	45 1066	5 1448	0 6384	1 9071	6 3756	.3503
.6498	47 8057	4 7601	0 3245	1 0523	6 7495	.3502
.6499	50 5051	4 3752	.4770 0104	.7600 1973	7 1233	.3501
.6500	.3853 2047	.3703 9900	.4769 6960	.7599 3421	.9367 4970	**.3500**

E⁻ⁱⁱ=   Eⁱⁱ= .0000,0000+   .0000,0000+   .0000,0000+   .0000,0000+   .0000,0000+
.0000,0000,1

TABLE I

p	x	z	$\sqrt{pq}$	$\sqrt{1-p^2}$	$\sqrt{1-q^2}$	q
.6500	.3853 2047	.3703 9900	.4769 6960	.7599 3421	.9367 4970	.3500
.6501	55 9046	3 6046	9 3814	8 4866	7 8706	.3499
.6502	58 6048	3 2189	9 0666	7 6309	8 2440	.3498
.6503	61 3053	2 8329	8 7515	6 7750	8 6173	.3497
.6504	64 0061	2 4466	8 4362	5 9189	8 9906	.3496
.6505	66 7072	2 0601	8 1207	5 0625	9 3636	.3495
.6506	69 4085	1 6733	7 8049	4 2059	.9369 7366	.3494
.6507	72 1101	1 2862	7 4890	3 3491	.9370 1094	.3493
.6508	74 8120	0 8988	7 1727	2 4921	0 4822	.3492
.6509	77 5142	0 5112	6 8563	1 6348	0 8548	.3491
.6510	.3880 2167	.3700 1233	.4766 5396	.7590 7773	.9371 2272	.3490
.6511	82 9195	.3699 7352	6 2227	.7589 9196	1 5996	.3489
.6512	85 6225	9 3468	5 9056	9 0616	1 9718	.3488
.6513	88 3258	8 9581	5 5882	8 2034	2 3439	.3487
.6514	91 0294	8 5691	5 2706	7 3450	2 7159	.3486
.6515	93 7333	8 1798	4 9528	6 4863	3 0878	.3485
.6516	96 4375	7 7903	4 6347	5 6275	3 4596	.3484
.6517	.3899 1419	7 4006	4 3164	4 7684	3 8312	.3483
.6518	.3901 8467	7 0105	3 9979	3 9090	4 2027	.3482
.6519	04 5517	6 6202	3 6791	3 0495	4 5741	.3481
.6520	.3907 2570	.3696 2296	.4763 3602	.7582 1897	.9374 9453	.3480
.6521	09 9626	5 8387	3 0409	1 3296	5 3165	.3479
.6522	12 6685	5 4476	2 7215	.7580 4694	5 6875	.3478
.6523	15 3747	5 0562	2 4018	.7579 6089	6 0584	.3477
.6524	18 0811	4 6645	2 0819	8 7482	6 4292	.3476
.6525	20 7879	4 2726	1 7618	7 8872	6 7998	.3475
.6526	23 4949	3 8804	1 4414	7 0261	7 1704	.3474
.6527	26 2022	3 4879	1 1208	6 1647	7 5408	.3473
.6528	28 9098	3 0951	0 7999	5 3030	7 9111	.3472
.6529	31 6177	2 7021	0 4789	4 4412	8 2812	.3471
.6530	.3934 3259	.3692 3088	.4760 1576	.7573 5791	.9378 6513	.3470
.6531	37 0344	1 9152	.4759 8360	2 7168	9 0212	.3469
.6532	39 7431	1 5214	9 5143	1 8542	9 3910	.3468
.6533	42 4522	1 1273	9 1923	0 9914	.9379 7607	.3467
.6534	45 1615	0 7329	8 8700	.7570 1284	.9380 1303	.3466
.6535	47 8712	.3690 3383	8 5476	.7569 2652	0 4997	.3465
.6536	50 5811	.3689 9433	8 2249	8 4017	0 8690	.3464
.6537	53 2913	9 5481	7 9020	7 5380	1 2382	.3463
.6538	56 0018	9 1527	7 5788	6 6740	1 6073	.3462
.6539	58 7126	8 7569	7 2554	5 8099	1 9763	.3461
.6540	.3961 4237	.3688 3609	.4756 9318	.7564 9455	.9382 3451	.3460
.6541	64 1351	7 9647	6 6079	4 0808	2 7138	.3459
.6542	66 8468	7 5681	6 2838	3 2160	3 0824	.3458
.6543	69 5587	7 1713	5 9595	2 3509	3 4509	.3457
.6544	72 2710	6 7742	5 6350	1 4856	3 8193	.3456
.6545	74 9835	6 3768	5 3102	.7560 6200	4 1875	.3455
.6546	77 6964	5 9792	4 9852	.7559 7542	4 5556	.3454
.6547	80 4095	5 5813	4 6599	8 8882	4 9236	.3453
.6548	83 1230	5 1831	4 3344	8 0220	5 2915	.3452
.6549	85 8367	4 7847	4 0087	7 1555	5 6592	.3451
.6550	.3988 5507	.3684 3860	.4753 6828	.7556 2888	.9386 0268	.3450

$E^{-ii} =$ $E^{ii} = .0000,0000+$ .0000,0000+   .0000,0000+   .0000,0000+   .0000,0000+
.0000,0000,1

TABLE I

.6550   .3450

p	x	z	√pq	√1-p²	√1-q²	q
.6550	.3988 5507	.3684 3860	.4753 6828	.7556 2888	.9386 0268	**.3450**
.6551	91 2650	3 9870	3 3566	5 4218	6 3944	.3449
.6552	93 9796	3 5877	3 0302	4 5547	6 7617	.3448
.6553	96 6945	3 1882	2 7035	3 6872	7 1290	.3447
.6554	.3999 4097	2 7884	2 3767	2 8196	7 4962	.3446
.6555	.4002 1252	2 3883	2 0496	1 9517	7 8632	.3445
.6556	04 8409	1 9879	1 7222	1 0836	8 2301	.3444
.6557	07 5570	1 5873	1 3946	.7550 2153	8 5969	.3443
.6558	10 2734	1 1864	1 0668	.7549 3467	8 9635	.3442
.6559	12 9900	0 7853	0 7388	8 4779	9 3301	.3441
.6560	.4015 7070	.3680 3838	.4750 4105	.7547 6089	.9389 6965	**.3440**
.6561	18 4243	.3679 9821	.4750 0820	6 7396	.9390 0628	.3439
.6562	21 1418	9 5801	.4749 7533	5 8701	0 4290	.3438
.6563	23 8596	9 1779	9 4243	5 0004	0 7950	.3437
.6564	26 5778	8 7754	9 0951	4 1304	1 1610	.3436
.6565	29 2962	8 3726	8 7656	3 2602	1 5268	.3435
.6566	32 0150	7 9695	8 4360	2 3898	1 8925	.3434
.6567	34 7340	7 5662	8 1060	1 5191	2 2580	.3433
.6568	37 4533	7 1626	7 7759	.7540 6482	2 6235	.3432
.6569	40 1730	6 7587	7 4455	.7539 7771	2 9888	.3431
.6570	.4042 8929	.3676 3545	.4747 1149	.7538 9058	.9393 3540	**.3430**
.6571	45 6131	5 9501	6 7841	8 0342	3 7191	.3429
.6572	48 3337	5 5454	6 4530	7 1623	4 0841	.3428
.6573	51 0545	5 1404	6 1217	6 2903	4 4489	.3427
.6574	53 7757	4 7352	5 7901	5 4180	4 8137	.3426
.6575	56 4971	4 3297	5 4584	4 5454	5 1783	.3425
.6576	59 2188	3 9239	5 1263	3 6727	5 5428	.3424
.6577	61 9409	3 5178	4 7941	2 7997	5 9071	.3423
.6578	64 6632	3 1115	4 4616	1 9264	6 2714	.3422
.6579	67 3859	2 7049	4 1289	1 0530	6 6355	.3421
.6580	.4070 1088	.3672 2980	.4743 7959	.7530 1793	.9396 9995	**.3420**
.6581	72 8320	1 8909	3 4628	.7529 3053	7 3634	.3419
.6582	75 5556	1 4835	3 1293	8 4312	7 7272	.3418
.6583	78 2794	1 0758	2 7957	7 5568	8 0908	.3417
.6584	81 0036	˙0 6678	2 4618	6 6821	8 4543	.3416
.6585	83 7280	.3670 2596	2 1277	5 8073	8 8177	.3415
.6586	86 4528	.3669 8511	1 7933	4 9322	9 1810	.3414
.6587	89 1778	9 4423	1 4587	4 0568	9 5442	.3413
.6588	91 9032	9 0332	1 1239	3 1812	.9399 9072	.3412
.6589	94 6288	8 6239	0 7889	2 3054	.9400 2702	.3411
.6590	.4097 3548	.3668 2143	.4740 4536	.7521 4294	.9400 6330	**.3410**
.6591	.4100 0811	7 8044	.4740 1180	.7520 5531	0 9956	.3409
.6592	02 8077	7 3943	.4739 7823	.7519 6766	1 3582	.3408
.6593	05 5345	6 9839	9 4463	8 7998	1 7206	.3407
.6594	08 2617	6 5732	9 1100	7 9229	2 0830	.3406
.6595	10 9892	6 1622	8 7736	7 0456	2 4452	.3405
.6596	13 7170	5 7510	8 4369	6 1682	2 8072	.3404
.6597	16 4451	5 3395	8 0999	5 2905	3 1692	.3403
.6598	19 1736	4 9277	7 7628	4 4126	3 5310	.3402
.6599	21 9023	4 5156	7 4254	3 5344	3 8928	.3401
.6600	.4124 6313	.3664 1033	.4737 0877	.7512 6560	.9404 2544	**.3400**

E^{-ii}=   E^{ii}=.0000,0000+   .0000,0000+   .0000,0000+   .0000,0000+   .0000,0000+
.0000,0000,1

TABLE I

p	x	z	$\sqrt{pq}$	$\sqrt{1-p^2}$	$\sqrt{1-q^2}$	q
.6600	.4124 6313	.3664 1033	.4737 0877	.7512 6560	.9404 2544	.3400
.6601	27 3606	3 6907	6 7498	1 7774	4 6158	.3399
.6602	30 0903	3 2778	6 4117	0 8985	4 9772	.3398
.6603	32 8202	2 8647	6 0734	.7510 0194	5 3384	.3397
.6604	35 5505	2 4513	5 7348	.7509 1400	5 6995	.3396
.6605	38 2810	2 0376	5 3960	8 2605	6 0605	.3395
.6606	41 0119	1 6236	5 0569	7 3806	6 4214	.3394
.6607	43 7431	1 2094	4 7176	6 5006	6 7822	.3393
.6608	46 4746	0 7949	4 3781	5 6203	7 1428	.3392
.6609	49 2064	.3660 3801	4 0383	4 7398	7 5033	.3391
.6610	.4151 9385	.3659 9650	.4733 6983	.7503 8590	.9407 8637	.3390
.6611	54 6709	9 5497	3 3581	2 9780	8 2240	.3389
.6612	57 4037	9 1341	3 0176	2 0968	8 5842	.3388
.6613	60 1367	8 7182	2 6769	1 2153	8 9442	.3387
.6614	62 8700	8 3021	2 3360	.7500 3336	9 3041	.3386
.6615	65 6037	7 8656	1 9948	.7499 4516	.9409 6639	.3385
.6616	68 3377	7 4689	1 6534	8 5695	.9410 0236	.3384
.6617	71 0720	7 0520	1 3118	7 6870	0 3831	.3383
.6618	73 8066	6 6347	0 9699	6 8044	0 7426	.3382
.6619	76 5415	6 2172	0 6278	5 9215	1 1019	.3381
.6620	.4179 2767	.3655 7994	.4730 2854	.7495 0384	.9411 4611	.3380
.6621	82 0122	5 3814	.4729 9428	4 1550	1 8202	.3379
.6622	84 7481	4 9630	9 6000	3 2714	2 1791	.3378
.6623	87 4843	⁻4 5444	9 2569	2 3875	2 5380	.3377
.6624	90 2207	4 1255	8 9136	1 5035	2 8967	.3376
.6625	92 9575	3 7064	8 5701	.7490 6191	3 2553	.3375
.6626	95 6946	3 2869	8 2263	.7489 7346	3 6138	.3374
.6627	.4198 4321	2 8672	7 8823	8 8498	3 9721	.3373
.6628	.4201 1698	2 4472	7 5380	7 9647	4 3304	.3372
.6629	03 9078	2 0270	7 1936	7 0795	4 6885	.3371
.6630	.4206 6462	.3651 6065	.4726 8488	.7486 1940	.9415 0465	.3370
.6631	09 3849	1 1857	6 5039	5 3082	5 4043	.3369
.6632	12 1239	0 7646	6 1587	4 4222	5 7621	.3368
.6633	14 8632	.3650 3432	5 8133	3 5360	6 1197	.3367
.6634	17 6028	.3649 9216	5 4676	2 6495	6 4773	.3366
.6635	20 3427	9 4997	5 1217	1 7628	6 8347	.3365
.6636	23 0830	9 0775	4 7755	.7480 8759	7 1919	.3364
.6637	25 8236	8 6551	4 4292	.7479 9887	7 5491	.3363
.6638	28 5645	8 2324	4 0826	9 1013	7 9061	.3362
.6639	31 3057	7 8094	3 7357	8 2136	8 2631	.3361
.6640	.4234 0472	.3647 3861	.4723 3886	.7477 3257	.9418 6199	.3360
.6641	36 7891	6 9626	3 0413	6 4376	8 9765	.3359
.6642	39 5312	6 5388	2 6937	5 5492	9 3331	.3358
.6643	42 2737	6 1147	2 3459	4 6606	.9419 6895	.3357
.6644	45 0165	5 6903	1 9979	3 7717	.9420 0459	.3356
.6645	47 7597	5 2657	1 6496	2 8826	0 4021	.3355
.6646	50 5031	4 8408	1 3011	1 9933	0 7581	.3354
.6647	53 2469	4 4156	0 9523	1 1037	1 1141	.3353
.6648	55 9910	3 9901	0 6034	.7470 2139	1 4699	.3352
.6649	58 7354	3 5644	.4720 2541	.7469 3239	1 8257	.3351
.6650	.4261 4801	.3643 1384	.4719 9047	.7468 4336	.9422 1813	.3350

$E^{-ii} =$ .0000,0000,1 $E^{ii} =$ .0000,0000+ .0000,0000+ .0000,0000+ .0000,0000+ .0000,0000+

TABLE I

.6650   **.3350**

p	x	z	√pq	√(1−p²)	√(1−q²)	q
.6650	.4261 4801	.3643 1384	.4719 9047	.7468 4336	.9422 1813	**.3350**
.6651	64 2251	2 7121	9 5550	7 5430	2 5368	.3349
.6652	66 9705	2 2855	9 2050	6 6523	2 8921	.3348
.6653	69 7162	1 8587	8 8548	5 7612	3 2474	.3347
.6654	72 4622	1 4316	8 5044	4 8700	3 6025	.3346
.6655	75 2085	1 0042	8 1538	3 9785	3 9575	.3345
.6656	77 9552	0 5765	7 8029	3 0868	4 3124	.3344
.6657	80 7021	.3640 1486	7 4517	2 1948	4 6672	.3343
.6658	83 4494	.3639 7204	7 1004	1 3026	5 0218	.3342
.6659	86 1971	9 2919	6 7488	.7460 4101	5 3763	.3341
.6660	.4288 9450	.3638 8632	.4716 3969	.7459 5174	.9425 7307	**.3340**
.6661	91 6933	8 4341	6 0448	8 6245	6 0850	.3339
.6662	94 4419	8 0048	5 6925	7 7313	6 4392	.3338
.6663	97 1909	7 5752	5 3400	6 8379	6 7933	.3337
.6664	.4299 9401	7 1454	4 9872	5 9442	7 1472	.3336
.6665	.4302 6896	6 7152	4 6341	5 0503	7 5010	.3335
.6666	05 4395	6 2848	4 2809	4 1562	7 8547	.3334
.6667	08 1898	5 8542	3 9273	3 2618	8 2083	.3333
.6668	10 9403	5 4232	3 5736	2 3671	8 5617	.3332
.6669	13 6912	4 9920	3 2196	1 4723	8 9150	.3331
.6670	.4316 4424	.3634 5605	.4712 8654	.7450 5772	.9429 2683	**.3330**
.6671	19 1939	4 1287	2 5109	.7449 6818	9 6214	.3329
.6672	21 9458	3 6966	2 1562	8 7862	.9429 9743	.3328
.6673	24 6980	3 2643	1 8012	7 8904	.9430 3272	.3327
.6674	27 4504	2 8317	1 4461	6 9943	0 6799	.3326
.6675	30 2033	2 3988	1 0906	6 0980	1 0326	.3325
.6676	32 9565	1 9656	0 7350	5 2014	1 3851	.3324
.6677	35 7100	1 5322	0 3791	4 3046	1 7374	.3323
.6678	38 4638	1 0985	.4710 0229	3 4076	2 0897	.3322
.6679	41 2179	0 6645	.4709 6665	2 5103	2 4418	.3321
.6680	.4343 9724	.3630 2303	.4709 3099	.7441 6127	.9432 7939	**.3320**
.6681	46 7272	.3629 7957	8 9531	.7440 7150	3 1458	.3319
.6682	49 4824	9 3609	8 5960	.7439 8169	3 4975	.3318
.6683	52 2378	8 9258	8 2386	8 9187	3 8492	.3317
.6684	54 9936	8 4905	7 8811	8 0202	4 2008	.3316
.6685	57 7498	8 0548	7 5232	7 1214	4 5522	.3315
.6686	60 5062	7 6189	7 1652	6 2224	4 9035	.3314
.6687	63 2630	7 1827	6 8069	5 3232	5 2547	.3313
.6688	66 0201	6 7463	6 4483	4 4237	5 6058	.3312
.6689	68 7776	6 3095	6 0896	3 5240	5 9567	.3311
.6690	.4371 5354	.3625 8725	.4705 7305	.7432 6240	.9436 3075	**.3310**
.6691	74 2935	5 4352	5 3713	1 7238	6 6583	.3309
.6692	77 0520	4 9976	5 0118	.7430 8234	7 0088	.3308
.6693	79 8108	4 5598	4 6521	.7429 9227	7 3593	.3307
.6694	82 5699	4 1217	4 2921	9 0217	7 7097	.3306
.6695	85 3294	3 6833	3 9319	8 1206	8 0599	.3305
.6696	88 0892	3 2446	3 5714	7 2191	8 4100	.3304
.6697	90 8493	2 8057	3 2107	6 3175	8 7600	.3303
.6698	93 6098	2 3664	2 8498	5 4155	9 1099	.3302
.6699	96 3706	1 9270	2 4886	4 5134	9 4597	.3301
.6700	.4399 1317	.3621 4872	.4702 1272	.7423 6110	.9439 8093	**.3300**

$E^{-ii}=$ .0000,0000,1 $E^{ii}=$ .0000,0000+ .0000,0000+ .0000,0000+ .0000,0000+ .0000,0000+

TABLE I

.6700   **.3300**

p	x	z	$\sqrt{pq}$	$\sqrt{1-p^2}$	$\sqrt{1-q^2}$	q
.6700	.4399 1317	.3621 4872	.4702 1272	.7423 6110	.9439 8093	.3300
.6701	.4401 8931	1 0471	1 7655	2 7083	.9440 1588	.3299
.6702	04 6549	0 6068	1 4036	1 8054	0 5082	.3298
.6703	07 4171	.3620 1662	1 0415	.7420 9023	0 8575	.3297
.6704	10 1795	.3619 7253	0 6791	.7419 9989	1 2067	.3296
.6705	12 9423	9 2842	.4700 3165	9 0953	1 5558	.3295
.6706	15 7055	8 8427	.4699 9536	8 1914	1 9047	.3294
.6707	18 4690	8 4010	9 5905	7 2873	2 2535	.3293
.6708	21 2328	7 9590	9 2272	6 3829	2 6022	.3292
.6709	23 9969	7 5168	8 8636	5 4783	2 9508	.3291
.6710	.4426 7614	.3617 0742	.4698 4998	.7414 5735	.9443 2992	.3290
.6711	29 5262	6 6314	8 1357	3 6684	3 6475	.3289
.6712	32 2914	6 1883	7 7714	2 7630	3 9958	.3288
.6713	35 0569	5 7450	7 4068	1 8575	4 3439	.3287
.6714	37 8228	5 3013	7 0420	0 9516	4 6918	.3286
.6715	40 5890	4 8574	6 6770	.7410 0455	5 0397	.3285
.6716	43 3555	4 4132	6 3117	.7409 1392	5 3874	.3284
.6717	46 1224	3 9687	5 9462	8 2327	5 7351	.3283
.6718	48 8896	3 5240	5 5805	7 3258	6 0826	.3282
.6719	51 6572	3 0789	5 2145	6 4188	6 4300	.3281
.6720	.4454 4251	.3612 6336	.4694 8482	.7405 5115	.9446 7772	.3280
.6721	57 1933	2 1881	4 4818	4 6039	7 1244	.3279
.6722	59 9619	1 7422	4 1150	3 6961	7 4714	.3278
.6723	62 7308	1 2961	3 7481	2 7881	7 8183	.3277
.6724	65 5001	0 8497	3 3809	1 8798	8 1651	.3276
.6725	68 2697	.3610 4030	3 0134	0 9712	8 5118	.3275
.6726	71 0396	.3609 9560	2 6457	.7400 0624	8 8583	.3274
.6727	73 8099	9 5088	2 2778	.7399 1534	9 2048	.3273
.6728	76 5805	9 0612	1 9096	8 2441	9 5511	.3272
.6729	79 3515	8 6134	1 5412	7 3346	.9449 8973	.3271
.6730	.4482 1228	.3608 1654	.4691 1726	.7396 4248	.9450 2434	.3270
.6731	84 8945	7 7170	0 8037	5 5148	0 5893	.3269
.6732	87 6665	7 2684	0 4345	4 6045	0 9352	.3268
.6733	90 4388	6 8195	.4690 0651	3 6940	1 2809	.3267
.6734	93 2115	6 3703	.4689 6955	2 7832	1 6265	.3266
.6735	95 9846	5 9208	9 3256	1 8722	1 9720	.3265
.6736	.4498 7580	5 4711	8 9555	0 9610	2 3174	.3264
.6737	.4501 5317	5 0211	8 5852	.7390 0495	2 6626	.3263
.6738	04 3058	4 5708	8 2146	.7389 1377	3 0078	.3262
.6739	07 0802	4 1202	7 8437	8 2257	3 3528	.3261
.6740	.4509 8550	.3603 6694	.4687 4727	.7387 3134	.9453 6977	.3260
.6741	12 6301	3 2183	7 1013	6 4010	4 0425	.3259
.6742	15 4056	2 7669	6 7298	5 4882	4 3871	.3258
.6743	18 1814	2 3152	6 3580	4 5752	4 7317	.3257
.6744	20 9576	1 8632	5 9859	3 6620	5 0761	.3256
.6745	23 7341	1 4110	5 6136	2 7485	5 4204	.3255
.6746	26 5110	0 9585	5 2411	1 8347	5 7646	.3254
.6747	29 2882	0 5057	4 8683	0 9207	6 1087	.3253
.6748	32 0657	.3600 0526	4 4953	.7380 0065	6 4526	.3252
.6749	34 8436	.3599 5993	4 1220	.7379 0920	6 7964	.3251
.6750	.4537 6219	.3599 1456	.4683 7485	.7378 1773	.9457 1402	.3250

$E^{-ii}=$ $E^{ii}=.0000,0000+$ .0000,0000+ .0000,0000+ .0000,0000+ .0000,0000+
.0000,0000,
1,0000,0000,1

TABLE I

p	x	z	$\sqrt{pq}$	$\sqrt{1-p^2}$	$\sqrt{1-q^2}$	q
.6750	.4537 6219	.3599 1456	.4683 7485	.7378 1773	.9457 1402	**.3250**
.6751	40 4005	8 6917	3 3747	7 2623	7 4838	.3249
.6752	43 1795	8 2376	3 0007	6 3471	7 8272	.3248
.6753	45 9588	7 7831	2. 6265	5 4316	8 1706	.3247
.6754	48 7385	7 3284	2 2520	4 5158	8 5138	.3246
.6755	51 5185	6 8734	1 8773	3 5999	8 8570	.3245
.6756	54 2988	6 4181	1 5023	2 6836	9 2000	.3244
.6757	57 0795	5 9625	1 1271	1 7672	9 5429	.3243
.6758	59 8606	5 5067	0 7516	.7370 8504	.9459 8856	.3242
.6759	62 6420	5 0505	0 3759	.7369 9334	.9460 2283	.3241
.6760	.4565 4238	.3594 5941	.4680 0000	.7369 0162	.9460 5708	**.3240**
.6761	68 2059	4 1374	.4679 6238	8 0987	0 9132	.3239
.6762	70 9884	3 6805	9 2474	7 1810	1 2555	.3238
.6763	73 7713	3 2232	8 8707	6 2630	1 5977	.3237
.6764	76 5545	2 7657	8 4938	5 3448	1 9398	.3236
.6765	79 3380	2 3079	8 1166	4 4263	2 2817	.3235
.6766	82 1219	1 8499	7 7392	3 5076	2 6235	.3234
.6767	84 9062	1 3915	7 3615	2 5886	2 9652	.3233
.6768	87 6908	0 9329	6 9836	1 6694	3 3068	.3232
.6769	90 4758	0 4740	6 6055	.7360 7499	3 6483	.3231
.6770	.4593 2611	.3590 0148	.4676 2271	.7359 8302	.9463 9896	**.3230**
.6771	96 0468	.3589 5553	5 8485	8 9102	4 3309	.3229
.6772	70 9884	9 0956	5 4696	7 9899	4 6720	.3228
.6773	.4601 6192	8 6356	5 0905	7 0695	5 0130	.3227
.6774	04 4060	8 1753	4 7111	6 1487	5 3539	.3226
.6775	07 1931	7 7147	4 3315	5 2277	5 6946	.3225
.6776	09 9805	7 2538	3 9516	4 3065	6 0353	.3224
.6777	12 7684	6 7927	3 5715	3 3850	6 3758	.3223
.6778	15 5566	6 3313	3 1912	2 4633	6 7162	.3222
.6779	18 3451	5 8696	2 8106	1 5413	7 0565	.3221
.6780	.4621 1340	.3585 4076	.4672 4298	.7350 6190	.9467 3967	**.3220**
.6781	23 9233	4 9453	2 0487	.7349 6965	7 7367	.3219
.6782	26 7129	4 4828	1 6674	8 7738	8 0767	.3218
.6783	29 5029	4 0200	1 2858	7 8508	8 4165	.3217
.6784	32 2932	3 5569	0 9040	6 9275	8 7562	.3216
.6785	35 0839	3 0935	0 5219	6 0040	9 0958	.3215
.6786	37 8750	2 6299	.4670 1396	5 0803	9 4353	.3214
.6787	40 6664	2 1660	.4669 7571	4 1562	.9469 7746	.3213
.6788	43 4582	1 7018	9 3743	3 2320	.9470 1138	.3212
.6789	46 2504	1 2373	8 9912	2 3075	0 4529	.3211
.6790	.4649 0429	.3580 7725	.4668 6079	.7341 3827	.9470 7919	**.3210**
.6791	51 8358	.3580 3075	8 2244	.7340 4577	1 1308	.3209
.6792	54 6290	.3579 8421	7 8406	.7339 5324	1 4696	.3208
.6793	57 4226	9 3765	7 4566	8 6069	1 8082	.3207
.6794	60 2166	8 9107	7 0723	7 6811	2 1467	.3206
.6795	63 0109	8 4445	6 6878	6 7551	2 4852	.3205
.6796	65 8056	7 9781	6 3030	5 8288	2 8234	.3204
.6797	68 6006	7 5113	5 9180	4 9022	3 1616	.3203
.6798	71 3961	7 0443	5 5328	3 9755	3 4997	.3202
.6799	74 1919	6 5771	5 1473	3 0484	3 8376	.3201
.6800	.4676 9880	.3576 1095	.4664 7615	.7332 1211	.9474 1754	**.3200**

$E^{-ii}=$ $E^{ii}=$.0000,0000+ .0000,0000+ .0000,0000+ .0000,0000+ .0000,0000+
.0000,0000,1

TABLE I

p	x	z	$\sqrt{pq}$	$\sqrt{1-p^2}$	$\sqrt{1-q^2}$	q
.6800	.4676 9880	.3576 1095	.4664 7615	.7332 1211	.9474 1754	**.3200**
.6801	79 7845	5 6417	4 3755	1 1936	4 5131	.3199
.6802	82 5814	5 1735	3 9893	.7330 2658	4 8507	.3198
.6803	85 3787	4 7051	3 6028	.7329 3377	5 1882	.3197
.6804	88 1763	4 2365	3 2161	8 4094	5 5255	.3196
.6805	90 9743	3 7675	2 8291	7 4808	5 8628	.3195
.6806	93 7726	3 2983	2 4418	6 5520	6 1999	.3194
.6807	96 5713	2 8288	2 0544	5 6229	6 5369	.3193
.6808	.4699 3704	2 3590	1 6667	4 6936	6 8737	.3192
.6809	.4702 1699	1 8889	1 2787	3 7640	7 2105	.3191
.6810	.4704 9697	.3571 4185	.4660 8905	.7322 8342	.9477 5472	**.3190**
.6811	07 7699	0 9479	0 5020	1 9041	7 8837	.3189
.6812	10 5705	0 4770	.4660 1133	0 9737	8 2201	.3188
.6813	13 3714	.3570 0058	.4659 7243	.7320 0431	8 5564	.3187
.6814	16 1727	.3569 5343	9 3351	.7319 1122	8 8926	.3186
.6815	18 9744	9 0625	8 9457	8 1811	9 2286	.3185
.6816	21 7764	8 5905	8 5560	7 2498	9 5645	.3184
.6817	24 5788	8 1182	8 1661	6 3181	.9479 9004	.3183
.6818	27 3816	7 6456	7 7759	5 3863	.9480 2361	.3182
.6819	30 1848	7 1727	7 3854	4 4541	0 5717	.3181
.6820	.4732 9883	.3566 6995	.4656 9947	.7313 5217	.9480 9071	**.3180**
.6821	35 7922	6 2261	6 6038	2 5891	1 2425	.3179
.6822	38 5964	5 7524	6 2126	1 6562	1 5777	.3178
.6823	41 4011	5 2784	5 8212	.7310 7230	1 9128	.3177
.6824	44 2061	4 8041	5 4295	.7309 7896	2 2478	.3176
.6825	47 0115	4 3295	5 0376	8 8559	2 5827	.3175
.6826	49 8172	3 8547	4 6454	7 9220	2 9175	.3174
.6827	52 6234	3 3796	4 2530	6 9878	3 2521	.3173
.6828	55 4299	2 9042	3 8603	6 0534	3 5867	.3172
.6829	58 2367	2 4285	3 4674	5 1187	3 9211	.3171
.6830	.4761 0440	.3561 9525	.4653 0743	.7304 1837	.9484 2554	**.3170**
.6831	63 8516	1 4763	2 6808	3 2485	4 5896	.3169
.6832	66 6596	0 9998	2 2872	2 3131	4 9236	.3168
.6833	69 4680	0 5230	1 8933	1 3773	5 2576	.3167
.6834	72 2768	.3560 0459	1 4991	.7300 4414	5 5914	.3166
.6835	75 0860	.3559 5685	1 1047	.7299 5051	5 9251	.3165
.6836	77 8955	9 0908	0 7101	8 5686	6 2587	.3164
.6837	80 7053	8 6129	.4650 3152	7 6319	6 5922	.3163
.6838	83 5156	8 1347	.4649 9200	6 6949	6 9255	.3162
.6839	86 3263	7 6562	9 5246	5 7576	7 2588	.3161
.6840	.4789 1373	.3557 1774	.4649 1290	.7294 8201	.9487 5919	**.3160**
.6841	91 9487	6 6984	8 7331	3 8823	7 9249	.3159
.6842	94 7605	6 2191	8 3369	2 9443	8 2578	.3158
.6843	.4797 5727	5 7394	7 9405	2 0060	8 5906	.3157
.6844	.4800 3852	5 2595	7 5439	1 0674	8 9232	.3156
.6845	03 1982	4 7794	7 1470	.7290 1286	9 2558	.3155
.6846	06 0115	4 2989	6 7498	.7289 1895	9 5882	.3154
.6847	08 8251	3 8182	6 3524	8 2502	.9489 9205	.3153
.6848	11 6392	3 3371	5 9548	7 3106	.9490 2527	.3152
.6849	14 4537	2 8558	5 5569	6 3708	0 5848	.3151
.6850	.4817 2685	.3552 3742	.4645 1588	.7285 4307	.9490 9167	**.3150**

$E^{-ii}=$ $E^{ii}=$.0000,0000+ .0000,0000+ .0000,0000+ .0000,0000+ .0000,0000+
.0000,0000,1

TABLE I

.6850   **.3150**

p	x	z	√pq	√1−p²	√1−q²	q
.6850	.4817 2685	.3552 3742	.4645 1588	.7285 4307	.9490 9167	**.3150**
.6851	20 0837	1 8924	4 7604	4 4903	1 2485	.3149
.6852	22 8993	1 4102	4 3617	3 5497	1 5803	.3148
.6853	25 7153	0 9278	3 9629	2 6088	1 9119	.3147
.6854	28 5316	.3550 4451	3 5637	1 6677	2 2434	.3146
.6855	31 3484	.3549 9621	3 1643	.7280 7263	2 5747	.3145
.6856	34 1655	9 4788	2 7647	.7279 7846	2 9060	.3144
.6857	36 9830	8 9953	2 3648	8 8427	3 2371	.3143
.6858	39 8009	8 5114	1 9647	7 9005	3 5681	.3142
.6859	42 6191	8 0273	1 5643	6 9581	3 8990	.3141
.6860	.4845 4378	.3547 5429	.4641 1636	.7276 0154	.9494 2298	**.3140**
.6861	48 2568	7 0582	0 7628	5 0724	4 5605	.3139
.6862	51 0763	6 5732	.4640 3616	4 1292	4 8910	.3138
.6863	53 8961	6 0880	.4639 9602	3 1858	5 2215	.3137
.6864	56 7163	5 6025	9 5586	2 2420	5 5518	.3136
.6865	59 5369	5 1167	9 1567	1 2980	5 8820	.3135
.6866	62 3579	4 6306	8 7546	.7270 3538	6 2121	.3134
.6867	65 1793	4 1442	8 3522	.7269 4093	6 5421	.3133
.6868	68 0010	3 6575	7 9495	8 4645	6 8719	.3132
.6869	70 8232	3 1706	7 5467	7 5195	7 2016	.3131
.6870	.4873 6457	.3542 6834	.4637 1435	.7266 5742	.9497 5313	**.3130**
.6871	76 4686	2 1958	6 7401	5 6286	7 8608	.3129
.6872	79 2919	1 7081	6 3365	4 6828	8 1901	.3128
.6873	82 1156	1 2200	5 9326	3 7367	8 5194	.3127
.6874	84 9397	0 7316	5 5284	2 7904	8 8486	.3126
.6875	87 7641	.3540 2430	5 1241	1 8438	9 1776	.3125
.6876	90 5890	.3539 7541	4 7194	.7260 8969	9 5065	.3124
.6877	93 4142	9 2649	4 3145	.7259 9498	.9499 8353	.3123
.6878	96 2399	8 7754	3 9094	9 0024	.9500 1640	.3122
.6879	.4899 0659	8 2856	3 5040	8 0548	0 4926	.3121
.6880	.4901 8923	.3537 7956	.4633 0983	.7257 1069	.9500 8210	**.3120**
.6881	04 7191	7 3053	2 6924	6 1587	1 1494	.3119
.6882	07 5463	6 8146	2 2863	5 2103	1 4776	.3118
.6883	10 3739	6 3237	1 8799	4 2616	1 8057	.3117
.6884	13 2019	5 8326	1 4732	3 3126	2 1337	.3116
.6885	16 0303	5 3411	1 0663	2 3634	2 4615	.3115
.6886	18 8591	4 8494	0 6591	1 4139	2 7893	.3114
.6887	21 6882	4 3573	.4630 2517	.7250 4642	3 1169	.3113
.6888	24 5178	3 8650	.4629 8441	.7249 5142	3 4444	.3112
.6889	27 3477	3 3724	9 4361	8 5639	3 7718	.3111
.6890	.4930 1781	.3532 8796	.4629 0280	.7247 6134	.9504 0991	**.3110**
.6891	33 0089	2 3864	8 6196	6 6626	4 4263	.3109
.6892	35 8400	1 8930	8 2109	5 7116	4 7533	.3108
.6893	38 6716	1 3992	7 8020	4 7602	5 0803	.3107
.6894	41 5035	0 9052	7 3928	3 8087	5 4071	.3106
.6895	44 3358	.3530 4109	6 9834	2 8568	5 7338	.3105
.6896	47 1686	.3529 9164	6 5737	1 9047	6 0604	.3104
.6897	50 0017	9 4215	6 1637	.7240 9524	6 3869	.3103
.6898	52 8352	8 9264	5 7536	.7239 9997	6 7132	.3102
.6899	55 6692	8 4309	5 3431	9 0468	7 0394	.3101
.6900	.4958 5035	.3527 9352	.4624 9324	.7238 0937	.9507 3656	**.3100**

$E^{-ii} =$ .0000,0000,2 $E^{ii} =$ .0000,0001   .0000,0000+   .0000,0000+   .0000,0000+   .0000,0000+

TABLE I

p	x	z	√pq	√1−p²	√1−q²	q
.6900	.4958 5035	.3527 9352	.4624 9324	.7238 0937	.9507 3656	.3100
.6901	61 3382	7 4392	4 5215	7 1403	7 6916	.3099
.6902	64 1733	6 9430	4 1103	6 1866	8 0175	.3098
.6903	67 0088	6 4464	3 6988	5 2326	8 3432	.3097
.6904	69 8447	5 9496	3 2871	4 2784	8 6689	.3096
.6905	72 6810	5 4524	2 8752	3 3239	8 9944	.3095
.6906	75 5178	4 9550	2 4630	2 3692	9 3198	.3094
.6907	78 3549	4 4573	2 0505	1 4142	9 6452	.3093
.6908	81 1924	3 9593	1 6378	.7230 4589	.9509 9703	.3092
.6909	84 0303	3 4611	1 2248	.7229 5034	.9510 2954	.3091
.6910	.4986 8686	.3522 9625	.4620 8116	.7228 5476	.9510 6204	.3090
.6911	89 7073	2 4637	.4620 3981	7 5915	0 9452	.3089
.6912	92 5464	1 9646	.4619 9844	6 6352	1 2699	.3088
.6913	95 3860	1 4652	9 5704	5 6786	1 5946	.3087
.6914	.4998 2259	0 9655	9 1562	4 7217	1 9190	.3086
.6915	.5001 0662	.3520 4656	8 7417	3 7646	2 2434	.3085
.6916	03 9070	.3519 9653	8 3270	2 8072	2 5677	.3084
.6917	06 7481	9 4648	7 9120	1 8496	2 8918	.3083
.6918	09 5897	8 9640	7 4967	.7220 8916	3 2159	.3082
.6919	12 4316	8 4629	7 0812	.7219 9334	3 5398	.3081
.6920	.5015 2740	.3517 9615	.4616 6655	.7218 9750	.9513 8636	.3080
.6921	18 1168	7 4598	6 2495	8 0163	4 1872	.3079
.6922	20 9599	6 9578	5 8332	7 0573	4 5108	.3078
.6923	23 8035	6 4556	5 4167	6 0980	4 8343	.3077
.6924	26 6475	5 9531	4 9999	5 1385	5 1576	.3076
.6925	29 4919	5 4503	4 5829	4 1787	5 4808	.3075
.6926	32 3367	4 9472	4 1656	3 2187	5 8039	.3074
.6927	35 1819	4 4438	3 7480	2 2584	6 1269	.3073
.6928	38 0275	3 9401	3 3303	1 2978	6 4498	.3072
.6929	40 8735	3 4362	2 9122	.7210 3370	6 7725	.3071
.6930	.5043 7199	.3512 9320	.4612 4939	.7209 3758	.9517 0951	.3070
.6931	46 5667	2 4275	2 0753	8 4145	7 4177	.3069
.6932	49 4140	1 9227	1 6565	7 4528	7 7401	.3068
.6933	52 2616	1 4176	1 2375	6 4909	8 0624	.3067
.6934	55 1097	0 9122	0 8181	5 5287	8 3845	.3066
.6935	57 9581	.3510 4066	.4610 3986	4 5663	8 7066	.3065
.6936	60 8070	.3509 9006	.4609 9787	3 6035	9 0285	.3064
.6937	63 6563	9 3944	9 5587	2 6406	9 3503	.3063
.6938	66 5060	8 8879	9 1383	1 6773	9 6721	.3062
.6939	69 3561	8 3811	8 7177	.7200 7138	.9519 9936	.3061
.6940	.5072 2066	.3507 8740	.4608 2969	.7199 7500	.9520 3151	.3060
.6941	75 0575	7 3667	7 8758	8 7859	0 6365	.3059
.6942	77 9089	6 8590	7 4544	7 8216	0 9577	.3058
.6943	80 7607	6 3511	7 0328	6 8570	1 2789	.3057
.6944	83 6128	5 8429	6 6109	5 8922	1 5999	.3056
.6945	86 4654	5 3343	6 1888	4 9270	1 9208	.3055
.6946	89 3184	4 8256	5 7664	3 9616	2 2415	.3054
.6947	92 1719	4 3165	5 3437	2 9960	2 5622	.3053
.6948	95 0257	3 8071	4 9208	2 0300	2 8828	.3052
.6949	.5097 8799	3 2975	4 4977	1 0638	3 2032	.3051
.6950	.5100 7346	.3502 7875	.4604 0743	.7190 0974	.9523 5235	.3050

E⁻ⁱᴸ= E ⁱᴸ .0000,0001 .0000,0000+ .0000,0000+ .0000,0000+ .0000,0000+
.0000,0000,2

TABLE I

.6950 | | | | | | .3050

p	x	z	$\sqrt{pq}$	$\sqrt{1-p^2}$	$\sqrt{1-q^2}$	q
.6950	.5100 7346	.3502 7875	.4604 0743	.7190 0974	.9523 5235	**.3050**
.6951	03 5897	· 2 2773	3 6506	.7189 1306	3 8437	.3049
.6952	06 4452	1 7668	3 2267	8 1636	4 1638	.3048
.6953	09 3011	1 2560	2 8025	7 1963	4 4838	.3047
.6954	12 1574	0 7450	2 3781	6 2288	4 8036	.3046
.6955	15 0141	.3500 2336	1 9534	5 2610	5 1234	.3045
.6956	17 8713	.3499 7220	1 5284	4 2929	5 4430	.3044
.6957	20 7289	9 2100	1 1032	3 3245	5 7625	.3043
.6958	23 5869	8 6978	0 6778	2 3559	6 0819	.3042
.6959	26 4453	8 1853	.4600 2521	1 3870	6 4012	.3041
.6960	.5129 3041	.3497 6725	.4599 8261	.7180 4178	.9526 7203	**.3040**
.6961	32 1634	7 1595	9 3999	.7179 4484	7 0394	.3039
.6962	35 0230	6 6461	8 9734	8 4787	7 3583	.3038
.6963	37 8831	6 1325	8 5466	7 5087	7 6771	.3037
.6964	40 7436	5 6185	8 1196	6 5384	7 9958	.3036
.6965	43 6046	5 1043	7 6924	5 5679	8 3144	.3035
.6966	46 4659	4 5898	7 2648	4 5971	8 6329	.3034
.6967	49 3277	4 0750	6 8371	3 6261	8 9512	.3033
.6968	52 1899	3 5599	6 4090	2 6547	9 2694	.3032
.6969	55 0526	3 0446	5 9807	1 6831	9 5876	.3031
.6970	.5157 9156	.3492 5289	.4595 5522	.7170 7113	.9529 9056	**.3030**
.6971	60 7790	2 0130	5 1234	.7169 7391	.9530 2234	.3029
.6972	63 6429	1 4968	4 6943	8 7667	0 5412	.3028
.6973	66 5073	0 9803	4 2650	7 7940	0 8589	.3027
.6974	69 3720	.3490 4635	3 8354	6 8211	1 1764	.3026
.6975	72 2371	.3489 9464	3 4056	5 8478	1 4938	.3025
.6976	75 1027	9 4290	2 9755	4 8743	1 8112	.3024
.6977	77 9687	8 9114	2 5452	3 9005	2 1284	.3023
.6978	80 8352	8 3934	2 1145	2 9265	2 4454	.3022
.6979	83 7020	7 8752	1 6837	1 9522	2 7624	.3021
.6980	.5186 5693	.3487 3567	.4591 2526	.7160 9776	.9533 0793	**.3020**
.6981	89 4370	6 8379	0 8212	.7160 0027	3 3960	.3019
.6982	92 3052	6 3188	.4590 3895	.7159 0276	3 7126	.3018
.6983	95 1737	5 7994	.4589 9576	8 0522	4 0291	.3017
.6984	.5198 0427	5 2798	9 5255	7 0765	4 3455	.3016
.6985	.5200 9121	4 7598	9 0930	6 1005	4 6618	.3015
.6986	03 7820	4 2396	8 6604	5 1243	4 9779	.3014
.6987	06 6523	3 7191	8 2274	4 1478	5 2940	.3013
.6988	09 5230	3 1983	7 7942	3 1710	5 6099	.3012
.6989	12 3941	2 6772	7 3608	2 1940	5 9257	.3011
.6990	.5215 2657	.3482 1558	.4586 9271	.7151 2167	.9536 2414	**.3010**
.6991	18 1377	1 6341	6 4931	.7150 2391	6 5570	.3009
.6992	21 0101	1 1121	6 0589	.7149 2612	6 8724	.3008
.6993	23 8830	0 5899	5 6244	8 2831	7 1878	.3007
.6994	26 7563	.3480 0674	5 1896	7 3047	7 5030	.3006
.6995	29 6300	.3479 5446	4 7546	6 3260	7 8181	.3005
.6996	32 5042	9 0214	4 3194	5 3470	8 1332	.3004
.6997	35 3788	8 4980	3 8838	4 3678	8 4480	.3003
.6998	38 2538	7 9744	3 4480	3 3883	8 7628	.3002
.6999	41 1292	7 4504	3 0120	2 4085	9 0775	.3001
.7000	.5244 0051	.3476 9261	.4582 5757	.7141 4284	.9539 3920	**.3000**

E^{-iL} E^{iL}=.0000,0001   .0000,0000+   .0000,0000+   .0000,0000+   .0000,0000+
.0000,0000,2

TABLE I

p	x	z	$\sqrt{pq}$	$\sqrt{1-p^2}$	$\sqrt{1-q^2}$	q
.7000	.5244 0051	.3476 9261	.4582 5757	.7141 4284	.9539 3920	**.3000**
.7001	46 8814	6 4016	2 1391	.7140 4481	.9539 7064	.2999
.7002	49 7582	5 8768	1 7023	.7139 4675	.9540 0208	.2998
.7003	52 6354	5 3516	1 2652	8 4866	0 3350	.2997
.7004	55 5130	4 8262	0 8279	7 5054	0 6490	.2996
.7005	58 3911	4 3005	.4580 3903	6 5240	0 9630	.2995
.7006	61 2696	3 7746	.4579 9524	5 5423	1 2769	.2994
.7007	64 1485	3 2483	9 5143	4 5603	1 5906	.2993
.7008	67 0279	2 7217	9 0759	3 5781	1 9042	.2992
.7009	69 9077	2 1949	8 6372	2 5955	2 2177	.2991
.7010	.5272 7879	.3471 6677	.4578 1983	.7131 6127	.9542 5311	**.2990**
.7011	75 6686	1 1403	7 7592	.7130 6296	2 8444	.2989
.7012	78 5497	0 6126	7 3197	.7129 6463	3 1575	.2988
.7013	81 4313	.3470 0846	6 8801	8 6626	3 4706	.2987
.7014	84 3133	.3469 5563	6 4401	7 6787	3 7835	.2986
.7015	87 1957	9 0278	5 9999	6 6945	4 0963	.2985
.7016	90 0786	8 4989	5 5594	5 7101	4 4090	.2984
.7017	92 9619	·7 9697	5 1187	4 7253	4 7216	.2983
.7018	95 8457	7 4403	4 6777	3 7403	5 0341	.2982
.7019	.5298 7299	6 9106	4 2364	2 7550	5 3465	.2981
.7020	.5301 6145	.3466 3806	.4573 7949	.7121 7694	.9545 6587	**.2980**
.7021	04 4996	5 8502	3 3531	.7120 7836	5 9708	.2979
.7022	07 3851	5 3197	2 9111	.7119 7975	6 2828	.2978
.7023	10 2710	4 7888	2 4688	.8 8111	6 5947	.2977
.7024	13 1574	4 2576	2 0262	7 8244	6 9065	.2976
.7025	16 0442	3 7261	1 5834	6 8374	7 2182	.2975
.7026	18 9315	3 1944	1 1403	5 8502	7 5297	.2974
.7027	21 8193	2 6624	0 6970	4 8627	7 8412	.2973
.7028	24 7074	2 1300	.4570 2534	3 8749	8 1525	.2972
.7029	27 5960	1 5974	.4569 8095	2 8868	8 4637	.2971
.7030	.5330 4851	.3461 0645	.4569 3654	.7111 8985	.9548 7748	**.2970**
.7031	33 3746	.3460 5313	8 9210	.7110 9099	9 0858	.2969
.7032	36 2646	.3459 9978	8 4763	.7109 9210	9 3966	.2968
.7033	39 1550	9 4641	8 0314	8 9318	.9549 7074	.2967
.7034	42 0458	8 9300	7 5862	7 9423	.9550 0180	.2966
.7035	44 9371	8 3957	7 1408	6 9526	0 3285	.2965
.7036	47 8288	7 8610	6 6951	5 9626	0 6389	.2964
.7037	50 7210	7 3261	6 2491	4 9723	0 9492	.2963
.7038	53 6137	6 7909	5 8029	3 9817	1 2594	.2962
.7039	56 5068	6 2554	5 3564	2 9908	1 5695	.2961
.7040	.5359 4003	.3455 7196	.4564 9096	.7101 9997	.9551 8794	**.2960**
.7041	62 2943	5 1835	4 4626	1 0083	2 1892	.2959
.7042	65 1887	4 6471	4 0153	.7100 0166	2 4989	.2958
.7043	68 0836	4 1104	3 5678	.7099 0247	2 8085	.2957
.7044	70 9789	3 5735	3 1200	8 0324	3 1180	.2956
.7045	73 8747	3 0362	2 6719	7 0399	3 4274	.2955
.7046	76 7709	2 4987	2 2236	6 0471	3 7367	.2954
.7047	79 6676	1 9609	1 7750	5 0540	4 0458	.2953
.7048	82 5647	1 4228	1 3261	4 0606	4 3548	.2952
.7049	85 4623	0 8844	0 8770	3 0670	4 6637	.2951
.7050	.5388 3603	.3450 3457	.4560 4276	.7092 0730	.9554 9725	**.2950**

$E^{-ii} =$ $E^{ii} = .0000,0001$ .0000,0000+ .0000,0000+ .0000,0000+ .0000,0000+
.0000,0000,2

TABLE I

.7050   .2950

p	x	z	$\sqrt{pq}$	$\sqrt{1-p^2}$	$\sqrt{1-q^2}$	q
.7050	.5388 3603	.3450 3457	.4560 4276	.7092 0730	.9554 9725	**.2950**
.7051	91 2588	.3449 8067	.4559 9780	1 0788	5 2812	.2949
.7052	94 1577	9 2674	9 5280	.7090 0843	5 5898	.2948
.7053	97 0571	8 7279	9 0779	.7089 0896	5 8982	.2947
.7054	.5399 9570	8 1880	8 6274	8 0945	6 2066	.2946
.7055	.5402 8573	7 6479	8 1767	7 0992	6 5148	.2945
.7056	05 7580	7 1075	7 7257	6 1036	6 8229	.2944
.7057	08 6592	6 5667	7 2745	5 1077	7 1309	.2943
.7058	11 5609	6 0257	6 8230	4 1115	7 4388	.2942
.7059	14 4630	5 4844	6 3713	3 1151	7 7465	.2941
.7060	.5417 3656	.3444 9428	.4555 9192	.7082 1183	.9558 0542	**.2940**
.7061	20 2686	4 4009	5 4669	1 1213	8 3617	.2939
.7062	23 1721	3 8588	5 0144	.7080 1240	8 6692	.2938
.7063	26 0761	3 3163	4 5616	.7079 1264	8 9765	.2937
.7064	28 9805	2 7736	4 1085	8 1286	9 2837	.2936
.7065	31 8853	2 2305	3 6551	7 1304	9 5907	.2935
.7066	34 7906	1 6872	3 2015	6 1320	.9559 8977	.2934
.7067	37 6964	1 1436	2 7476	5 1333	.9560 2045	.2933
.7068	40 6027	0 5996	2 2935	4 1343	0 5113	.2932
.7069	43 5094	.3440 0554	1 8391	3 1350	0 8179	.2931
.7070	.5446 4165	.3439 5109	.4551 3844	.7072 1355	.9561 1244	**.2930**
.7071	49 3241	8 9662	0 9295	1 1356	1 4308	.2929
.7072	52 2322	8 4211	0 4743	.7070 1355	1 7371	.2928
.7073	55 1408	7 8757	.4550 0188	.7069 1351	2 0432	.2927
.7074	58 0498	7 3301	.4549 5631	8 1344	2 3493	.2926
.7075	60 9592	6 7841	9 1071	7 1334	2 6552	.2925
.7076	63 8691	6 2379	8 6508	6 1322	2 9610	.2924
.7077	66 7795	5 6913	8 1943	5 1306	3 2668	.2923
.7078	69 6904	5 1445	7 7375	4 1288	3 5723	.2922
.7079	72 6017	4 5974	7 2804	3 1267	3 8778	.2921
.7080	.5475 5135	.3434 0500	.4546 8231	.7062 1243	.9564 1832	**.2920**
.7081	78 4258	3 5023	6 3655	1 1217	4 4884	.2919
.7082	81 3385	2 9543	5 9076	.7060 1187	4 7936	.2918
.7083	84 2517	2 4060	5 4495	.7059 1155	5 0986	.2917
.7084	87 1653	1 8574	4 9911	8 1119	5 4035	.2916
.7085	90 0794	1 3086	4 5324	7 1081	5 7083	.2915
.7086	92 9940	0 7594	4 0735	6 1040	6 0130	.2914
.7087	95 9091	.3430 2100	3 6143	5 0996	6 3175	.2913
.7088	.5498 8246	.3429 6602	3 1549	4 0950	6 6220	.2912
.7089	.5501 7406	9 1102	2 6951	3 0900	6 9263	.2911
.7090	.5504 6570	.3428 5599	.4542 2351	.7052 0848	.9567 2305	**.2910**
.7091	07 5739	8 0093	1 7749	1 0793	7 5346	.2909
.7092	10 4913	7 4584	1 3143	.7050 0735	7 8386	.2908
.7093	13 4091	6 9072	0 8536	.7049 0674	8 1425	.2907
.7094	16 3274	6 3557	.4540 3925	8 0610	8 4463	.2906
.7095	19 2462	5 8039	.4539 9312	7 0543	8 7499	.2905
.7096	22 1655	5 2519	9 4696	6 0474	9 0535	.2904
.7097	25 0852	4 6995	9 0077	5 0402	9 3569	.2903
.7098	28 0054	4 1468	8 5456	4 0327	9 6602	.2902
.7099	30 9261	3 5939	8 0832	3 0248	.9569 9634	.2901
.7100	.5533 8472	.3423 0407	.4537 6205	.7042 0168	.9570 2665	**.2900**

E^{-iL} = .0000,0000,2 E^{iL} .0000,0001   .0000,0000+   .0000,0000+   .0000,0000+   .0000,0000+

TABLE I
.7100

.2900

p	x	z	$\sqrt{pq}$	$\sqrt{1-p^2}$	$\sqrt{1-q^2}$	q
.7100	.5533 8472	.3423 0407	.4537 6205	.7042 0168	.9570 2665	**.2900**
.7101	36 7688	2 4871	7 1576	.7041 0084	0 5694	.2899
.7102	39 6909	1 9333	6 6944	.7039 9997	0 8723	.2898
.7103	42 6135	1 3792	6 2309	8 9908	1 1750	.2897
.7104	45 5365	0 8248	5 7672	7 9815	1 4776	.2896
.7105	48 4600	.3420 2701	5 3032	6 9720	1 7801	.2895
.7106	51 3840	.3419 7151	4 8389	5 9622	2 0825	.2894
.7107	54 3085	9 1598	4 3744	4 9521	2 3848	.2893
.7108	57 2334	8 6042	3 9096	3 9417	2 6870	.2892
.7109	60 1588	8 0484	3 4445	2 9310	2 9890	.2891
.7110	.5563 0847	.3417 4922	.4532 9792	.7031 9201	.9573 2910	**.2890**
.7111	66 0111	6 9357	2 5135	.7030 9088	3 5928	.2889
.7112	68 9379	6 3790	2 0477	.7029 8973	3 8945	.2888
.7113	71 8652	5 8219	1 5815	8 8855	4 1961	.2887
.7114	74 7930	5 2646	1 1151	7 8734	4 4976	.2886
.7115	77 7213	4 7070	0 6484	6 8610	4 7990	.2885
.7116	80 6500	4 1491	.4530 1815	5 8483	5 1002	.2884
.7117	83 5792	3 5909	.4529 7142	4 8353	5 4013	.2883
.7118	86 5089	3 0324	9 2467	3 8220	5 7024	.2882
.7119	89 4391	2 4736	8 7790	2 8085	6 0033	.2881
.7120	.5592 3698	.3411 9145	.4528 3109	.7021 7946	.9576 3041	**.2880**
.7121	95 3009	1 3551	7 8426	.7020 7805	6 6048	.2879
.7122	.5598 2326	0 7954	7 3741	.7019 7661	6 9053	.2878
.7123	.5601 1647	.3410 2354	6 9052	8 7514	7 2058	.2877
.7124	04 0973	.3409 6752	6 4361	7 7364	7 5061	.2876
.7125	07 0303	9 1146	5 9667	6 7211	7 8064	.2875
.7126	09 9639	8 5538	5 4971	5 7055	8 1065	.2874
.7127	12 8979	7 9926	5 0272	4 6897	8 4065	.2873
.7128	15 8324	7 4312	4 5570	3 6735	8 7064	.2872
.7129	18 7674	6 8695	4 0865	2 6571	9 0062	.2871
.7130	.5621 7029	.3406 3074	.4523 6158	.7011 6403	.9579 3058	**.2870**
.7131	24 6389	5 7451	3 1448	.7010 6233	9 6054	.2869
.7132	27 5753	5 1825	2 6735	.7009 6060	.9579 9048	.2868
.7133	30 5123	4 6196	2 2020	8 5884	.9580 2041	.2867
.7134	33 4497	4 0564	1 7302	7 5705	0 5033	.2866
.7135	36 3876	3 4929	1 2581	6 5523	0 8024	.2865
.7136	39 3260	2 9291	0 7858	5 5338	1 1014	.2864
.7137	42 2649	2 3650	.4520 3132	4 5150	1 4003	.2863
.7138	45 2043	1 8007	.4519 8403	3 4960	1 6990	.2862
.7139	48 1441	1 2360	9 3671	2 4766	1 9977	.2861
.7140	.5651 0845	.3400 6710	.4518 8937	.7001 4570	.9582 2962	**.2860**
.7141	54 0253	.3400 1058	8 4200	.7000 4371	2 5946	.2859
.7142	56 9667	.3399 5402	7 9460	.6999 4168	2 8929	.2858
.7143	59 9085	8 9744	7 4717	8 3963	3 1911	.2857
.7144	62 8508	8 4083	7 0972	7 3755	3 4891	.2856
.7145	65 7936	7 8418	6 5224	6 3544	3 7871	.2855
.7146	68 7369	7 2751	6 0474	5 3330	4 0849	.2854
.7147	71 6807	6 7081	5 5721	4 3113	4 3827	.2853
.7148	74 6250	6 1408	5 0965	3 2894	4 6803	.2852
.7149	77 5697	5 5732	4 6206	2 2671	4 9778	.2851
.7150	.5680 5150	.3395 0053	.4514 1444	.6991 2445	.9585 2752	**.2850**

$E^{-ii}=$ .0000,0000,2 $E^{ii}=$.0000,0001   .0000,0000+   .0000,0000+   .0000,0000+   .0000,0000+

TABLE I

p	x	z	$\sqrt{pq}$	$\sqrt{1-p^2}$	$\sqrt{1-q^2}$	q
.7150	.5680 5150	.3395 0053	.4514 1444	.6991 2445	.9585 2752	**.2850**
.7151	83 4607	4 4371	3 6680	.6990 2217	5 5724	.2849
.7152	86 4070	3 8686	3 1913	.6989 1985	5 8696	.2848
.7153	89 3537	3 2998	2 7144	8 1751	6 1666	.2847
.7154	92 3009	2 7307	2 2371	7 1514	6 4636	.2846
.7155	95 2487	2 1613	1 7596	6 1273	6 7604	.2845
.7156	.5698 1969	1 5916	1 2819	5 1030	7 0571	.2844
.7157	.5701 1456	1 0217	0 8038	4 0784	7 3537	.2843
.7158	04 0948	.3390 4514	.4510 3255	3 0535	7 6502	.2842
.7159	07 0445	.3389 8809	.4509 8469	2 0283	7 9465	.2841
.7160	.5709 9947	.3389 3100	.4509 3680	.6981 0028	.9588 2428	**.2840**
.7161	12 9454	8 7389	8 8889	.6979 9770	8 5389	.2839
.7162	15 8966	8 1674	8 4095	8 9509	8 8350	.2838
.7163	18 8483	7 5957	7 9298	7 9245	9 1309	.2837
.7164	21 8005	7 0236	7 4498	6 8979	9 4267	.2836
.7165	24 7532	6 4513	6 9696	5 8709	.9589 7224	.2835
.7166	27 7064	5 8787	6 4891	4 8437	.9590 0179	.2834
.7167	30 6601	5 3058	6 0083	3 8161	0 3134	.2833
.7168	33 6143	4 7326	5 5273	2 7883	0 6087	.2832
.7169	36 5690	4 1590	5 0459	1 7601	0 9040	.2831
.7170	.5739 5242	.3383 5852	.4504 5644	.6970 7317	.9591 1991	**.2830**
.7171	42 4799	3 0111	4 0825	.6969 7029	1 4941	.2829
.7172	45 4361	2 4367	3 6003	8 6739	1 7890	.2828
.7173	48 3928	1 8621	3 1179	7 6446	2 0838	.2827
.7174	51 3500	1 2871	2 6352	6 6150	2 3784	.2826
.7175	54 3077	0 7118	2 1523	5 5850	2 6730	.2825
.7176	57 2659	.3380 1362	1 6690	4 5548	2 9674	.2824
.7177	60 2246	.3379 5603	1 1855	3 5243	3 2617	.2823
.7178	63 1839	8 9842	0 7017	2 4935	3 5560	.2822
.7179	66 1436	8 4077	.4500 2177	1 4624	3 8501	.2821
.7180	.5769 1038	.3377 8309	.4499 7333	.6960 4310	.9594 1440	**.2820**
.7181	72 0645	7 2539	9 2487	.6959 3993	4 4379	.2819
.7182	75 0258	6 6765	8 7638	8 3673	4 7317	.2818
.7183	77 9875	6 0989	8 2787	7 3351	5 0253	.2817
.7184	80 9497	5 5209	7 7932	6 3025	5 3189	.2816
.7185	83 9125	4 9427	7 3075	5 2696	5 6123	.2815
.7186	86 8758	4 3641	6 8215	4 2364	5 9056	.2814
.7187	89 8395	3 7853	6 3353	3 2029	6 1988	.2813
.7188	92 8038	3 2062	5 8488	2 1692	6 4919	.2812
.7189	95 7686	2 6267	5 3619	1 1351	6 7848	.2811
.7190	.5798 7339	.3372 0470	.4494 8749	.6950 1007	.9597 0777	**.2810**
.7191	.5801 6997	1 4670	4 3875	.6949 0661	7 3704	.2809
.7192	04 6660	0 8867	3 8999	8 0311	7 6630	.2808
.7193	07 6329	.3370 3061	3 4120	6 9958	7 9556	.2807
.7194	10 6002	.3369 7252	2 9238	5 9603	8 2480	.2806
.7195	13 5681	9 1439	2 4353	4 9244	8 5403	.2805
.7196	16 5365	8 5624	1 9466	3 8882	8 8324	.2804
.7197	19 5054	7 9806	1 4576	2 8518	9 1245	.2803
.7198	22 4748	7 3985	0 9683	1 8150	9 4164	.2802
.7199	25 4447	6 8161	.4490 4787	.6940 7780	.9599 7083	.2801
.7200	.5828 4151	.3366 2334	.4489 9889	.6939 7406	.9600 0000	**.2800**

$E^{-ii} = \ $ $E^{ii} = .0000,0001$ .0000,0000+   .0000,0000+   .0000,0000+   .0000,0000+
.0000,0002

TABLE I

.7200 　　　　　　　　　　　　　　　　　　　　　　　　　　　　　　　　**.2800**

p	x	z	$\sqrt{pq}$	$\sqrt{1-p^2}$	$\sqrt{1-q^2}$	q
.7200	.5828 4151	.3366 2334	.4489 9889	.6939 7406	.9600 0000	**.2800**
.7201	31 3860	5 6505	9 4987	8 7030	0 2916	.2799
.7202	34 3575	5 0672	9 0084	7 6650	0 5831	.2798
.7203	37 3295	4 4836	8 5177	6 6268	0 8745	.2797
.7204	40 3019	3 8997	8 0267	5 5882	1 1658	.2796
.7205	43 2749	3 3155	7 5355	4 5494	1 4569	.2795
.7206	46 2485	2 7311	7 0440	3 5102	1 7480	.2794
.7207	49 2225	2 1463	6 5522	2 4708	2 0389	.2793
.7208	52 1970	1 5612	6 0602	1 4310	2 3297	.2792
.7209	55 1721	0 9758	5 5679	.6930 3910	2 6204	.2791
.7210	.5858 1477	.3360 3902	.4485 0753	.6929 3506	.9602 9110	**.2790**
.7211	61 1238	.3359 8042	4 5824	8 3100	3 2015	.2789
.7212	64 1004	9 2179	4 0892	7 2690	3 4919	.2788
.7213	67 0776	8 6314	3 5958	6 2278	3 7821	.2787
.7214	70 0552	8 0445	3 1021	5 1862	4 0723	.2786
.7215	73 0334	7 4574	2 6081	4 1444	4 3623	.2785
.7216	76 0121	6 8699	2 1138	3 1022	4 6522	.2784
.7217	78 9913	6 2822	1 6192	2 0597	4 9420	.2783
.7218	81 9710	5 6941	1 1244	.6921 0170	5 2317	.2782
.7219	84 9513	5 1058	0 6293	.6919 9739	5 5213	.2781
.7220	.5887 9321	.3354 5171	.4480 1339	.6918 9306	.9605 8107	**.2780**
.7221	90 9134	3 9282	.4479 6383	7 8869	6 1001	.2779
.7222	93 8953	3 3390	9 1423	6 8429	6 3893	.2778
.7223	96 8776	2 7494	8 6461	5 7987	6 6785	.2777
.7224	.5899 8605	2 1596	8 1496	4 7541	6 9675	.2776
.7225	.5902 8439	1 5694	7 6528	3 7092	7 2564	.2775
.7226	05 8279	0 9790	7 1558	2 6640	7 5452	.2774
.7227	08 8123	.3350 3883	6 6585	1 6186	7 8338	.2773
.7228	11 7973	.3349 7972	6 1609	.6910 5728	8 1224	.2772
.7229	14 7829	9 2059	5 6630	.6909 5267	8 4108	.2771
.7230	.5917 7689	.3348 6143	.4475 1648	.6908 4803	.9608 6992	**.2770**
.7231	20 7555	8 0224	4 6664	7 4336	8 9874	.2769
.7232	23 7426	7 4301	4 1676	6 3866	9 2755	.2768
.7233	26 7302	6 8376	3 6686	5 3393	9 5635	.2767
.7234	29 7184	6 2448	3 1693	4 2917	.9609 8514	.2766
.7235	32 7071	5 6517	2 6698	3 2438	.9610 1392	.2765
.7236	35 6963	5 0583	2 1699	2 1956	0 4268	.2764
.7237	38 6860	4 4645	1 6698	1 1471	0 7144	.2763
.7238	41 6763	3 8705	1 1694	.6900 0983	1 0018	.2762
.7239	44 6672	3 2762	0 6687	.6899 0491	1 2891	.2761
.7240	.5947 6585	.3342 6816	.4470 1678	.6897 9997	.9611 5764	**.2760**
.7241	50 6504	2 0867	.4469 6665	6 9500	1 8635	.2759
.7242	53 6428	1 4915	9 1650	5 8999	2 1504	.2758
.7243	56 6357	0 8959	8 6632	4 8496	2 4373	.2757
.7244	59 6292	.3340 3001	8 1611	3 7990	2 7241	.2756
.7245	62 6232	.3339 7040	7 6588	2 7480	3 0107	.2755
.7246	65 6177	9 1076	7 1561	1 6967	3 2972	.2754
.7247	68 6128	8 5109	6 6532	.6890 6452	3 5837	.2753
.7248	71 6085	7 9139	6 1500	.6889 5933	3 8700	.2752
.7249	74 6046	7 3166	5 6465	8 5411	4 1562	.2751
.7250	.5977 6013	.3336 7190	.4465 1428	.6887 4887	.9614 4423	**.2750**

$E^{-ii} = $ .0000,0000,2　　$E^{ii} = $ .0000,0001　　.0000,0000+　　.0000,0000+　　.0000,0000+　　.0000,0000+

TABLE I

p	x	z	$\sqrt{pq}$	$\sqrt{1-p^2}$	$\sqrt{1-q^2}$	q
.7250	.5977 6013	.3336 7190	.4465 1428	.6887 4887	.9614 4423	.2750
.7251	80 5985	6 1210	4 6387	6 4359	4 7282	.2749
.7252	83 5963	5 5228	4 1344	5 3828	5 0141	.2748
.7253	86 5946	4 9243	3 6298	4 3294	5 2998	.2747
.7254	89 5934	4 3255	3 1249	3 2757	5 5855	.2746
.7255	92 5928	3 7264	2 6197	2 2217	5 8710	.2745
.7256	95 5927	3 1270	2 1143	1 1673	6 1564	.2744
.7257	.5998 5931	2 5273	1 6086	.6880 1127	6 4417	.2743
.7258	.6001 5941	1 9273	1 1026	.6879 0578	6 7269	.2742
.7259	04 5956	1 3270	0 5963	8 0025	7 0120	.2741
.7260	.6007 5977	.3330 7264	.4460 0897	.6876 9470	.9617 2969	.2740
.7261	10 6003	.3330 1255	.4459 5828	5 8911	7 5818	.2739
.7262	13 6035	.3329 5242	9 0757	4 8350	7 8665	.2738
.7263	16 6072	8 9227	8 5683	3 7785	8 1511	.2737
.7264	19 6115	8 3209	8 0606	2 7217	8 4356	.2736
.7265	22 6163	7 7188	7 5526	1 6646	8 7200	.2735
.7266	25 6216	7 1164	7 0443	.6870 6073	9 0043	.2734
.7267	28 6275	6 5137	6 5358	.6869 5495	9 2885	.2733
.7268	31 6339	5 9107	6 0269	8 4915	9 5725	.2732
.7269	34 6409	5 3074	5 5178	7 4332	.9619 8565	.2731
.7270	.6037 6484	.3324 7037	.4455 0084	.6866 3746	.9620 1403	.2730
.7271	40 6565	4 0998	4 4987	5 3157	0 4241	.2729
.7272	43 6651	3 4956	3 9888	4 2564	0 7077	.2728
.7273	46 6742	2 8911	3 4785	3 1968	0 9912	.2727
.7274	49 6839	2 2863	2 9680	2 1370	1 2746	.2726
.7275	52 6941	1 6812	2 4572	1 0768	1 5578	.2725
.7276	55 7049	1 0757	1 9461	.6860 0163	1 8410	.2724
.7277	58 7163	.3320 4700	1 4347	.6858 9555	2 1240	.2723
.7278	61 7282	.3319 8640	0 9231	7 8944	2 4070	.2722
.7279	64 7406	9 2577	.4450 4111	6 8330	2 6898	.2721
.7280	.6067 7536	.3318 6510	.4449 8989	.6855 7713	.9622 9725	.2720
.7281	70 7672	8 0441	9 3864	4 7093	3 2551	.2719
.7282	73 7813	7 4369	8 8736	3 6469	3 5376	.2718
.7283	76 7959	6 8294	8 3605	2 5843	3 8200	.2717
.7284	79 8111	6 2215	7 8471	1 5213	4 1022	.2716
.7285	82 8269	5 6134	7 3335	.6850 4580	4 3844	.2715
.7286	85 8432	5 0050	6 8195	.6849 3944	4 6664	.2714
.7287	88 8601	4 3962	6 3053	8 3305	4 9484	.2713
.7288	91 8775	3 7872	5 7908	7 2663	5 2302	.2712
.7289	94 8955	3 1779	5 2760	6 2018	5 5119	.2711
.7290	.6097 9140	.3312 5682	.4444 7610	.6845 1370	.9625 7935	.2710
.7291	.6100 9331	1 9583	4 2456	4 0718	6 0750	.2709
.7292	03 9527	1 3480	3 7300	3 0064	6 3563	.2708
.7293	06 9729	0 7375	3 2140	1 9406	6 6376	.2707
.7294	09 9937	.3310 1266	2 6978	.6840 8745	6 9187	.2706
.7295	13 0150	.3309 5155	2 1813	.6839 8081	7 1997	.2705
.7296	16 0368	8 9040	1 6646	8 7414	7 4807	.2704
.7297	19 0593	8 2923	1 1475	7 6744	7 7615	.2703
.7298	22 0823	7 6802	0 6301	6 6071	8 0422	.2702
.7299	25 1058	7 0679	.4440 1125	5 5394	8 3228	.2701
.7300	.6128 1299	.3306 4552	.4439 5946	.6834 4714	.9628 6032	.2700

$E^{-ii} =$ .0000,0000,2 $E^{ii} =$ .0000,0001   .0000,0000+   .0000,0000+   .0000,0000+   .0000,0000+

TABLE I

p	x	z	$\sqrt{pq}$	$\sqrt{1-p^2}$	$\sqrt{1-q^2}$	q
.7300	.6128 1299	.3306 4552	.4439 5946	.6834 4714	.9628 6032	**.2700**
.7301	31 1546	5 8422	9 0764	3 4032	8 8836	.2699
.7302	34 1798	5 2290	8 5579	2 3346	9 1638	.2698
.7303	37 2056	4 6154	8 0391	1 2657	9 4440	.2697
.7304	40 2319	4 0015	7 5200	.6830 1965	.9629 7240	.2696
.7305	43 2589	3 3874	7 0007	.6829 1270	.9630 0039	.2695
.7306	46 2863	2 7729	6 4810	8 0571	0 2837	.2694
.7307	49 3144	2 1581	5 9611	6 9870	0 5634	.2693
.7308	52 3430	1 5430	5 4409	5 9165	0 8430	.2692
.7309	55 3722	0 9276	4 9204	4 8457	1 1224	.2691
.7310	.6158 4019	.3300 3119	.4434 3996	.6823 7746	.9631 4018	**.2690**
.7311	61 4322	.3299 6959	3 8786	2 7032	1 6810	.2689
.7312	64 4631	9 0797	3 3572	1 6315	1 9601	.2688
.7313	67 4945	8 4631	2 8355	.6820 5594	2 2391	.2687
.7314	70 5265	7 8462	2 3136	.6819 4871	2 5181	.2686
.7315	73 5591	7 2289	1 7914	8 4144	2 7968	.2685
.7316	76 5922	6 6114	1 2689	7 3414	3 0755	.2684
.7317	79 6259	5 9936	0 7461	6 2681	3 3541	.2683
.7318	82 6602	5 3755	.4430 2230	5 1945	3 6325	.2682
.7319	85 6950	4 7571	.4429 6997	4 1206	3 9109	.2681
.7320	.6188 7304	.3294 1384	.4429 1760	.6813 0463	.9634 1891	**.2680**
.7321	91 7664	3 5194	8 6521	1 9717	4 4672	.2679
.7322	94 8029	2 9000	8 1278	.6810 8969	4 7452	.2678
.7323	.6197 8401	2 2804	7 6033	.6809 8217	5 0231	.2677
.7324	.6200 8778	1 6605	7 0785	8 7461	5 3009	.2676
.7325	03 9160	1 0402	6 5534	7 6703	5 5786	.2675
.7326	06 9549	.3290 4197	6 0280	6 5942	5 8562	.2674
.7327	09 9943	.3289 7988	5 5023	5 5177	6 1336	.2673
.7328	13 0343	9 1777	4 9764	4 4409	6 4110	.2672
.7329	16 0749	8 5562	4 4501	3 3638	6 6882	.2671
.7330	.6219 1160	.3287 9345	.4423 9236	.6802 2864	.9636 9653	**.2670**
.7331	22 1577	7 3124	3 3968	1 2086	7 2423	.2669
.7332	25 2000	6 6900	2 8697	.6800 1306	7 5192	.2668
.7333	28 2428	6 0674	2 3423	.6799 0522	7 7960	.2667
.7334	31 2863	5 4444	1 8146	7 9735	8 0726	.2666
.7335	34 3303	4 8211	1 2866	6 8945	8 3492	.2665
.7336	37 3749	4 1975	0 7583	5 8152	8 6256	.2664
.7337	40 4200	3 5736	.4420 2297	4 7355	8 9020	.2663
.7338	43 4658	2 9494	.4419 7009	3 6556	9 1782	.2662
.7339	46 5121	2 3249	9 1718	2 5753	9 4543	.2661
.7340	.6249 5590	.3281 7001	.4418 6423	.6791 4947	.9639 7303	**.2660**
.7341	52 6065	1 0750	8 1126	.6790 4138	.9640 0062	.2659
.7342	55 6546	.3280 4496	7 5826	.6789 3325	0 2819	.2658
.7343	58 7032	.3279 8239	7 0523	8 2510	0 5576	.2657
.7344	61 7524	9 1979	6 5217	8 1691	0 8332	.2656
.7345	64 8023	8 5715	5 9908	6 0869	1 1086	.2655
.7346	67 8527	7 9449	5 4597	5 0043	1 3839	.2654
.7347	70 9036	7 3180	4 9282	3 9215	1 6591	.2653
.7348	73 9552	6 6907	4 3964	2 8383	1 9342	.2652
.7349	77 0074	6 0632	3 8644	1 7549	2 2092	.2651
.7350	.6280 0601	.3275 4353	.4413 3321	.6780 6711	.9642 4841	**.2650**

E^{-iL} .0000,0000 2 E^{iL} .0000,0001   .0000,0000+   .0000,0000+   .0000,0000+   .0000,0000+

TABLE I

p	x	z	$\sqrt{pq}$	$\sqrt{1-p^2}$	$\sqrt{1-q^2}$	q
.7350	.6280 0601	.3275 4353	.4413 3321	.6780 6711	.9642 4841	**.2650**
.7351	83 1134	4 8072	2 7995	.6779 5869	2 7589	.2649
.7352	86 1673	4 1787	2 2665	8 5025	3 0335	.2648
.7353	89 2218	3 5499	1 7333	7 4177	3 3081	.2647
.7354	92 2769	2 9209	1 1998	6 3326	3 5825	.2646
.7355	95 3326	2 2915	0 6660	5 2472	3 8569	.2645
.7356	.6298 3889	1 6618	.4410 1320	4 1615	4 1311	.2644
.7357	.6301 4457	1 0318	.4409 5976	3 0754	4 4052	.2643
.7358	04 5032	.3270 4015	9 0629	1 9891	4 6792	.2642
.7359	07 5612	.3269 7709	8 5280	.6770 9024	4 9530	.2641
.7360	.6310 6198	.3269 1400	.4407 9927	.6769 8154	.9645 2268	**.2640**
.7361	13 6790	8 5088	7 4572	8 7280	5 5005	.2639
.7362	16 7388	7 8773	6 9214	7 6404	5 7740	.2638
.7363	19 7992	7 2454	6 3853	6 5524	6 0474	.2637
.7364	22 8602	6 6133	5 8488	5 4641	6 3207	.2636
.7365	25 9217	5 9809	5 3121	4 3754	6 5940	.2635
.7366	28 9839	5 3481	4 7751	3 2865	6 8671	.2634
.7367	32 0467	4 7151	4 2378	2 1972	7 1400	.2633
.7368	35 1100	4 0817	3 7003	1 1076	7 4129	.2632
.7369	38 1740	3 4480	3 1624	.6760 0177	7 6857	.2631
.7370	.6341 2385	.3262 8141	.4402 6242	.6758 9274	.9647 9583	**.2630**
.7371	44 3036	2 1798	2 0858	7 8369	8 2309	.2629
.7372	47 3694	1 5452	1 5470	6 7460	8 5033	.2628
.7373	50 4357	0 9103	1 0080	5 6547	8 7756	.2627
.7374	53 5026	.3260 2751	.4400 4686	4 5632	9 0478	.2626
.7375	56 5701	.3259 6396	.4399 9290	3 4713	9 3199	.2625
.7376	59 6382	9 0038	9 3890	2 3791	9 5919	.2624
.7377	62 7070	8 3677	8 8488	1 2866	.9649 8638	.2623
.7378	65 7763	7 7313	8 3083	.6750 1938	.9650 1355	.2622
.7379	68 8462	7 0945	7 7675	.6749 1006	0 4072	.2621
.7380	.6371 9167	.3256 4575	.4397 2264	.6748 0071	.9650 6787	**.2620**
.7381	74 9878	5 8201	6 6850	6 9133	0 9502	.2619
.7382	78 0596	5 1825	6 1433	5 8191	1 2215	.2618
.7383	81 1319	4 5445	5 6013	4 7247	1 4927	.2617
.7384	84 2048	3 9063	5 0590	3 6299	1 7638	.2616
.7385	87 2784	3 2677	4 5165	2 5348	2 0348	.2615
.7386	90 3525	2 6288	3 9736	1 4393	2 3056	.2614
.7387	93 4272	1 9896	3 4304	.6740 3435	2 5764	.2613
.7388	96 5026	1 3501	2 8870	.6739 2474	2 8470	.2612
.7389	.6399 5785	0 7103	2 3432	8 1510	3 1176	.2611
.7390	.6402 6551	.3250 0702	.4391 7992	.6737 0543	.9653 3880	**.2610**
.7391	05 7323	.3249 4298	1 2548	5 9572	3 6583	.2609
.7392	08 8100	8 7891	0 7102	4 8598	3 9285	.2608
.7393	11 8884	8 1480	.4390 1653	3 7620	4 1986	.2607
.7394	14 9674	7 5067	.4389 6200	2 6640	4 4886	.2606
.7395	18 0470	6 8650	9 0745	1 5656	4 7385	.2605
.7396	21 1272	6 2231	8 5287	.6730 4668	5 0082	.2604
.7397	24 2080	5 5808	7 9826	.6729 3678	5 2779	.2603
.7398	27 2894	4 9382	7 4362	8 2684	5 5474	.2602
.7399	30 3715	4 2954	6 8894	7 1687	5 8168	.2601
.7400	.6433 4541	.3243 6522	.4386 3424	.6726 0687	.9656 0862	**.2600**

$E^{-i\iota}=$ $E^{i\iota}=$.0000,0001   .0000,0000+   .0000,0000+   .0000,0000+   .0000,0000+
.0000,0000,2

TABLE I

p	x	z	$\sqrt{pq}$	$\sqrt{1-p^2}$	$\sqrt{1-q^2}$	q
.7400	.6433 4541	.3243 6522	.4386 3424	.6726 0687	.9656 0862	.2600
.7401	36 5374	3 0087	5 7951	4 9683	6 3554	.2599
.7402	39 6212	2 3649	5 2475	3 8676	6 6245	.2598
.7403	42 7057	1 7207	4 6996	2 7666	6 8934	.2597
.7404	45 7908	1 0763	4 1515	1 6653	7 1623	.2596
.7405	48 8765	.3240 4316	3 6030	.6720 5636	7 4311	.2595
.7406	51 9628	.3239 7865	3 0542	.6719 4616	7 6997	.2594
.7407	55 0497	9 1412	2 5051	8 3592	7 9683	.2593
.7408	58 1373	8 4955	1 9557	7 2566	8 2367	.2592
.7409	61 2254	7 8496	1 4061	6 1536	8 5050	.2591
.7410	.6464 3142	.3237 2033	.4380 8561	.6715 0503	.9658 7732	.2590
.7411	67 4036	6 5567	.4380 3058	3 9466	9 0413	.2589
.7412	70 4936	5 9098	.4379 7552	2 8426	9 3093	.2588
.7413	73 5842	5 2626	9 2044	1 7383	9 5772	.2587
.7414	76 6755	4 6151	8 6532	.6710 6337	.9659 8449	.2586
.7415	79 7673	3 9673	8 1018	.6709 5287	.9660 1126	.2585
.7416	82 8598	3 3191	7 5500	8 4234	0 3801	.2584
.7417	85 9530	2 6707	6 9979	7 3177	0 6475	.2583
.7418	89 0467	2 0219	6 4456	6 2117	0 9149	.2582
.7419	92 1410	1 3729	5 8929	5 1054	1 1821	.2581
.7420	.6495 2360	.3230 7235	.4375 3400	.6703 9988	.9661 4492	.2580
.7421	.6498 3316	.3230 0738	4 7867	2 8918	1 7162	.2579
.7422	.6501 4278	.3229 4238	4 2332	1 7845	1 9830	.2578
.7423	04 5246	8 7735	3 6793	.6700 6769	2 2498	.2577
.7424	07 6221	8 1229	3 1252	.6699 5689	2 5164	.2576
.7425	10 7202	7 4720	2 5708	8 4606	2 7830	.2575
.7426	13 8189	6 8208	2 0160	7 3520	3 0494	.2574
.7427	16 9182	6 1693	1 4610	6 2431	3 3157	.2573
.7428	20 0182	5 5174	0 9056	5 1338	3 5819	.2572
.7429	23 1188	4 8653	.4370 3500	4 0241	3 8480	.2571
.7430	.6526 2200	.3224 2128	.4369 7940	.6692 9142	.9664 1140	.2570
.7431	29 3218	3 5600	9 2378	1 8039	4 3799	.2569
.7432	32 4243	2 9069	8 6813	.6690 6932	4 6457	.2568
.7433	35 5274	2 2535	8 1244	.6689 5823	4 9113	.2567
.7434	38 6312	1 5998	7 5673	8 4710	5 1769	.2566
.7435	41 7355	0 9458	7 0098	7 3593	5 4423	.2565
.7436	44 8405	.3220 2915	6 4521	6 2474	5 7076	.2564
.7437	47 9461	.3219 6368	5 8941	5 1351	5 9728	.2563
.7438	51 0524	8 9819	5 3357	4 0224	6 2379	.2562
.7439	54 1593	8 3266	4 7771	2 9095	6 5029	.2561
.7440	.6557 2668	.3217 6710	.4364 2181	.6681 7962	.9666 7678	.2560
.7441	60 3749	7 0152	3 6589	.6680 6825	7 0326	.2559
.7442	63 4837	6 3590	3 0994	.6679 5685	7 2972	.2558
.7443	66 5932	5 7025	2 5395	8 4542	7 5618	.2557
.7444	69 7032	5 0457	1 9794	7 3396	7 8262	.2556
.7445	72 8139	4 3885	1 4189	6 2246	8 0906	.2555
.7446	75 9252	3 7311	0 8582	5 1093	8 3548	.2554
.7447	79 0372	3 0733	.4360 2971	3 9936	8 6189	.2553
.7448	82 1498	2 4153	.4359 7358	2 8776	8 8829	.2552
.7449	85 2630	1 7569	9 1741	1 7613	9 1468	.2551
.7450	.6588 3769	.3211 0982	.4358 6122	.6670 6446	.9669 4105	.2550

$E^{-ii} \underline{\llcorner}$ E^{ii}=.0000,0001   .0000,0000+   .0000,0000+   .0000,0000+   .0000,0000+
.0000,0000,2

TABLE I

.7450   .2550

p	x	z	$\sqrt{pq}$	$\sqrt{1-p^2}$	$\sqrt{1-q^2}$	q
.7450	.6588 3769	.3211 0982	.4358 6122	.6670 6446	.9669 4105	**.2550**
.7451	91 4914	.3210 4392	8 0499	.6669 5276	9 6742	.2549
.7452	94 6066	.3209 7799	7 4873	8 4103	.9669 9377	.2548
.7453	.6597 7224	9 1203	6 9245	7 2926	.9670 2012	.2547
.7454	.6600 8388	8 4604	6 3613	6 1746	0 4645	.2546
.7455	03 9559	7 8001	5 7979	5 0563	0 7277	.2545
.7456	07 0737	7 1396	5 2341	3 9376	0 9908	.2544
.7457	10 1920	6 4787	4 6700	2 8185	1 2538	.2543
.7458	13 3110	5 8176	4 1056	1 6992	1 5167	.2542
.7459	16 4307	5 1561	3 5410	.6660 5795	1 7795	.2541
.7460	.6619 5510	.3204 4943	.4352 9760	.6659 4594	.9672 0422	**.2540**
.7461	22 6719	3 8322	2 4107	8 3391	2 3047	.2539
.7462	25 7935	3 1697	1 8451	7 2183	2 5672	.2538
.7463	28 9158	2 5070	1 2792	6 0973	2 8295	.2537
.7464	32 0386	1 8440	0 7130	4 9759	3 0917	.2536
.7465	35 1621	1 1806	.4350 1465	3 8541	3 3539	.2535
.7466	38 2863	.3200 5169	.4349 5797	2 7321	3 6159	.2534
.7467	41 4111	.3199 8529	9 0126	1 6097	3 8778	.2533
.7468	44 5366	9 1886	8 4452	.6650 4869	4 1395	.2532
.7569	47 6627	8 5240	7 8775	.6649 3638	4 4012	.2531
.7470	.6650 7895	.3197 8591	.4347 3095	.6648 2404	.9674 6628	**.2530**
.7471	53 9169	7 1939	6 7412	7 1166	4 9242	.2529
.7472	57 0450	6 5283	6 1726	5 9925	5 1856	.2528
.7473	60 1737	5 8625	5 6036	4 8680	5 4468	.2527
.7474	63 3031	5 1963	5 0344	3 7432	5 7079	.2526
.7475	66 4331	4 5298	4 4649	2 6181	5 9689	.2525
.7476	69 5638	3 8630	3 8950	1 4926	6 2298	.2524
.7477	72 6951	3 1959	3 3249	.6640 3668	6 4906	.2523
.7478	75 8271	2 5285	2 7544	.6639 2406	6 7513	.2522
.7479	78 9597	1 8607	2 1837	8 1141	7 0119	.2521
.7480	.6682 0930	.3191 1927	.4341 6126	.6636 9873	.9677 2723	**.2520**
.7481	85 2269	.3190 5243	1 0412	5 8601	7 5327	.2519
.7482	88 3615	.3189 8556	.4340 4696	4 7325	7 7929	.2518
.7483	91 4968	9 1866	.4339 8976	3 6047	8 0531	.2517
.7484	94 6327	8 5173	9 3253	2 4765	8 3131	.2516
.7485	.6697 7693	7 8477	8 7527	1 3479	8 5730	.2515
.7486	.6700 9065	7 1778	8 1798	.6630 2190	8 8328	.2514
.7487	04 0444	6 5075	7 6066	.6629 0898	9 0925	.2513
.7488	07 1830	5 8370	7 0331	7 9602	9 3520	.2512
.7489	10 3222	5 1661	6 4593	6 8302	9 6115	.2511
.7490	.6713 4621	.3184 4949	.4335 8851	.6625 7000	.9679 8709	**.2510**
.7491	16 6027	3 8234	5 3107	4 5693	.9680 1301	.2509
.7492	19 7439	3 1516	4 7360	3 4384	0 3892	.2508
.7493	22 8857	2 4794	4 1609	2 3071	0 6483	.2507
.7494	26 0283	1 8070	3 5856	1 1754	0 9072	.2506
.7495	29 1715	1 1342	3 0099	.6620 0434	1 1660	.2505
.7496	32 3154	.3180 4612	2 4340	.6618 9111	1 4247	.2504
.7497	35 4599	.3179 7878	1 8577	7 7784	1 6833	.2503
.7498	38 6051	9 1141	1 2811	6 6454	1 9417	.2502
.7499	41 7510	8 4401	0 7042	5 5120	2 2001	.2501
.7500	.6744 8975	.3177 7657	.4330 1270	.6614 3783	.9682 4584	**.2500**

$E^{-ii}=$ .0000,0000,2 $E^{ii}=$ .0000,0001   .0000,0000+   .0000,0000+   .0000,0000+   .0000,0000+

TABLE I

.7500   **.2500**

p	x	z	$\sqrt{pq}$	$\sqrt{1-p^2}$	$\sqrt{1-q^2}$	q
.7500	.6744 8975	.3177 7657	.4330 1270	.6614 3783	.9682 4584	**.2500**
.7501	48 0447	7 0911	.4329 5495	3 2442	2 7165	.2499
.7502	51 1926	6 4161	8 9717	2 1098	2 9745	.2498
.7503	54 3411	5 7408	8 3936	.6610 9750	3 2325	.2497
.7504	57 4903	5 0652	7 8152	.6609 8399	3 4903	.2496
.7505	60 6402	4 3893	7 2364	8 7045	3 7480	.2495
.7506	63 7907	3 7131	6 6574	7 5687	4 0056	.2494
.7507	66 9419	3 0366	6 0780	6 4325	4 2631	.2493
.7508	70 0938	2 3597	5 4984	5 2961	4 5204	.2492
.7509	73 2464	1 6826	4 9184	4 1592	4 7777	.2491
.7510	.6776 3996	.3171 0051	.4324 3381	.6603 0220	.9685 0348	**.2490**
.7511	79 5535	.3170 3273	3 7575	1 8845	5 2919	.2489
.7512	82 7081	.3169 6492	3 1766	.6600 7466	5 5488	.2488
.7513	85 8634	8 9707	2 5954	.6599 6084	5 8056	.2487
.7514	89 0193	8 2920	2 0139	8 4698	6 0624	.2486
.7515	92 1759	7 6129	1 4321	7 3309	6 3190	.2485
.7516	95 3332	6 9336	0 8499	6 1916	6 5755	.2484
.7517	.6798 4912	6 2539	.4320 2675	5 0520	6 8318	.2483
.7518	.6801 6499	5 5739	.4319 6847	3 9120	7 0881	.2482
.7519	04 8092	4 8935	9 1016	2 7717	7 3443	.2481
.7520	.6807 9692	.3164 2129	.4318 5183	.6591 6311	.9687 6003	**.2480**
.7521	11 1299	3 5320	7 9346	.6590 4900	7 8563	.2479
.7522	14 2912	2 8507	7 3506	.6589 3487	8 1121	.2478
.7523	17 4533	2 1691	6 7663	8 2070	8 3678	.2477
.7524	20 6160	1 4872	6 1816	7 0649	8 6234	.2476
.7525	23 7794	0 8050	5 5967	5 9225	8 8789	.2475
.7526	26 9435	.3160 1224	5 0115	4 7797	9 1343	.2474
.7527	30 1083	.3159 4396	4 4259	3 6366	9 3896	.2473
.7528	33 2737	8 7564	3 8401	2 4931	9 6448	.2472
.7529	36 4399	8 0729	3 2539	1 3493	.9689 8998	.2471
.7530	.6839 6067	.3157 3891	.4312 6674	.6580 2052	.9690 1548	**.2470**
.7531	42 7742	6 7050	2 0806	.6579 0606	0 4096	.2469
.7532	45 9424	6 0206	1 4935	7 9158	0 6644	.2468
.7533	49 1113	5 3358	0 9061	6 7706	0 9190	.2467
.7534	52 2809	4 6507	.4310 3183	5 6250	1 1735	.2466
.7535	55 4512	3 9654	.4309 7303	4 4791	1 4279	.2465
.7536	58 6221	3 2797	9 1419	3 3328	1 6822	.2464
.7537	61 7938	2 5936	8 5532	2 1862	1 9364	.2463
.7538	64 9661	1 9073	7 9643	.6571 0392	2 1905	.2462
.7539	68 1392	1 2206	7 3750	.6569 8919	2 4444	.2461
.7540	.6871 3129	.3150 5337	.4306 7853	.6568 7442	.9692 6983	**.2460**
.7541	74 4873	.3149 8464	6 1954	7 5961	2 9520	.2459
.7542	77 6624	9 1588	5 6052	6 4477	3 2057	.2458
.7543	80 8382	8 4708	5 0146	5 2990	3 4592	.2457
.7544	84 0147	7 7826	4 4238	4 1499	3 7126	.2456
.7545	87 1919	7 0940	3 8326	3 0005	3 9659	.2455
.7546	90 3697	6 4052	3 2411	1 8507	4 2191	.2454
.7547	93 5483	5 7160	2 6493	.6560 7005	4 4722	.2453
.7548	96 7276	5 0265	2 0572	.6559 5500	4 7252	.2452
.7549	.6899 9075	4 3366	1 4648	8 3991	4 9780	.2451
.7550	.6903 0882	.3143 6465	.4300 8720	.6557 2479	.9695 2308	**.2450**

$E^{-ii}=$ .0000,0002 $E^{ii}=$ .0000,0001   .0000,0000+   .0000,0000+   .0000,0000+   .0000,0000+

TABLE I

.7550 .2450

p	x	z	$\sqrt{pq}$	$\sqrt{1-p^2}$	$\sqrt{1-q^2}$	q
.7550	.6903 0882	.3143 6465	.4300 8720	.6557 2479	.9695 2308	**.2450**
.7551	06 2696	2 9560	.4300 2789	6 0963	5 4834	.2449
.7552	09 4517	2 2652	.4299 6856	4 9444	5 7360	.2448
.7553	12 6344	1 5741	9 0919	3 7921	5 9884	.2447
.7554	15 8179	0 8827	8 4979	2 6395	6 2407	.2446
.7555	19 0021	.3140 1910	7 9036	1 4865	6 4929	.2445
.7556	22 1870	.3139 4989	7 3089	.6550 3331	6 7450	.2444
.7557	25 3725	8 8065	6 7140	.6549 1794	6 9970	.2443
.7558	28 5588	8 1138	6 1187	8 0254	7 2489	.2442
.7559	31 7458	7 4208	5 5231	6 8709	7 5007	.2441
.7560	.6934 9335	.3136 7275	.4294 9272	.6545 7162	.9697 7523	**.2440**
.7561	38 1219	6 0338	4 3310	4 5610	8 0039	.2439
.7562	41 3110	5 3398	3 7345	3 4055	8 2553	.2438
.7563	44 5008	4 6456	3 1377	2 2497	8 5066	.2437
.7564	47 6913	3 9509	2 5405	.6541 0935	8 7579	.2436
.7565	50 8825	3 2560	1 9430	.6539 9369	9 0090	.2435
.7566	54 0744	2 5608	1 3452	8 7800	9 2600	.2434
.7567	57 2671	1 8652	0 7471	7 6227	9 5109	.2433
.7568	60 4604	1 1693	.4290 1487	6 4651	.9699 7616	.2432
.7569	63 6544	.3130 4731	.4289 5500	5 3071	.9700 0123	.2431
.7570	.6966 8492	.3129 7766	.4288 9509	.6534 1488	.9700 2629	**.2430**
.7571	70 0447	9 0797	8 3515	2 9901	0 5133	.2429
.7572	73 2408	8 3826	7 7519	1 8310	0 7637	.2428
.7573	76 4377	7 6851	7 1519	.6530 6716	1 0139	.2427
.7574	79 6353	6 9873	6 5515	.6529 5118	1 2640	.2426
.7575	82 8337	6 2892	5 9509	8 3516	1 5141	.2425
.7576	86 0327	5 5907	5 3499	7 1911	1 7640	.2424
.7577	89 2325	4 8920	4 7487	6 0303	2 0138	.2423
.7578	92 4329	4 1929	4 1471	4 8690	2 2634	.2422
.7579	95 6341	3 4935	3 5451	3 7075	2 5130	.2421
.7580	.6998 8360	.3122 7937	.4282 9429	.6522 5455	.9702 7625	**.2420**
.7581	.7002 0386	2 0937	2 3404	1 3832	3 0119	.2419
.7582	05 2420	1 3933	1 7375	.6520 2205	3 2611	.2418
.7583	08 4460	.3120 6927	1 1343	.6519 0575	3 5102	.2417
.7584	11 6508	.3119 9917	.4280 5308	7 8941	3 7593	.2416
.7585	14 8563	9 2903	.4279 9270	6 7304	4 0082	.2415
.7586	18 0625	8 5887	9 3228	5 5663	4 2570	.2414
.7587	21 2694	7 8867	8 7184	4 4018	4 5057	.2413
.7588	24 4771	7 1844	8 1136	3 2370	4 7543	.2412
.7589	27 6855	6 4818	7 5085	2 0718	5 0028	.2411
.7590	.7030 8946	.3115 7789	.4276 9031	.6510 9062	.9705 2512	**.2410**
.7591	34 1044	5 0756	6 2973	.6509 7403	5 4994	.2409
.7592	37 3150	4 3721	5 6913	8 5740	5 7476	.2408
.7593	40 5263	3 6682	5 0849	7 4074	5 9956	.2407
.7594	43 7383	2 9640	4 4782	6 2404	6 2436	.2406
.7595	46 9510	2 2594	3 8712	5 0730	6 4914	.2405
.7596	50 1645	1 5546	3 2639	3 9053	6 7391	.2404
.7597	53 3787	0 8494	2 6562	2 7372	6 9867	.2403
.7598	56 5936	.3110 1439	2 0482	1 5687	7 2342	.2402
.7599	59 8092	.3109 4381	1 4399	.6500 3999	7 4816	.2401
.7600	.7063 0256	.3108 7319	.4270 8313	.6499 2307	.9707 7289	**.2400**

$E^{-ii}=$ .0000,0000,3 $E^{ii}=$.0000,0001   .0000,0000+   .0000,0000+   .0000,0000+   .0000,0000+

TABLE I

p	x	z	$\sqrt{pq}$	$\sqrt{1-p^2}$	$\sqrt{1-q^2}$	q
.7600	.7063 0256	.3108 7319	.4270 8313	.6499 2307	.9707 7289	.2400
.7601	66 2427	8 0255	.4270 2224	8 0612	7 9761	.2399
.7602	69 4605	7 3187	.4269 6131	6 8913	8 2231	.2398
.7603	72 6791	6 6116	9 0035	5 7210	8 4701	.2397
.7604	75 8984	5 9041	8 3936	4 5503	8 7169	.2396
.7605	79 1185	5 1964	7 7834	3 3793	8 9636	.2395
.7606	82 3393	4 4883	7 1728	2 2079	9 2103	.2394
.7607	85 5608	3 7799	6 5620	.6491 0362	9 4568	.2393
.7608	88 7830	3 0712	5 9508	.6489 8641	9 7032	.2392
.7609	92 0060	2 3622	5 3393	8 6916	.9709 9495	.2391
.7610	.7095 2297	.3101 6528	.4264 7274	.6487 5188	.9710 1957	.2390
.7611	.7098 4542	0 9431	4 1153	6 3456	0 4418	.2389
.7612	.7101 6794	.3100 2331	3 5028	5 1720	0 6877	.2388
.7613	04 9053	.3099 5228	2 8900	3 9981	0 9336	.2387
.7614	08 1320	8 8121	2 2769	2 8238	1 1793	.2386
.7615	11 3594	8 1012	1 6634	1 6491	1 4250	.2385
.7616	14 5875	7 3899	1 0496	.6480 4741	1 6705	.2384
.7617	17 8164	6 6782	.4260 4355	.6479 2987	1 9159	.2383
.7618	21 0461	5 9663	.4259 8211	8 1229	2 1612	.2382
.7619	24 2765	5 2540	9 2064	6 9467	2 4064	.2381
.7620	.7127 5076	.3094 5414	.4258 5913	.6475 7702	.9712 6515	.2380
.7621	30 7395	3 8285	7 9759	4 5933	2 8965	.2379
.7622	33 9721	3 1153	7 3602	3 4161	3 1414	.2378
.7623	37 2054	2 4017	6 7442	2 2385	3 3862	.2377
.7624	40 4395	1 6879	6 1278	.6471 0605	3 6308	.2376
.7625	43 6744	0 9737	5 5111	.6469 8821	3 8754	.2375
.7626	46 9100	.3090 2591	4 8941	8 7034	4 1198	.2374
.7627	50 1464	.3089 5443	4 2768	7 5243	4 3642	.2373
.7628	53 3834	8 8291	3 6591	6 3449	4 6084	.2372
.7629	56 6213	8 1136	3 0411	5 1650	4 8525	.2371
.7630	.7159 8599	.3087 3978	.4252 4228	.6463 9848	.9715 0965	.2370
.7631	63 0992	6 6816	1 8042	2 8043	5 3404	.2369
.7632	66 3393	5 9652	1 1852	1 6233	5 5842	.2368
.7633	69 5802	5 2484	.4250 5660	.6460 4420	5 8279	.2367
.7634	72 8218	4 5312	.4249 9464	.6459 2603	6 0714	.2366
.7635	76 0642	3 8138	9 3264	8 0783	6 3149	.2365
.7636	79 3073	3 0960	8 7062	6 8958	6 5582	.2364
.7637	82 5512	2 3779	8 0856	5 7131	6 8015	.2363
.7638	85 7958	1 6595	7 4647	4 5299	7 0446	.2362
.7639	89 0412	0 9408	6 8434	3 3463	7 2876	.2361
.7640	.7192 2873	.3080 2217	.4246 2219	.6452 1624	.9717 5306	.2360
.7641	95 5342	.3079 5023	5 6000	.6450 9781	7 7734	.2359
.7642	.7198 7819	8 7826	4 9777	.6449 7935	8 0161	.2358
.7643	.7202 0303	8 0626	4 3552	8 6085	8 2586	.2357
.7644	05 2794	7 3422	3 7323	7 4231	8 5011	.2356
.7645	08 5294	6 6215	3 1091	6 2373	8 7435	.2355
.7646	11 7801	5 9005	2 4856	5 0511	8 9857	.2354
.7647	15 0316	5 1791	1 8617	3 8646	9 2279	.2353
.7648	18 2838	4 4575	1 2376	2 6777	9 4699	.2352
.7649	21 5367	3 7355	.4240 6130	1 4904	9 7119	.2351
.7650	.7224 7905	.3073 0132	.4239 9882	.6440 3028	.9719 9537	.2350

$E^{-ii} =$ .0000,0000,3 $E^{ii} =$ .0000,0001   .0000,0000+   .0000,0000+   .0000,0000+   .0000,0000+

TABLE I

p	x	z	√pq	√1−p²	√1−q²	q
.7650	.7224 7905	.3073 0132	.4239 9882	.6440 3028	.9719 9537	**.2350**
.7651	28 0450	2 2905	9 3630	.6439 1148	.9720 1954	.2349
.7652	31 3003	1 5676	8 7375	7 9264	0 4370	.2348
.7653	34 5564	0 8443	8 1117	6 7376	0 6785	.2347
.7654	37 8132	.3070 1207	7 4856	5 5485	0 9199	.2346
.7655	41 0708	.3069 3967	6 8591	4 3589	1 1612	.2345
.7656	44 3291	8 6724	6 2323	3 1690	1 4024	.2344
.7657	47 5882	7 9478	5 6052	1 9788	1 6434	.2343
.7658	50 8481	7 2229	4 9777	.6430 7881	1 8844	.2342
.7659	54 1088	6 4977	4 3499	.6429 5971	2 1252	.2341
.7660	.7257 3702	.3065 7721	.4233 7218	.6428 4057	.9722 3660	**.2340**
.7661	60 6324	5 0462	3 0933	7 2139	2 6066	.2339
.7662	63 8954	4 3200	2 4645	6 0218	2 8471	.2338
.7663	67 1592	3 5934	1 8354	4 8293	3 0875	.2337
.7664	70 4237	2 8665	1 2060	3 6364	3 3278	.2336
.7665	73 6890	2 1393	.4230 5762	2 4431	3 5680	.2335
.7666	76 9551	1 4118	.4229 9461	1 2494	3 8081	.2334
.7667	80 2219	.3060 6839	9 3157	.6420 0554	4 0481	.2333
.7668	83 4896	.3059 9558	8 6849	.6418 8610	4 2879	.2332
.7669	86 7580	9 2272	8 0538	7 6662	4 5277	.2331
.7670	.7290 0272	.3058 4984	.4227 4224	.6416 4710	.9724 7673	**.2330**
.7671	93 2972	7 7692	6 7906	5 2754	5 0069	.2329
.7672	96 5679	7 0397	6 1585	4 0795	5 2463	.2328
.7673	.7299 8394	6 3099	5 5261	2 8832	5 4856	.2327
.7674	.7303 1117	5 5798	4 8934	1 6865	5 7249	.2326
.7675	06 3848	4 8493	4 2603	.6410 4895	5 9640	.2325
.7676	09 6587	4 1185	3 6269	.6409 2920	6 2030	.2324
.7677	12 9333	3 3874	2 9931	8 0942	6 4418	.2323
.7678	16 2088	2 6559	2 3591	6 8960	6 6806	.2322
.7679	19 4850	1 9241	1 7246	5 6974	6 9193	.2321
.7680	.7322 7620	.3051 1920	.4221 0899	.6404 4984	.9727 1579	**.2320**
.7681	26 0398	.3050 4596	.4220 4548	3 2991	7 3963	.2319
.7682	29 3184	.3049 7268	.4219 8194	2 0993	7 6347	.2318
.7683	32 5978	8 9937	9 1837	.6400 8992	7 8729	.2317
.7684	35 8780	8 2603	8 5476	.6399 6987	8 1110	.2316
.7685	39 1589	7 5265	7 9112	8 4979	8 3490	.2315
.7686	42 4407	6 7925	7 2745	7 2966	8 5869	.2314
.7687	45 7232	6 0581	6 6374	6 0950	8 8247	.2313
.7688	49 0066	5 3233	6 0000	4 8930	9 0624	.2312
.7689	52 2907	4 5883	5 3623	3 6906	9 3000	.2311
.7690	.7355 5756	.3043 8529	.4214 7242	.6392 4878	.9729 5375	**.2310**
.7691	58 8613	3 1171	4 0858	1 2846	.9729 7749	.2309
.7692	62 1478	2 3811	3 4470	.6390 0811	.9730 0121	.2308
.7693	65 4351	1 6447	2 8080	.6388 8771	0 2493	.2307
.7694	68 7232	0 9080	2 1686	7 6728	0 4863	.2306
.7695	72 0121	.3040 1710	1 5288	6 4681	0 7233	.2305
.7696	75 3018	.3039 4336	0 8887	5 2630	0 9601	.2304
.7697	78 5922	8 6959	.4210 2483	4 0576	1 1968	.2303
.7698	81 8835	7 9579	.4209 6076	2 8517	1 4334	.2302
.7699	85 1756	7 2195	8 9665	1 6455	1 6699	.2301
.7700	.7388 4685	.3036 4808	.4208 3251	.6380 4389	.9731 9063	**.2300**

$E^{-ii}=$ .0000,0000,3 $E^{ii}=$.0000,0001   .0000,0000+   .0000,0000+   .0000,0000+   .0000,0000+

TABLE I

p	x	z	$\sqrt{pq}$	$\sqrt{1-p^2}$	$\sqrt{1-q^2}$	q
.7700	.7388 4685	.3036 4808	.4208 3251	.6380 4389	.9731 9063	**.2300**
.7701	91 7622	5 7418	7 6833	.6379 2319	2 1426	.2299
.7702	95 0567	5 0025	7 0412	8 0245	2 3787	.2298
.7703	.7398 3520	4 2628	6 3988	6 8167	2 6148	.2297
.7704	.7401 6480	3 5228	5 7561	5 6085	2 8508	.2296
.7705	04 9449	2 7825	5 1130	4 4000	3 0866	.2295
.7706	08 2426	2 0418	4 4695	3 1910	3 3224	.2294
.7707	11 5411	1 3008	3 8258	1 9817	3 5580	.2293
.7708	14 8405	.3030 5595	3 1817	.6370 7720	3 7935	.2292
.7709	18 1406	.3029 8179	2 5372	.6369 5619	4 0289	.2291
.7710	.7421 4415	.3029 0759	.4201 8924	.6368 3514	.9734 2642	**.2290**
.7711	24 7432	8 3336	1 2473	7 1406	4 4994	.2289
.7712	28 0458	7 5909	.4200 6019	5 9293	4 7345	.2288
.7713	31 3492	6 8480	.4199 9561	4 7177	4 9695	.2287
.7714	34 6534	6 1047	9 3099	3 5056	5 2044	.2286
.7715	37 9583	5 3610	8 6635	2 2932	5 4391	.2285
.7716	41 2641	4 6171	8 0167	.6361 0804	5 6738	.2284
.7717	44 5708	3 8728	7 3695	.6359 8672	5 9083	.2283
.7718	47 8782	3 1282	6 7221	8 6536	6 1428	.2282
.7719	51 1864	2 3832	6 0742	7 4397	6 3771	.2281
.7720	.7454 4955	.3021 6379	.4195 4261	.6356 2253	.9736 6113	**.2280**
.7721	57 8054	0 8923	4 7776	5 0105	6 8454	.2279
.7722	61 1161	.3020 1464	4 1288	3 7954	7 0794	.2278
.7723	64 4276	.3019 4001	3 4796	2 5799	7 3133	.2277
.7724	67 7399	8 6535	2 8301	1 3639	7 5471	.2276
.7725	71 0530	7 9065	2 1802	.6350 1476	7 7808	.2275
.7726	74 3670	7 1593	1 5300	.6348 9309	8 0144	.2274
.7727	77 6818	6 4117	0 8795	7 7138	8 2478	.2273
.7728	80 9974	5 6637	.4190 2286	6 4964	8 4812	.2272
.7729	84 3138	4 9155	.4189 5774	5 2785	8 7144	.2271
.7730	.7487 6311	.3014 1669	.4188 9259	.6344 0602	.9738 9476	**.2270**
.7731	90 9492	3 4179	8 2740	2 8416	9 1806	.2269
.7732	94 2681	2 6687	7 6218	1 6225	9 4135	.2268
.7733	.7497 5878	1 9191	6 9692	.6340 4031	9 6463	.2267
.7734	.7500 9084	1 1692	6 3163	.6339 1832	.9739 8791	.2266
.7735	04 2298	.3010 4189	5 6630	7 9630	.9740 1117	.2265
.7736	07 5520	.3009 6683	5 0094	6 7424	0 3441	.2264
.7737	10 8750	8 9174	4 3555	5 5214	0 5765	.2263
.7738	14 1989	8 1661	3 7012	4 3000	0 8088	.2262
.7739	17 5236	7 4146	3 0466	3 0782	1 0410	.2261
.7740	.7520 8491	.3006 6626	.4182 3917	.6331 8560	.9741 2730	**.2260**
.7741	24 1755	5 9104	1 7364	.6330 6334	1 5050	.2259
.7742	27 5027	5 1578	1 0807	.6329 4104	1 7368	.2258
.7743	30 8307	4 4049	.4180 4247	8 1870	1 9685	.2257
.7744	34 1595	3 6516	.4179 7684	6 9633	2 2002	.2256
.7745	37 4892	2 8981	9 1117	5 7391	2 4317	.2255
.7746	40 8198	2 1441	8 4547	4 5145	2 6631	.2254
.7747	44 1512	1 3899	7 7974	3 2896	2 8944	.2253
.7748	47 4834	.3000 6353	7 1397	2 0642	3 1256	.2252
.7749	50 8164	.2999 8804	6 4817	.6320 8385	3 3567	.2251
.7750	.7554 1503	.2999 1251	.4175 8233	.6319 6123	.9743 5876	**.2250**

$E^{-ii}=$ $E^{iL}=.0000,0001$ .0000,0000+ .0000,0000+ .0000,0001 .0000,0000+
.0000,00003

TABLE I

p	x	z	$\sqrt{pq}$	$\sqrt{1-p^2}$	$\sqrt{1-q^2}$	q
.7750	.7554 1503	.2999 1251	.4175 8233	.6319 6123	.9743 5876	**.2250**
.7751	57 4850	8 3696	5 1645	8 3858	3 8185	.2249
.7752	60 8206	7 6137	4 5055	7 1589	4 0493	.2248
.7753	64 1570	6 8574	3 8461	5 9315	4 2799	.2247
.7754	67 4942	6 1008	3 1863	4 7038	4 5105	.2246
.7755	70 8323	5 3439	2 5262	3 4757	4 7409	.2245
.7756	74 1712	4 5867	1 8658	2 2471	4 9712	.2244
.7757	77 5110	3 8291	1 2050	.6311 0182	5 2014	.2243
.7758	80 8516	3 0712	.4170 5438	.6309 7889	5 4315	.2242
.7759	84 1931	2 3129	.4169 8824	8 5592	5 6615	.2241
.7760	.7587 5354	.2991 5543	.4169 2206	.6307 3291	.9745 8914	**.2240**
.7761	90 8786	0 7954	8 5584	6 0986	6 1212	.2239
.7762	94 2226	.2990 0361	7 8959	4 8676	6 3509	.2238
.7763	.7597 5675	.2989 2765	7 2330	3 6363	6 5805	.2237
.7764	.7600 9132	8 5166	6 5698	2 4046	6 8099	.2236
.7765	04 2598	7 7564	5 9063	.6301 1725	7 0393	.2235
.7766	07 6072	6 9958	5 2424	.6299 9400	7 2685	.2234
.7767	10 9554	6 2348	4 5781	8 7071	7 4977	.2233
.7768	14 3046	5 4736	3 9135	7 4738	7 7267	.2232
.7769	17 6546	4 7120	3 2486	6 2401	7 9556	.2231
.7770	.7621 0054	.2983 9500	.4162 5833	.6295 0060	.9748 1844	**.2230**
.7771	24 3571	3 1878	1 9177	3 7714	8 4132	.2229
.7772	27 7096	2 4252	1 2517	2 5365	8 6418	.2228
.7773	31 0631	1 6622	.4160 5854	1 3012	8 8702	.2227
.7774	34 4173	0 8990	.4159 9187	.6290 0655	9 0986	.2226
.7775	37 7724	.2980 1354	9 2517	.6288 8294	9 3269	.2225
.7776	41 1284	.2979 3714	8 5844	7 5929	9 5551	.2224
.7777	44 4853	8 6071	7 9167	6 3559	.9749 7831	.2223
.7778	47 8430	7 8425	7 2486	5 1186	.9750 0111	.2222
.7779	51 2016	7 0776	6 5802	3 8809	0 2389	.2221
.7780	.7654 5610	.2976 3123	.4155 9115	.6282 6428	.9750 4667	**.2220**
.7781	57 9213	5 5466	5 2424	1 4042	0 6943	.2219
.7782	61 2824	4 7807	4 5729	.6280 1653	0 9218	.2218
.7783	64 6444	4 0144	3 9031	.6278 9259	1 1492	.2217
.7784	68 0073	3 2478	3 2330	7 6862	1 3765	.2216
.7785	71 3711	2 4808	2 5625	6 4460	1 6037	.2215
.7786	74 7357	1 7135	1 8916	5 2055	1 8308	.2214
.7787	78 1012	0 9458	1 2204	3 9645	2 0578	.2213
.7788	81 4676	.2970 1779	.4150 5489	2 7232	2 2847	.2212
.7789	84 8348	.2969 4095	.4149 8770	1 4814	2 5114	.2211
.7790	.7688 2029	.2968 6409	.4149 2047	.6270 2392	.9752 7381	**.2210**
.7791	91 5719	7 8719	8 5322	.6268 9967	2 9646	.2209
.7792	94 9417	7 1026	7 8592	7 7537	3 1911	.2208
.7793	.7698 3125	6 3329	7 1859	6 5103	3 4174	.2207
.7794	.7701 6841	5 5629	6 5123	5 2665	3 6436	.2206
.7795	05 0565	4 7926	5 8383	4 0223	3 8697	.2205
.7796	08 4299	4 0219	5 1639	2 7777	4 0958	.2204
.7797	11 8041	3 2509	4 4892	1 5326	4 3217	.2203
.7798	15 1793	2 4795	3 8142	.6260 2872	4 5475	.2202
.7799	18 5553	1 7079	3 1388	.6259 0414	4 7731	.2201
.7800	.7721 9321	.2960 9358	.4142 4630	.6257 7951	.9754 9987	**.2200**

$E^{-ii}=$ .0000,0000,3 $E^{ii}=$ .0000,0001 .0000,0000+ .0000,0000+ .0000,0001 .0000,0000+

TABLE I

p	x	z	$\sqrt{pq}$	$\sqrt{1-p^2}$	$\sqrt{1-q^2}$	q
.7800	.7721 9321	.2960 9358	.4142 4630	.6257 7951	.9754 9987	.2200
.7801	25 3099	.2960 1635	1 7869	6 5485	5 2242	.2199
.7802	28 6885	.2959 3908	1 1105	5 3014	5 4496	.2198
.7803	32 0680	8 6177	.4140 4337	4 0540	5 6748	.2197
.7804	35 4484	7 8444	.4139 7565	2 8061	5 9000	.2196
.7805	38 8297	7 0706	9 0790	1 5578	6 1250	.2195
.7806	42 2119	6 2966	8 4011	.6250 3091	6 3499	.2194
.7807	45 5949	5 5222	7 7229	.6249 0600	6 5748	.2193
.7808	48 9789	4 7475	7 0444	7 8105	6 7995	.2192
.7809	52 3637	3 9724	6 3654	6 5606	7 0214	.2191
.7810	.7755 7494	.2953 1970	.4135 6862	.6245 3102	.9757 2486	.2190
.7811	59 1360	2 4213	5 0065	4 0595	7 4730	.2189
.7812	62 5235	1 6452	4 3265	2 8083	7 6973	.2188
.7813	65 9119	0 8688	3 6462	1 5568	7 9214	.2187
.7814	69 3012	.2950 0920	2 9655	.6240 3048	8 1455	.2186
.7815	72 6913	.2949 3149	2 2845	.6239 0524	8 3695	.2185
.7816	76 0824	8 5375	1 6031	7 7996	8 5933	.2184
.7817	79 4744	7 7597	0 9213	6 5464	8 8171	.2183
.7818	82 8672	6 9816	.4130 2392	5 2928	9 0407	.2182
.7819	86 2610	6 2031	.4129 5568	4 0387	9 2643	.2181
.7820	.7789 6556	.2945 4243	.4128 8739	.6232 7843	.9759 4877	.2180
.7821	93 0511	4 6452	8 1908	1 5294	9 7110	.2179
.7822	96 4476	3 8657	7 5072	.6230 2742	.9759 9342	.2178
.7823	.7799 8449	3 0859	6 8234	.6229 0185	.9760 1573	.2177
.7824	.7803 2432	2 3057	6 1391	7 7624	0 3803	.2176
.7825	06 6423	1 5252	5 4545	6 5058	0 6032	.2175
.7826	10 0424	.2940 7444	4 7696	5 2489	0 8260	.2174
.7827	13 4433	.2939 9632	4 0843	3 9916	1 0487	.2173
.7828	16 8452	9 1817	3 3986	2 7338	1 2712	.2172
.7829	20 2479	8 3999	2 7126	1 4756	1 4937	.2171
.7830	.7823 6516	.2937 6177	.4122 0262	.6220 2170	.9761 7160	.2170
.7831	27 0562	6 8351	1 3395	.6218 9580	1 9383	.2169
.7832	30 4617	6 0523	.4120 6524	7 6986	2 1604	.2168
.7833	33 8681	5 2690	.4119 9649	6 4388	2 3824	.2167
.7834	37 2754	4 4855	9 2771	5 1785	2 6044	.2166
.7835	40 6836	3 7016	8 5890	3 9178	2 8262	.2165
.7836	44 0927	2 9173	7 9004	2 6568	3 0479	.2164
.7837	47 5027	2 1328	7 2116	1 3953	3 2695	.2163
.7838	50 9137	1 3478	6 5223	.6210 1333	3 4910	.2162
.7839	54 3255	.2930 5626	5 8327	.6208 8710	3 7124	.2161
.7840	.7857 7383	.2929 7770	.4115 1428	.6207 6082	.9763 9336	.2160
.7841	61 1520	8 9910	4 4525	6 3451	4 1548	.2159
.7842	64 5666	8 2047	3 7618	5 0815	4 3759	.2158
.7843	67 9821	7 4181	3 0708	3 8175	4 5968	.2157
.7844	71 3985	6 6311	2 3794	2 5530	4 8177	.2156
.7845	74 8159	5 8438	1 6876	1 2882	5 0384	.2155
.7846	78 2342	5 0562	0 9955	.6200 0229	5 2590	.2154
.7847	81 6534	4 2682	.4110 3030	.6198 7572	5 4796	.2153
.7848	85 0735	3 4799	.4109 6102	7 4911	5 7000	.2152
.7849	88 4945	2 6912	8 9170	6 2246	5 9203	.2151
.7850	.7891 9165	.2921 9022	.4108 2235	.6194 9576	.9766 1405	.2150

$\epsilon^{-i\iota}=$ .0000,0003 $\epsilon^{i\iota}=$ .0000,0001   .0000,0000+   .0000,0000+   .0000,0001   .0000,0000+

TABLE I

.7850   .2150

p	x	z	$\sqrt{pq}$	$\sqrt{1-p^2}$	$\sqrt{1-q^2}$	q
.7850	.7891 9165	.2921 9022	.4108 2235	.6194 9576	.9766 1405	.2150
.7851	95 3394	1 1128	7 5295	3 6903	6 3606	.2149
.7852	.7898 7632	.2920 3231	6 8353	2 4225	6 5806	.2148
.7853	.7902 1880	.2919 5330	6 1406	.6191 1543	6 8004	.2147
.7854	05 6136	8 7427	5 4457	.6189 8856	7 0202	.2146
.7855	09 0402	7 9519	4 7503	8 6166	7 2399	.2145
.7856	12 4678	7 1608	4 0546	7 3471	7 4594	.2144
.7857	15 8962	6 3694	3 3585	6 0772	7 6789	.2143
.7858	19 3256	5 5777	2 6621	4 8069	7 8982	.2142
.7859	22 7559	4 7856	1 9653	3 5361	8 1175	.2141
.7860	.7926 1872	.2913 9931	.4101 2681	.6182 2650	.9768 3366	.2140
.7861	29 6194	3 2003	.4100 5706	.6180 9934	8 5556	.2139
.7862	33 0525	2 4072	.4099 8727	.6179 7214	8 7745	.2138
.7863	36 4866	1 6137	9 1744	8 4489	8 9933	.2137
.7864	39 9215	0 8199	8 4758	7 1761	9 2120	.2136
.7865	43 3574	.2910 0257	7 7768	5 9028	9 4306	.2135
.7866	46 7943	.2909 2312	7 0775	4 6291	9 6491	.2134
.7867	50 2321	8 4364	6 3778	3 3549	.9769 8675	.2133
.7868	53 6709	7 6412	5 6777	2 0804	.9770 0858	.2132
.7869	57 1106	6 8456	4 9773	.6170 8054	0 3039	.2131
.7870	.7960 5512	.2906 0498	.4094 2765	.6169 5300	.9770 5220	.2130
.7871	63 9928	5 2535	3 5753	8 2541	0 7399	.2129
.7872	67 4353	4 4570	2 8738	6 9779	0 9578	.2128
.7873	70 8787	3 6600	2 1719	5 7012	1 1755	.2127
.7874	74 3231	2 8628	1 4697	4 3241	1 3931	.2126
.7875	77 7685	2 0652	0 7670	3 1465	1 6107	.2125
.7876	81 2148	1 2672	.4090 0641	1 8685	1 8281	.2124
.7877	84 6620	.2900 4689	.4089 3607	.6160 5902	2 0454	.2123
.7878	88 1102	.2899 6703	8 6570	.6159 3113	2 2626	.2122
.7879	91 5593	8 8713	7 9529	8 0321	2 4797	.2121
.7880	.7995 0094	.2898 0720	.4087 2485	.6156 7524	.9772 6967	.2120
.7881	.7998 4605	7 2723	6 5436	5 4723	2 9135	.2119
.7882	.8001 9125	6 4723	5 8385	4 1917	3 1303	.2118
.7883	05 3654	5 6719	5 1329	2 9108	3 3470	.2117
.7884	08 8193	4 8712	4 4270	1 6294	3 5635	.2116
.7885	12 2742	4 0702	3 7207	.6150 3476	3 7800	.2115
.7886	15 7300	3 2688	3 0141	.6149 0653	3 9963	.2114
.7887	19 1868	2 4670	2 3071	7 7826	4 2126	.2113
.7888	22 6445	1 6649	1 5997	6 4995	4 4287	.2112
.7889	26 1032	0 8625	0 8919	5 2159	4 6447	.2111
.7890	.8029 5629	.2890 0597	.4080 1838	.6143 9320	.9774 8606	.2110
.7891	33 0235	.2889 2566	.4079 4753	2 6476	5 0764	.2109
.7892	36 4851	8 4531	8 7665	1 3627	5 2921	.2108
.7893	39 9476	7 6493	8 0573	.6140 0774	5 5077	.2107
.7894	43 4111	6 8451	7 3477	.6138 7917	5 7232	.2106
.7895	46 8756	6 0406	6 6377	7 5056	5 9386	.2105
.7896	50 3411	5 2357	5 9274	6 2190	6 1538	.2104
.7897	53 8075	4 4305	5 2167	4 9320	6 3690	.2103
.7898	57 2748	3 6250	4 5056	3 6446	6 5841	.2102
.7899	60 7432	2 8191	3 7942	2 3567	6 7990	.2101
.7900	.8064 2125	.2882 0128	.4073 0824	.6131 0684	.9777 0139	.2100

$E^{-ii}=$ $E^{ii}=.0000,0001$ .0000,0000+   .0000,0000+   .0000,0001   .0000,0000+
.0000,0000,3

TABLE I

p	x	z	√pq	√1−p²	√1−q²	q
.7900	.8064 2125	.2882 0128	.4073 0824	.6131 0684	.9777 0139	**.2100**
.7901	67 6828	1 2062	2 3702	.6129 7797	7 2286	.2099
.7902	71 1540	.2880 3993	1 6576	8 4905	7 4432	.2098
.7903	74 6263	.2879 5920	0 9447	7 2009	7 6577	.2097
.7904	78 0995	8 7844	.4070 2314	5 9109	7 8722	.2096
.7905	81 5737	7 9764	.4069 5178	4 6204	8 0865	.2095
.7906	85 0488	7 1680	8 8038	3 3295	8 3007	.2094
.7907	88 5249	6 3594	8 0894	2 0381	8 5148	.2093
.7908	92 0020	5 5503	7 3746	.6120 7464	8 7288	.2092
.7909	95 4801	4 7410	6 6594	.6119 4541	8 9426	.2091
.7910	.8098 9592	.2873 9312	.4065 9439	.6118 1615	.9779 1564	**.2090**
.7911	.8102 4392	3 1212	5 2280	6 8684	9 3701	.2089
.7912	05 9203	2 3108	4 5118	5 5749	9 5836	.2088
.7913	09 4023	1 5000	3 7951	4 2809	.9779 7971	.2087
.7914	12 8853	.2870 6889	3 0781	2 9865	.9780 0104	.2086
.7915	16 3692	.2869 8774	2 3608	1 6917	0 2237	.2085
.7916	19 8542	9 0656	1 6430	.6110 3964	0 4368	.2084
.7917	23 3401	8 2534	0 9249	.6109 1007	0 6498	.2083
.7918	26 8271	7 4409	.4060 2064	7 8045	0 8627	.2082
.7919	30 3150	6 6281	.4059 4875	6 5079	1 0756	.2081
.7920	.8133 8039	.2865 8149	.4058 7683	.6105 2109	.9781 2883	**.2080**
.7921	37 2938	5 0013	8 0487	3 9134	1 5009	.2079
.7922	40 7847	4 1874	7 3287	2 6155	1 7133	.2078
.7923	44 2766	3 3732	6 6083	1 3172	1 9257	.2077
.7924	47 7695	2 5586	5 8876	.6100 0184	2 1380	.2076
.7925	51 2634	1 7436	5 1665	.6098 7191	2 3502	.2075
.7926	54 7582	0 9283	4 4450	7 4195	2 5622	.2074
.7927	58 2541	.2860 1127	3 7231	6 1193	2 7742	.2073
.7928	61 7510	.2859 2967	3 0009	4 8188	2 9860	.2072
.7929	65 2488	8 4803	2 2782	3 5178	3 1978	.2071
.7930	.8168 7477	.2857 6636	.4051 5553	.6092 2163	.9783 4094	**.2070**
.7931	72 2476	6 8466	0 8319	.6090 9145	3 6210	.2069
.7932	75 7484	6 0292	.4050 1081	.6089 6121	3 8324	.2068
.7933	79 2503	5 2114	.4049 3840	8 3094	4 0437	.2067
.7934	82 7532	4 3933	8 6595	7 0062	4 2549	.2066
.7935	86 2570	3 5749	7 9347	5 7025	4 4660	.2065
.7936	89 7619	2 7561	7 2094	4 3984	4 6770	.2064
.7937	93 2678	1 9369	6 4838	3 0939	4 8879	.2063
.7938	.8196 7747	1 1174	5 7578	1 7889	5 0987	.2062
.7939	.8200 2826	.2850 2975	5 0314	.6080 4835	5 3093	.2061
.7940	.8203 7915	.2849 4773	.4044 3046	.6079 1776	.9785 5199	**.2060**
.7941	07 3014	8 6568	3 5775	7 8713	5 7304	.2059
.7942	10 8123	7 8359	2 8500	6 5645	5 9407	.2058
.7943	14 3243	7 0146	2 1221	5 2573	6 1510	.2057
.7944	17 8372	6 1930	1 3938	3 9496	6 3611	.2056
.7945	21 3512	5 3711	.4040 6652	2 6415	6 5712	.2055
.7946	24 8662	4 5487	.4039 9361	1 3330	6 7811	.2054
.7947	28 3822	3 7261	9 2067	.6070 0240	6 9909	.2053
.7948	31 8992	2 9031	8 4769	.6068 7145	7 2006	.2052
.7949	35 4172	2 0797	7 7468	7 4046	7 4102	.2051
.7950	.8238 9363	.2841 2560	.4037 0162	.6066 0943	.9787 6197	**.2050**

E⁻ⁱⁱ=
.0000,0000,+   Eⁱⁱ=.0000,0001   .0000,0000+   .0000,0000+   .0000,0001   .0000,0000+

TABLE I

p	x	z	$\sqrt{pq}$	$\sqrt{1-p^2}$	$\sqrt{1-q^2}$	q
.7950	.8238 9363	.2841 2560	.4037 0162	.6066 0943	.9787 6197	.2050
.7951	42 4564	.2840 4319	6 2853	4 7835	7 8291	.2049
.7952	45 9775	.2839 6075	5 5540	3 4723	8 0384	.2048
.7953	49 4996	8 7827	4 8223	2 1606	8 2476	.2047
.7954	53 0228	7 9576	4 0902	.6060 8485	8 4567	.2046
.7955	56 5469	7 1321	3 3578	.6059 5359	8 6656	.2045
.7956	60 0721	6 3063	2 6250	8 2228	8 8745	.2044
.7957	63 5984	5 4801	1 8917	6 9094	9 0833	.2043
.7958	67 1256	4 6536	1 1581	5 5954	9 2919	.2042
.7959	70 6539	3 8267	.4030 4242	4 2810	9 5004	.2041
.7960	.8274 1832	.2832 9994	.4029 6898	.6052 9662	.9789 7089	.2040
.7961	77 7135	2 1718	8 9551	1 6509	.9789 9172	.2039
.7962	81 2449	1 3439	8 2200	.6050 3352	.9790 1254	.2038
.7963	84 7773	.2830 5156	7 4845	.6049 0190	0 3335	.2037
.7964	88 3108	.2829 6869	6 7486	7 7024	0 5416	.2036
.7965	91 8453	8 8579	6 0123	6 3853	0 7495	.2035
.7966	95 3808	8 0286	5 2756	5 0677	0 9573	.2034
.7967	.8298 9173	7 1989	4 5386	3 7497	1 1649	.2033
.7968	.8302 4549	6 3688	3 8012	2 4313	1 3725	.2032
.7969	05 9935	5 5384	3 0634	.6041 1124	1 5800	.2031
.7970	.8309 5332	.2824 7076	.4022 3252	.6039 7930	.9791 7874	.2030
.7971	13 0739	3 8765	1 5866	8 4732	1 9946	.2029
.7972	16 6157	3 0450	0 8477	7 1530	2 2018	.2028
.7973	20 1585	2 2131	.4020 1083	5 8323	2 4088	.2027
.7974	23 7023	1 3809	.4019 3686	4 5111	2 6158	.2026
.7975	27 2472	.2820 5484	8 6285	3 1895	2 8226	.2025
.7976	30 7931	.2819 7155	7 8880	1 8674	3 0294	.2024
.7977	34 3401	8 8822	7 1471	.6030 5448	3 2360	.2023
.7978	37 8881	8 0486	6 4059	.6029 2218	3 4425	.2022
.7979	41 4372	7 2147	5 6642	7 8984	3 6489	.2021
.7980	.8344 9873	.2816 3803	.4014 9222	.6026 5745	.9793 8552	.2020
.7981	48 5385	5 5457	4 1797	5 2501	4 0614	.2019
.7982	52 0907	4 7106	3 4369	3 9253	4 2675	.2018
.7983	55 6440	3 8752	2 6937	2 6000	4 4735	.2017
.7984	59 1984	3 0395	1 9501	.6021 2743	4 6794	.2016
.7985	62 7538	2 2034	1 2062	.6019 9481	4 8851	.2015
.7986	66 3103	1 3669	.4010 4618	8 6214	5 0908	.2014
.7987	69 8678	.2810 5301	.4009 7171	7 2943	5 2964	.2013
.7988	73 4264	.2809 6930	8 9719	5 9668	5 5018	.2012
.7989	76 9860	8 8555	8 2264	4 6387	5 7072	.2011
.7990	.8380 5467	.2808 0176	.4007 4805	.6013 3102	.9795 9124	.2010
.7991	84 1085	7 1793	6 7342	1 9813	6 1175	.2009
.7992	87 6713	6 3408	5 9875	.6010 6519	6 3226	.2008
.7993	91 2352	5 5018	5 2404	.6009 3220	6 5275	.2007
.7994	94 8001	4 6625	4 4930	7 9917	6 7323	.2006
.7995	.8398 3662	3 8228	3 7451	6 6609	6 9370	.2005
.7996	.8401 9333	2 9828	2 9969	5 3296	7 1416	.2004
.7997	05 5014	2 1425	2 2482	3 9979	7 3461	.2003
.7998	09 0706	1 3017	1 4992	2 6657	7 5505	.2002
.7999	12 6409	.2800 4606	0 7498	1 3331	7 7548	.2001
.8000	.8416 2123	.2799 6192	.4000 0000	.6000 0000	.9797 9590	.2000

$E^{-ii}=$ $E^{ii}=.0000,0001$ .0000,0000+   .0000,0000+   .0000,0001   .0000,0000+
.0000,0000,4

TABLE I

Page 98

.8000 .2000

p	x	z	$\sqrt{pq}$	$\sqrt{1-p^2}$	$\sqrt{1-q^2}$	q
.8000	.8416 2123	.2799 6192	.4000 0000	.6000 0000	.9797 9590	**.2000**
.8001	19 7848	8 7774	.3999 2498	.5998 6664	8 1630	.1999
.8002	23 3583	7 9352	8 4992	7 3324	8 3670	.1998
.8003	26 9329	7 0927	7 7482	5 9979	8 5709	.1997
.8004	30 5086	6 2499	6 9969	4 6630	8 7746	.1996
.8005	34 0854	5 4066	6 2451	3 3275	8 9783	.1995
.8006	37 6632	4 5630	5 4930	1 9917	9 1818	.1994
.8007	41 2421	3 7191	4 7404	.5990 6553	9 3852	.1993
.8008	44 8221	2 8748	3 9875	.5989 3185	9 5886	.1992
.8009	48 4032	2 0301	3 2342	7 9812	9 7918	.1991
.8010	.8451 9854	.2791 1851	.3992 4804	.5986 6435	.9799 9949	**.1990**
.8011	55 5686	.2790 3397	1 7263	5 3053	.9800 1979	.1989
.8012	59 1530	.2789 4940	0 9718	3 9666	0 4008	.1988
.8013	62 7384	8 6479	.3990 2169	2 6274	0 6036	.1987
.8014	66 3249	7 8015	.3989 4616	.5981 2878	0 8063	.1986
.8015	69 9125	6 9546	8 7059	.5979 9477	1 0089	.1985
.8016	73 5012	6 1075	7 9498	8 6072	1 2114	.1984
.8017	77 0910	5 2599	7 1934	7 2662	1 4137	.1983
.8018	80 6819	4 4121	6 4365	5 9247	1 6160	.1982
.8019	84 2738	3 5638	5 6792	4 5827	1 8181	.1981
.8020	.8487 8669	.2782 7152	.3984 9216	.5973 2403	.9802 0202	**.1980**
.8021	91 4611	1 8662	4 1635	1 8974	2 2221	.1979
.8022	95 0563	1 0169	3 4051	.5970 5541	2 4240	.1978
.8023	.8498 6527	.2780 1672	2 6462	.5969 2102	2 6257	.1977
.8024	.8502 2501	.2779 3172	1 8870	7 8660	2 8273	.1976
.8025	05 8487	8 4668	1 1274	6 5212	3 0289	.1975
.8026	09 4483	7 6160	.3980 3673	5 1759	3 2303	.1974
.8027	13 0491	6 7649	.3979 6069	3 8302	3 4316	.1973
.8028	16 6510	5 9134	8 8461	2 4840	3 6328	.1972
.8029	20 2539	5 0615	8 0848	.5961 1374	3 8339	.1971
.8030	.8523 8580	.2774 2093	.3977 3232	.5959 7903	.9804 0349	**.1970**
.8031	27 4632	3 3568	6 5612	8 4427	4 2358	.1969
.8032	31 0695	2 5039	5 7988	7 0946	4 4365	.1968
.8033	34 6769	1 6506	5 0360	5 7460	4 6372	.1967
.8034	38 2854	.2770 7969	4 2728	4 3970	4 8378	.1966
.8035	41 8950	.2769 9429	3 5092	3 0475	5 0382	.1965
.8036	45 5058	9 0885	2 7451	1 6976	5 2386	.1964
.8037	49 1176	8 2338	1 9807	.5950 3471	5 4388	.1963
.8038	52 7306	7 3787	1 2159	.5948 9962	5 6390	.1962
.8039	56 3447	6 5233	.3970 4507	7 6448	5 8390	.1961
.8040	.8559 9599	.2765 6674	.3969 6851	.5946 2930	.9806 0390	**.1960**
.8041	63 5762	4 8113	8 9191	4 9406	6 2388	.1959
.8042	67 1936	3 9547	8 1527	3 5878	6 4385	.1958
.8043	70 8122	3 0978	7 3859	2 2345	6 6381	.1957
.8044	74 4319	2 2406	6 6187	.5940 8807	6 8376	.1956
.8045	78 0527	1 3829	5 8511	.5939 5265	7 0370	.1955
.8046	81 6746	.2760 5250	5 0831	8 1718	7 2363	.1954
.8047	85 2977	.2759 6666	4 3147	6 8166	7 4355	.1953
.8048	88 9219	8 8079	3 5459	5 4609	7 6346	.1952
.8049	92 5472	7 9488	2 7767	4 1047	7 8336	.1951
.8050	.8596 1736	.2757 0894	.3962 0071	.5932 7481	.9808 0324	**.1950**

$E^{-ii}=$ $E^{ii}=$.0000,0002 .0000,0000+ .0000,0000+ .0000,0001 .0000,0000+
.0000,0000,4

TABLE I

.8050   **.1950**

p	x	z	$\sqrt{pq}$	$\sqrt{1-p^2}$	$\sqrt{1-q^2}$	q
.8050	.8596 1736	.2757 0894	.3962 0071	.5932 7481	.9808 0324	**.1950**
.8051	.8599 8012	6 2296	1 2371	1 3910	8 2312	.1949
.8052	.8603 4299	5 3694	.3960 4666	.5930 0334	8 4298	.1948
.8053	07 0597	4 5089	.3959 6958	.5928 6753	8 6284	.1947
.8054	10 6907	3 6480	8 9246	7 3168	8 8268	.1946
.8055	14 3228	2 7868	8 1530	5 9577	9 0252	.1945
.8056	17 9561	1 9251	7 3810	4 5982	9 2234	.1944
.8057	21 5905	1 0632	6 6085	3 2382	9 4215	.1943
.8058	25 2260	.2750 2008	5 8357	1 8777	9 6196	.1942
.8059	28 8627	.2749 3381	5 0625	.5920 5168	.9809 8175	.1941
.8060	.8632 5005	.2748 4751	.3954 2888	.5919 1553	.9810 0153	**.1940**
.8061	36 1395	7 6116	3 5148	7 7934	0 2130	.1939
.8062	39 7796	6 7478	2 7403	6 4310	0 4106	.1938
.8063	43 4208	5 8837	1 9655	5 0681	0 6081	.1937
.8064	47 0632	5 0191	1 1902	3 7048	0 8055	.1936
.8065	50 7067	4 1543	.3950 4145	2 3409	1 0028	.1935
.8066	54 3514	3 2890	.3949 6385	.5910 9766	1 1999	.1934
.8067	57 9973	2 4234	8 8620	.5909 6117	1 3970	.1933
.8068	61 6443	1 5574	8 0851	8 2464	1 5940	.1932
.8069	65 2924	.2740 6911	7 3078	6 8806	1 7908	.1931
.8070	.8668 9417	.2739 8243	.3946 5301	.5905 5144	.9811 9876	**.1930**
.8071	72 5921	8 9573	5 7520	4 1476	2 1842	.1929
.8072	76 2437	8 0898	4 9735	2 7804	2 3808	.1928
.8073	79 8965	7 2220	4 1946	1 4126	2 5772	.1927
.8074	83 5504	6 3539	3 4153	.5900 0444	2 7735	.1926
.8075	87 2055	5 4853	2 6355	.5898 6757	2 9697	.1925
.8076	90 8617	4 6164	1 8554	7 3065	3 1659	.1924
.8077	94 5191	3 7471	1 0749	5 9368	3 3619	.1923
.8078	.8698 1777	2 8775	.3940 2939	4 5667	3 5578	.1922
.8079	.8701 8374	2 0075	.3939 5125	3 1960	3 7536	.1921
.8080	.8705 4983	.2731 1371	.3938 7308	.5891 8248	.9813 9493	**.1920**
.8081	09 1604	.2730 2664	7 9486	.5890 4532	4 1448	.1919
.8082	12 8236	.2729 3953	7 1660	.5889 0811	4 3403	.1918
.8083	16 4880	8 5238	6 3830	7 7085	4 5357	.1917
.8084	20 1535	·7 6520	5 5996	6 3354	4 7310	.1916
.8085	23 8203	6 7798	4 8158	4 9618	4 9261	.1915
.8086	27 4882	5 9072	4 0315	3 5877	5 1212	.1914
.8087	31 1573	5 0343	3 2469	2 2131	5 3161	.1913
.8088	34 8275	4 1610	2 4618	.5880 8380	5 5110	.1912
.8089	38 4990	3 2873	1 6764	.5879 4625	5 7057	.1911
.8090	.8742 1716	.2722 4133	.3930 8905	.5878 0864	.9815 9004	**.1910**
.8091	45 8454	1 5389	.3930 1042	6 7099	6 0949	.1909
.8092	49 5204	.2720 6641	.3929 3175	5 3328	6 2893	.1908
.8093	53 1966	.2719 7890	8 5304	3 9553	6 4836	.1907
.8094	56 8739	8 9135	7 7429	2 5773	6 6778	.1906
.8095	60 5525	8 0376	6 9549	.5871 1988	6 8720	.1905
.8096	64 2322	7 1614	6 1666	.5869 8198	7 0660	.1904
.8097	67 9131	6 2848	5 3778	8 4403	7 2599	.1903
.8098	71 5952	5 4078	4 5886	7 0603	7 4536	.1902
.8099	75 2785	4 5305	3 7991	5 6798	7 6473	.1901
.8100	.8778 9630	.2713 6528	.3923 0090	.5864 2988	.9817 8409	**.1900**

$E^{-ii}=$ $E^{ii}=$.0000,0002   .0000,0000+   .0000,0001   .0000,0001   .0000,0000+
.0000,0000,4

TABLE I

.8100   .1900

p	x	z	√pq	√1−p²	√1−q²	q
.8100	.8778 9630	.2713 6528	.3923 0090	.5864 2988	.9817 8409	**.1900**
.8101	82 6487	2 7747	2 2186	2 9173	8 0344	.1899
.8102	86 3355	1 8962	1 4278	1 5353	8 2277	.1898
.8103	90 0236	1 0174	.3920 6366	.5860 1528	8 4210	.1897
.8104	93 7128	.2710 1382	.3919 8449	.5858 7698	8 6142	.1896
.8105	.8797 4033	.2709 2587	9 0528	7 3864	8 8072	.1895
.8106	.8801 0949	8 3787	8 2603	6 0024	9 0002	.1894
.8107	04 7878	7 4984	7 4674	4 6179	9 1930	.1893
.8108	08 4818	6 6178	6 6741	3 2330	9 3857	.1892
.8109	12 1770	5 7367	5 8804	1 8475	9 5784	.1891
.8110	.8815 8735	.2704 8553	.3915 0862	.5850 4615	.9819 7709	**.1890**
.8111	19 5712	3 9736	4 2916	.5849 0751	.9819 9633	.1889
.8112	23 2700	3 0914	3 4966	7 6881	.9820 1556	.1888
.8113	26 9701	2 2089	2 7012	6 3006	0 3478	.1887
.8114	30 6714	1 3260	1 9054	4 9127	0 5399	.1886
.8115	34 3739	.2700 4428	1 1092	3 5242	0 7319	.1885
.8116	38 0776	.2699 5592	.3910 3125	2 1352	0 9238	.1884
.8117	41 7825	8 6752	.3909 5154	.5840 7458	1 1156	.1883
.8118	45 4886	7 7908	8 7179	.5839 3558	1 3072	.1882
.8119	49 1959	6 9061	7 9200	7 9653	1 4988	.1881
.8120	.8852 9045	.2696 0210	.3907 1217	.5836 5743	.9821 6903	**.1880**
.8121	56 6143	5 1355	6 3230	5 1829	1 8816	.1879
.8122	60 3253	4 2496	5 5238	3 7909	2 0729	.1878
.8123	64 0375	3 3634	4 7242	2 3984	2 2640	.1877
.8124	67 7509	2 4768	3 9242	.5831 0054	2 4551	.1876
.8125	71 4656	1 5899	3 1237	.5829 6119	2 6460	.1875
.8126	75 1815	.2690 7025	2 3229	8 2179	2 8369	.1874
.8127	78 8986	.2689 8148	1 5216	6 8234	3 0276	.1873
.8128	82 6169	8 9268	.3900 7199	5 4284	3 2182	.1872
.8129	86 3365	8 0383	.3899 9178	4 0329	3 4087	.1871
.8130	.8890 0573	.2687 1495	.3899 1153	.5822 6369	.9823 5991	**.1870**
.8131	93 7793	6 2603	8 3123	.5821 2403	3 7894	.1869
.8132	.8897 5026	5 3707	7 5089	.5819 8433	3 9796	.1868
.8133	.8901 2271	4 4808	6 7051	8 4458	4 1697	.1867
.8134	04 9528	3 5905	5 9009	7 0477	4 3597	.1866
.8135	08 6798	2 6998	5 0963	5 6491	4 5496	.1865
.8136	12 4080	1 8088	4 2912	4 2501	4 7394	.1864
.8137	16 1374	0 9173	3 4857	2 8505	4 9291	.1863
.8138	19 8681	.2680 0255	2 6798	1 4504	5 1186	.1862
.8139	23 6000	.2679 1334	1 8735	.5810 0498	5 3081	.1861
.8140	.8927 3332	.2678 2408	.3891 0667	.5808 6487	.9825 4974	**.1860**
.8141	31 0676	7 3479	.3890 2595	7 2471	5 6867	.1859
.8142	34 8033	6 4546	.3889 4519	5 8450	5 8758	.1858
.8143	38 5402	5 5609	8 6439	4 4424	6 0649	.1857
.8144	42 2784	4 6669	7 8354	3 0392	6 2538	.1856
.8145	46 0178	3 7725	7 0265	1 6355	6 4426	.1855
.8146	49 7584	2 8777	6 2172	.5800 2314	6 6314	.1854
.8147	53 5003	1 9825	5 4074	.5798 8267	6 8200	.1853
.8148	57 2435	1 0870	4 5973	7 4215	7 0085	.1852
.8149	60 9879	.2670 1911	3 7867	6 0158	7 1969	.1851
.8150	.8964 7336	.2669 2948	.3882 9757	.5794 6096	.9827 3852	**.1850**

$E^{-ii}=$ .0000,0000,4 E^{iL} .0000,0002   .0000,0000+   .0000,0001   .0000,0001   .0000,0000+

TABLE I

.8150 .1850

p	x	z	$\sqrt{pq}$	$\sqrt{1-p^2}$	$\sqrt{1-q^2}$	q
.8150	.8964 7336	.2669 2948	.3882 9757	.5794 6096	.9827 3852	.1850
.8151	68 4806	8 3981	2 1642	3 2028	7 5734	.1849
.8152	72 2288	7 5011	1 3523	1 7956	7 7615	.1848
.8153	75 9782	6 6037	.3880 5400	.5790 3878	7 9495	.1847
.8154	79 7289	5 7059	.3879 7273	.5788 9795	8 1374	.1846
.8155	83 4809	4 8077	8 9142	7 5707	8 3251	.1845
.8156	87 2342	3 9092	8 1006	6 1614	8 5128	.1844
.8157	90 9887	3 0103	7 2866	4 7516	8 7004	.1843
.8158	94 7445	2 1110	6 4721	3 3412	8 8878	.1842
.8159	.8998 5016	1 2113	5 6572	1 9304	9 0752	.1841
.8160	.9002 2599	.2660 3113	.3874 8419	.5780 5190	.9829 2624	.1840
.8161	06 0195	.2659 4109	4 0262	.5779 1071	9 4496	.1839
.8162	09 7804	8 5101	3 2100	7 6947	9 6366	.1838
.8163	13 5425	7 6089	2 3934	6 2818	.9829 8235	.1837
.8164	17 3059	6 7074	1 5764	4 8683	.9830 0104	.1836
.8165	21 0706	5 8055	.3870 7590	3 4543	0 1971	.1835
.8166	24 8366	4 9032	.3869 9411	2 0398	0 3837	.1834
.8167	28 6038	4 0005	9 1228	.5770 6248	0 5702	.1833
.8168	32 3724	3 0975	8 3040	.5769 2093	0 7566	.1832
.8169	36 1422	2 1940	7 4848	7 7933	0 9429	.1831
.8170	.9039 9133	.2651 2902	.3866 6652	.5766 3767	.9831 1291	.1830
.8171	43 6857	.2650 3860	5 8452	4 9596	1 3152	.1829
.8172	47 4594	.2649 4815	5 0247	3 5420	1 5012	.1828
.8173	51 2343	8 5766	4 2038	2 1238	1 6871	.1827
.8174	55 0106	7 6712	3 3825	.5760 7052	1 8729	.1826
.8175	58 7881	6 7656	2 5607	.5759 2860	2 0585	.1825
.8176	62 5670	5 8595	1 7385	7 8663	2 2441	.1824
.8177	66 3471	4 9530	0 9158	6 4460	2 4296	.1823
.8178	70 1285	4 0462	.3860 0927	5 0253	2 6149	.1822
.8179	73 9113	3 1390	.3859 2692	3 6040	2 8002	.1821
.8180	.9077 6953	.2642 2314	.3858 4453	.5752 1822	.9832 9853	.1820
.8181	81 4806	1 3235	7 6209	.5750 7599	3 1703	.1819
.8182	85 2673	.2640 4151	6 7961	.5749 3370	3 3553	.1818
.8183	89 0552	.2639 5064	5 9708	7 9136	3 5401	.1817
.8184	92 8445	8 5973	5 1451	6 4897	3 7248	.1816
.8185	.9096 6350	7 6879	4 3190	5 0653	3 9094	.1815
.8186	.9100 4269	6 7780	3 4924	3 6403	4 0940	.1814
.8187	04 2200	5 8678	2 6654	2 2148	4 2784	.1813
.8188	08 0145	4 9572	1 8380	.5740 7888	4 4627	.1812
.8189	11 8103	4 0462	1 0101	.5739 3622	4 6469	.1811
.8190	.9115 6074	.2633 1348	.3850 1818	.5737 9352	.9834 8310	.1810
.8191	19 4058	2 2230	.3849 3531	6 5076	5 0149	.1809
.8192	23 2055	1 3109	8 5239	5 0794	5 1988	.1808
.8193	27 0066	.2630 3984	7 6942	3 6508	5 3826	.1807
.8194	30 8089	.2629 4855	6 8642	2 2216	5 5663	.1806
.8195	34 6126	8 5722	6 0337	.5730 7918	5 7498	.1805
.8196	38 4176	7 6586	5 2027	.5729 3616	5 9333	.1804
.8197	42 2240	6 7446	4 3713	7 9308	6 1167	.1803
.8198	46 0316	5 8301	3 5395	6 4995	6 2999	.1802
.8199	49 8406	4 9153	2 7072	5 0676	6 4831	.1801
.8200	.9153 6509	.2624 0002	.3841 8745	.5723 6352	.9836 6661	.1800

$E^{-ii}=$ $E^{ii}=$.0000,0002 .0000,0000+ .0000,0001 .0000,0001 .0000,0000+
.0000,0000,4

TABLE I

p	x	z	$\sqrt{pq}$	$\sqrt{1-p^2}$	$\sqrt{1-q^2}$	q
.8200	.9153 6509	.2624 0002	.3841 8745	.5723 6352	.9836 6661	**.1800**
.8201	57 4625	3 0846	1 0414	2 2023	6 8490	.1799
.8202	61 2755	2 1687	.3840 2078	.5720 7688	7 0319	.1798
.8203	65 0898	1 2524	.3839 3738	.5719 3348	7• 2146	.1797
.8204	68 9054	.2620 3357	8 5393	7 9003	7 3972	.1796
.8205	72 7224	.2619 4186	7 7044	6 4653	7 5797	.1795
.8206	76 5408	8 5011	6 8690	5 0297	7 7621	.1794
.8207	80 3604	7 5833	6 0332	3 5935	7 9444	.1793
.8208	84 1814	6 6650	5 1970	2 1569	8 1267	.1792
.8209	88 0037	5 7464	4 3603	.5710 7197	8 3087	.1791
.8210	.9191 8274	.2614 8274	.3833 5232	.5709 2819	.9838 4907	**.1790**
.8211	95 6524	3 9081	2 6856	7 8436	8 6726	.1789
.8212	.9199 4787	2 9883	1 8476	6 4048	8 8544	.1788
.8213	.9203 3064	2 0682	1 0091	4 9655	9 0361	.1787
.8214	07 1355	1 1477	.3830 1702	3 5256	9 2177	.1786
.8215	10 9660	.2610 2267	.3829 3309	2 0851	9 3991	.1785
.8216	14 7977	.2609 3055	8 4911	.5700 6442	9 5805	.1784
.8217	18 6308	8 3838	7 6508	.5699 2027	9 7617	.1783
.8218	22 4653	7 4617	6 8102	7 7606	.9839 9429	.1782
.8219	26 3011	6 5393	5 9690	6 3180	.9840 1239	.1781
.8220	.9230 1383	.2605 6165	.3825 1274	.5694 8749	.9840 3049	**.1780**
.8221	33 9768	4 6933	4 2854	3 4312	0 4857	.1779
.8222	37 8167	3 7697	3 4430	1 9870	0 6664	.1778
.8223	41 6580	2 8457	2 6000	.5690 5422	0 8471	.1777
.8224	45 5006	1 9213	1 7567	.5689 0969	1 0276	.1776
.8225	49 3446	0 9966	0 9128	7 6511	1 2080	.1775
.8226	53 1900	.2600 0715	.3820 0686	6 2047	1 3883	.1774
.8227	57 0367	.2599 1460	.3819 2239	4 7578	1 5685	.1773
.8228	60 8848	8 2201	8 3787	3 3103	1 7486	.1772
.8229	64 7342	7 2938	7 5331	1 8623	1 9286	.1771
.8230	.9268 5851	.2596 3671	.3816 6870	.5680 4137	.9842 1085	**.1770**
.8231	72 4373	5 4401	5 8405	.5678 9646	2 2883	.1769
.8232	76 2909	4 5126	4 9936	7 5149	2 4680	.1768
.8233	80 1459	3 5848	4 1462	6 0647	2 6476	.1767
.8234	84 0023	2 6566	3 2983	4 6140	2 8270	.1766
.8235	87 8600	1 7280	2 4500	3 1627	3 0064	.1765
.8236	91 7191	.2590 7990	1 6012	1 7109	3 1857	.1764
.8237	95 5797	.2589 8697	.3810 7520	.5670 2585	3 3648	.1763
.8238	.9299 4415	8 9399	.3809 9024	.5668 8055	3 5439	.1762
.8239	.9303 3048	8 0098	9 0522	7 3520	3 7228	.1761
.8240	.9307 1695	.2587 0793	.3808 2017	.5665 8980	.9843 9017	**.1760**
.8241	11 0356	6 1483	7 3507	4 4434	4 0804	.1759
.8242	14 9030	5 2171	6 4992	2 9883	4 2590	.1758
.8243	18 7718	4 2854	5 6473	1 5326	4 4376	.1757
.8244	22 6421	3 3533	4 7949	.5660 0763	4 6160	.1756
.8245	26 5137	2 4208	3 9420	.5658 6195	4 7943	.1755
.8246	30 3868	1 4880	3 0887	7 1622	4 9725	.1754
.8247	34 2612	.2580 5548	2 2350	5 7043	5 1506	.1753
.8248	38 1370	.2579 6211	1 3808	4 2458	5 3286	.1752
.8249	42 0143	8 6871	.3800 5261	2 7868	5 5065	.1751
.8250	.9345 8929	.2577 7527	.3799 6710	.5651 3273	.9845 6843	**.1750**

$E^{-ii} =$ .0000,0000,4 $E^{ii} =$ .0000,0002   .0000,0000+   .0000,0001   .0000,0001   .0000,0000+

TABLE I

.8250   **.1750**

p	x	z	√pq	√1−p²	√1−q²	q
.8250	.9345 8929	.2577 7527	.3799 6710	.5651 3273	.9845 6843	**.1750**
.8251	49 7730	6 8180	8 8155	.5649 8672	5 8620	.1749
.8252	53 6544	5 8828	7 9595	8 4065	6 0396	.1748
.8253	57 5373	4 9472	7 1030	6 9453	6 2171	.1747
.8254	61 4216	4 0113	6 2460	5 4835	6 3945	.1746
.8255	65 3073	3 0749	5 3886	4 0212	6 5717	.1745
.8256	69 1944	2 1382	4 5308	2 5583	6 7489	.1744
.8257	73 0829	1 2011	3 6725	.5641 0948	6 9260	.1743
.8258	76 9728	.2570 2636	2 8137	.5639 6308	7 1029	.1742
.8259	80 8642	.2569 3257	1 9545	8 1663	7 2798	.1741
.8260	.9384 7570	.2568 3874	.3791 0948	.5636 7012	.9847 4565	**.1740**
.8261	88 6512	7 4488	.3790 2347	5 2355	7 6332	.1739
.8262	92 5468	6 5097	.3789 3741	3 7693	7 8097	.1738
.8263	.9396 4439	5 5702	8 5130	2 3025	7 9861	.1737
.8264	.9400 3424	4 6304	7 6515	.5630 8351	8 1625	.1736
.8265	04 2423	3 6902	6 7895	.5629 3672	8 3387	.1735
.8266	08 1436	2 7496	5 9271	7 8987	8 5148	.1734
.8267	12 0464	1 8085	5 0642	6 4297	8 6908	.1733
.8268	15 9506	.2560 8671	4 2008	4 9601	8 8667	.1732
.8269	19 8562	.2559 9254	3 3370	3 4899	9 0425	.1731
.8270	.9423 7633	.2558 9832	.3782 4727	.5622 0192	.9849 2182	**.1730**
.8271	27 6718	8 0406	1 6080	.5620 5479	9 3938	.1729
.8272	31 5818	7 0976	.3780 7428	.5619 0761	9 5693	.1728
.8273	35 4932	6 1543	.3779 8771	7 6037	9 7447	.1727
.8274	39 4061	5 2105	9 0110	6 1307	.9849 9200	.1726
.8275	43 3204	4 2664	8 1444	4 6572	.9850 0952	.1725
.8276	47 2361	3 3219	7 2773	3 1831	0 2703	.1724
.8277	51 1533	2 3770	6 4098	1 7084	0 4452	.1723
.8278	55 0719	1 4317	5 5418	.5610 2332	0 6201	.1722
.8279	58 9920	.2550 4859	4 6734	.5608 7573	0 7948	.1721
.8280	.9462 9136	.2549 5399	.3773 8044	.5607 2810	.9850 9695	**.1720**
.8281	66 8366	8 5934	2 9351	5 8040	1 1440	.1719
.8282	70 7611	7 6465	2 0652	4 3265	1 3185	.1718
.8283	74 6870	6 6992	1 1949	2 8485	1 4928	.1717
.8284	78 6144	5 7515	.3770 3241	.5601 3698	1 6671	.1716
.8285	82 5432	4 8035	.3769 4529	.5599 8906	1 8412	.1715
.8286	86 4735	3 8550	8 5812	8 4108	2 0152	.1714
.8287	90 4053	2 9062	7 7090	6 9305	2 1891	.1713
.8288	94 3385	1 9570	6 8363	5 4496	2 3630	.1712
.8289	.9498 2732	1 0073	5 9632	3 9681	2 5367	.1711
.8290	.9502 2094	.2540 0573	.3765 0896	.5592 4860	.9852 7103	**.1710**
.8291	06 1471	.2539 1069	4 2156	.5591 0034	2 8838	.1709
.8292	10 0862	8 1561	3 3411	.5589 5202	3 0572	.1708
.8293	14 0268	7 2049	2 4661	8 0364	3 2305	.1707
.8294	17 9689	6 2533	1 5906	6 5521	3 4037	.1706
.8295	21 9124	5 3013	.3760 7147	5 0671	3 5768	.1705
.8296	25 8575	4 3489	.3759 8383	3 5816	3 7497	.1704
.8297	29 8040	3 3961	8 9614	2 0956	3 9226	.1703
.8298	33 7520	2 4429	8 0841	.5580 6089	4 0954	.1702
.8299	37 7015	1 4894	7 2063	.5579 1217	4 2681	.1701
.8300	.9541 6525	.2530 5354	.3756 3280	.5577 6339	.9854 4406	**.1700**

E⁻ⁱᴸ=   Eⁱᴸ=.0000,0002   .0000,0000+   .0000,0001   .0000,0001   .0000,0000+
.0000,0000,+

TABLE I

p	x	z	$\sqrt{pq}$	$\sqrt{1-p^2}$	$\sqrt{1-q^2}$	q
.8300	.9541 6525	.2530 5354	.3756 3280	.5577 6339	.9854 4406	**.1700**
.8301	45 6050	.2529 5810	5 4492	6 1455	4 6131	.1699
.8302	49 5590	8 6263	4 5700	4 6566	4 7854	.1698
.8303	53 5144	7 6711	3 6903	3 1671	4 9577	.1697
.8304	57 4714	6 7156	2 8101	1 6769	5 1298	.1696
.8305	61 4299	5 7596	1 9295	.5570 1863	5 3019	.1695
.8306	65 3898	4 8033	1 0484	.5568 6950	5 4738	.1694
.8307	69 3513	3 8465	.3750 1668	7 2032	5 6456	.1693
.8308	73 3143	2 8894	.3749 2847	5 7107	5 8174	.1692
.8309	77 2787	1 9319	8 4022	4 2177	5 9890	.1691
.8310	.9581 2447	.2520 9739	.3747 5192	.5562 7242	.9856 1605	**.1690**
.8311	85 2122	.2520 0156	6 6357	.5561 2300	6 3319	.1689
.8312	89 1811	.2519 0569	5 7517	.5559 7352	6 5032	.1688
.8313	93 1516	8 0978	4 8673	8 2399	6 6744	.1687
.8314	.9597 1236	7 1383	3 9824	6 7440	6 8455	.1686
.8315	.9601 0971	6 1784	3 0970	5 2475	7 0165	.1685
.8316	05 0722	5 2181	2 2111	3 7504	7 1874	.1684
.8317	09 0487	4 2573	1 3248	2 2528	7 3582	.1683
.8318	13 0268	3 2962	.3740 4379	.5550 7545	7 5289	.1682
.8319	17 0064	2 3347	.3739 5506	.5549 2557	7 6995	.1681
.8320	.9620 9875	.2511 3728	.3738 6629	.5547 7563	.9857 8700	**.1680**
.8321	24 9702	.2510 4105	7 7746	6 2563	8 0403	.1679
.8322	28 9544	.2509 4478	6 8859	4 7557	8 2106	.1678
.8323	32 9401	8 4848	5 9967	3 2545	8 3807	.1677
.8324	36 9273	7 5213	5 1070	1 7528	8 5508	.1676
.8325	40 9161	6 5574	4 2168	.5540 2504	8 7208	.1675
.8326	44 9064	5 5931	3 3261	.5538 7475	8 8906	.1674
.8327	48 8982	4 6284	2 4350	7 2440	9 0604	.1673
.8328	52 8916	3 6633	1 5434	5 7399	9 2300	.1672
.8329	56 8865	2 6978	.3730 6513	4 2352	9 3995	.1671
.8330	.9660 8830	.2501 7319	.3729 7587	.5532 7299	.9859 5690	**.1670**
.8331	64 8810	.2500 7656	8 8656	.5531 2240	9 7383	.1669
.8332	68 8805	.2499 7989	7 9721	.5529 7175	.9859 9075	.1668
.8333	72 8816	8 8319	7 0781	8 2105	.9860 0766	.1667
.8334	76 8843	7 8644	6 1836	6 7028	0 2456	.1666
.8335	80 8885	6 8965	5 2886	5 1946	0 4145	.1665
.8336	84 8942	5 9282	4 3931	3 6857	0 5833	.1664
.8337	88 9015	4 9595	3 4971	2 1763	0 7521	.1663
.8338	92 9104	3 9904	2 6007	.5520 6663	0 9206	.1662
.8339	.9696 9208	3 0209	1 7038	.5519 1556	1 0891	.1661
.8340	.9700 9328	.2492 0510	.3720 8064	.5517 6444	.9861 2575	**.1660**
.8341	04 9463	1 0807	.3719 9085	6 1326	1 4258	.1659
.8342	08 9614	.2490 1100	9 0101	4 6202	1 5940	.1658
.8343	12 9781	.2489 1389	8 1112	3 1072	1 7621	.1657
.8344	16 9963	8 1674	7 2119	1 5936	1 9300	.1656
.8345	21 0161	7 1955	6 3120	.5510 0794	2 0979	.1655
.8346	25 0375	6 2232	5 4117	.5508 5646	2 2657	.1654
.8347	29 0605	5 2505	4 5109	7 0492	2 4333	.1653
.8348	33 0850	4 2774	3 6096	5 5332	2 6009	.1652
.8349	37 1111	3 3039	2 7078	4 0166	2 7683	.1651
.8350	.9741 1388	.2482 3300	.3711 8055	.5502 4994	.9862 9357	**.1650**

$E^{-ii}=$.0000,00005 $E^{ii}=$.0000,0002.   .0000,0000+   .0000,0001   .0000,0001   .0000,0000+

TABLE I

.8350   **.1650**

p	x	z	√pq	√1−p²	√1−q²	q
.8350	.9741 1388	.2482 3300	.3711 8055	.5502 4994	.9862 9357	**.1650**
.8351	45 1681	1 3557	.3710 9027	.5500 9816	3 1029	.1649
.8352	49 1989	.2480 3810	.3709 9995	.5499 4632	3 2700	.1648
.8353	53 2313	.2479 4058	9 0957	7 9443	3 4371	.1647
.8354	57 2654	8 4303	8 1915	6 4247	3 6040	.1646
.8355	61 3010	7 4544	7 2867	4 9045	3 7708	.1645
.8356	65 3382	6 4781	6 3815	3 3837	3 9376	.1644
.8357	69 3770	5 5013	5 4758	1 8623	4 1042	.1643
.8358	73 4173	4 5242	4 5696	.5490 3402	4 2707	.1642
.8359	77 4593	3 5466	3 6629	.5488 8176	4 4371	.1641
.8360	.9781 5029	.2472 5687	.3702 7557	.5487 2944	.9864 6034	**.1640**
.8361	85 5481	1 5903	1 8481	5 7706	4 7696	.1639
.8362	89 5949	.2470 6116	0 9399	4 2462	4 9357	.1638
.8363	93 6432	.2469 6324	.3700 0312	2 7211	5 1017	.1637
.8364	.9797 6932	8 6529	.3699 1221	.5481 1955	5 2676	.1636
.8365	.9801 7448	7 6729	8 2124	.5479 6692	5 4333	.1635
.8366	05 7980	6 6925	7 3023	8 1424	5 5990	.1634
.8367	09 8529	5 7117	6 3916	6 6149	5 7646	.1633
.8368	13 9093	4 7305	5 4805	5 0868	5 9301	.1632
.8369	17 9673	3 7489	4 5689	3 5582	6 0954	.1631
.8370	.9822 0270	.2462 7669	.3693 6567	.5472 0289	.9866 2607	**.1630**
.8371	26 0883	1 7845	2 7441	.5470 4990	6 4258	.1629
.8372	30 1512	.2460 8017	1 8310	.5468 9685	6 5909	.1628
.8373	34 2157	.2459 8185	0 9174	7 4373	6 7558	.1627
.8374	38 2819	8 8349	.3690 0033	5 9056	6 9207	.1626
.8375	42 3496	7 8509	.3689 0886	4 3732	7 0854	.1625
.8376	46 4190	6 8664	8 1735	2 8403	7 2501	.1624
.8377	50 4901	5 8816	7 2579	.5461 3067	7 4146	.1623
.8378	54 5627	4 8963	6 3418	.5459 7725	7 5790	.1622
.8379	58 6370	3 9107	5 4252	8 2377	7 7434	.1621
.8380	.9862 7130	.2452 9246	.3684 5081	.5456 7023	.9867 9076	**.1620**
.8381	66 7906	1 9381	3 5905	5 1663	8 0717	.1619
.8382	70 8698	.2450 9512	2 6724	3 6296	8 2357	.1618
.8383	74 9507	.2449 9639	1 7538	2 0924	8 3996	.1617
.8384	79 0332	8 9762	.3680 8347	.5450 5545	8 5634	.1616
.8385	83 1173	7 9881	.3679 9151	.5449 0160	8 7271	.1615
.8386	87 2031	6 9996	8 9950	7 4768	8 8907	.1614
.8387	91 2906	6 0107	8 0744	5 9371	9 0542	.1613
.8388	95 3797	5 0214	7 1532	4 3968	9 2176	.1612
.8389	.9899 4705	4 0316	6 2316	2 8558	9 3809	.1611
.8390	.9903 5629	.2443 0415	.3675 3095	.5441 3142	.9869 5441	**.1610**
.8391	07 6570	2 0509	4 3869	.5439 7720	9 7071	.1609
.8392	11 7528	1 0599	3 4638	8 2291	.9869 8701	.1608
.8393	15 8502	.2440 0686	2 5401	6 6857	.9870 0330	.1607
.8394	19 9492	.2439 0768	1 6160	5 1416	0 1957	.1606
.8395	24 0500	8 0846	.3670 6914	3 5969	0 3584	.1605
.8396	28 1524	7 0920	.3669 7662	2 0515	0 5210	.1604
.8397	32 2565	6 0989	8 8406	.5430 5056	0 6834	.1603
.8398	36 3622	5 1055	7 9144	.5428 9590	0 8458	.1602
.8399	40 4697	4 1117	6 9877	7 4118	1 0080	.1601
.8400	.9944 5788	.2433 1174	.3666 0606	.5425 8640	.9871 1701	**.1600**

E⁻ⁱⁱ=.0000,0000,5   Eⁱⁱ=.0000,0002   .0000,0000+   .0000,0001   .0000,0001   .0000,0000+

TABLE I

.8400   .1600

p	x	z	$\sqrt{pq}$	$\sqrt{1-p^2}$	$\sqrt{1-q^2}$	q
.8400	.9944 5788	.2433 1174	.3666 0606	.5425 8640	.9871 1701	**.1600**
.8401	48 6896	2 1227	5 1329	4 3155	1 3322	.1599
.8402	52 8021	1 1277	4 2047	2 7665	1 4941	.1598
.8403	56 9162	.2430 1322	3 2760	.5421 2167	1 6559	.1597
.8404	61 0321	.2429 1363	2 3468	.5419 6664	1 8177	.1596
.8405	65 1496	8 1400	1 4171	8 1154	1 9793	.1595
.8406	69 2689	7 1433	.3660 4869	6 5639	2 1408	.1594
.8407	73 3898	6 1461	.3659 5561	5 0116	2 3022	.1593
.8408	77 5124	5 1486	8 6249	3 4588	2 4635	.1592
.8409	81 6367	4 1506	7 6931	1 9053	2 6247	.1591
.8410	.9985 7627	.2423 1523	.3656 7609	.5410 3512	.9872 7858	**.1590**
.8411	89 8904	2 1535	5 8281	5408 7964	2 9468	.1589
.8412	94 0198	1 1543	4 8948	7 2411	3 1077	.1588
.8413	.9998 1509	.2420 1547	3 9610	5 6851	3 2685	.1587
.8414	1.0002 2838	.2419 1546	3 0267	4 1284	3 4292	.1586
.8415	06 4183	8 1542	2 0919	2 5711	3 5898	.1585
.8416	10 5546	7 1534	1 1565	.5401 0132	3 7503	.1584
.8417	14 6925	6 1521	.3650 2207	.5399 4547	3 9106	.1583
.8418	18 8322	5 1504	.3649 2843	7 8955	4 0709	.1582
.8419	22 9736	4 1483	8 3474	6 3357	4 2311	.1581
.8420	1.0027 1167	.2413 1458	.3647 4100	.5394 7753	.9874 3911	**.1580**
.8421	31 2615	2 1429	6 4721	3 2142	4 5511	.1579
.8422	35 4080	1 1396	5 5337	1 6524	4 7109	.1578
.8423	39 5563	.2410 1358	4 5948	.5390 0901	4 8707	.1577
.8424	43 7064	.2409 1317	3 6553	.5388 5271	5 0303	.1576
.8425	47 8581	8 1271	2 7153	6 9634	5 1899	.1575
.8426	52 0116	7 1221	1 7748	5 3991	5 3493	.1574
.8427	56 1668	6 1167	.3640 8338	3 8342	5 5086	.1573
.8428	60 3237	5 1109	.3639 8923	2 2687	5 6679	.1572
.8429	64 4824	4 1046	8 9503	.5380 7025	5 8270	.1571
.8430	1.0068 6428	.2403 0980	.3638 0077	.5379 1356	.9875 9860	**.1570**
.8431	72 8050	2 0909	7 0646	7 5681	6 1449	.1569
.8432	76 9689	1 0834	6 1210	6 0000	6 3038	.1568
.8433	81 1345	.2400 0755	5 1769	4 4312	6 4625	.1567
.8434	85 3019	.2399 0672	4 2322	2 8618	6 6211	.1566
.8435	89 4711	8 0584	3 2871	.5371 2917	6 7796	.1565
.8436	93 6420	7 0493	2 3414	.5369 7210	6 9380	.1564
.8437	1.0097 8147	6 0397	1 3952	8 1497	7 0963	.1563
.8438	1.0101 9891	5 0297	.3630 4485	6 5777	7 2545	.1562
.8439	06 1653	4 0193	.3629 5012	5 0050	7 4126	.1561
.8440	1.0110 3433	.2393 0085	.3628 5534	.5363 4317	.9877 5706	**.1560**
.8441	14 5230	1 9972	7 6051	1 8578	7 7284	.1559
.8442	18 7045	.2390 9856	6 6563	.5360 2832	7 8862	.1558
.8443	22 8878	.2389 9735	5 7070	.5358 7080	8 0439	.1557
.8444	27 0728	8 9610	4 7571	7 1321	8 2015	.1556
.8445	31 2596	7 9481	3 8067	5 5555	8 3589	.1555
.8446	35 4482	6 9347	2 8558	3 9783	8 5163	.1554
.8447	39 6385	5 9210	1 9043	2 4005	8 6735	.1553
.8448	43 8307	4 9068	.3620 9524	.5350 8220	8 8307	.1552
.8449	48 0246	3 8922	.3619 9999	.5349 2428	8 9878	.1551
.8450	1.0152 2203	.2382 8772	.3619 0468	.5347 6630	.9879 1447	**.1550**

$E^{-ii}=$ $E^{ii}=$.0000,0002   .0000,0001   .0000,0001   .0000,0001   .0000,0000+
.0000,0000,5

TABLE I
.8450   .1550

p	x	z	$\sqrt{pq}$	$\sqrt{1-p^2}$	$\sqrt{1-q^2}$	q
.8450	1.0152 2203	.2382 8772	.3619 0468	.5347 6630	.9879 1447	**.1550**
.8451	56 4178	1 8618	8 0933	6 0826	9 3015	.1549
.8452	60 6171	.2380 8459	7 1392	4 5015	9 4583	.1548
.8453	64 8182	.2379 8297	6 1846	2 9197	9 6149	.1547
.8454	69 0211	8 8130	5 2295	.5341 3373	9 7715	.1546
.8455	73 2258	7 7959	4 2738	.5339 7542	.9879 9279	.1545
.8456	77 4322	6 7783	3 3176	8 1705	.9880 0842	.1544
.8457	81 6405	5 7604	2 3609	6 5861	0 2404	.1543
.8458	85 8506	4 7420	1 4036	5 0010	0 3966	.1542
.8459	90 0625	3 7232	.3610 4458	3 4153	0 5526	.1541
.8460	1.0194 2762	.2372 7040	.3609 4875	.5331 8290	.9880 7085	**.1540**
.8461	1.0198 4917	1 7843	8 5286	.5330 2419	0 8643	.1539
.8462	1.0202 7090	.2370 7643	7 5693	.5328 6542	1 0200	.1538
.8463	06 9282	.2369 7438	6 6093	7 0659	1 1756	.1537
.8464	11 1491	8 7229	5 6489	5 4769	1 3311	.1536
.8465	15 3719	7 6016	4 6879	3 8872	1 4865	.1535
.8466	19 5965	6 5798	3 7264	2 2969	1 6418	.1534
.8467	23 8229	5 5577	2 7644	.5320 7059	1 7970	.1533
.8468	28 0511	4 5351	1 8018	.5319 1142	1 9520	.1532
.8469	32 2812	3 5120	.3600 8387	7 5219	2 1070	.1531
.8470	1.0236 5131	.2362 4886	.3599 8750	.5315 9289	.9882 2619	**.1530**
.8471	40 7468	1 4647	8 9108	4 3352	2 4167	.1529
.8472	44 9824	.2360 4405	7 9461	2 7409	2 5713	.1528
.8473	49 2199	.2359 4157	6 9808	.5311 1459	2 7259	.1527
.8474	53 4591	8 3906	6 0150	.5309 5503	2 8803	.1526
.8475	57 7002	7 3651	5 0487	7 9539	3 0347	.1525
.8476	61 9432	6 3391	4 0818	6 3569	3 1890	.1524
.8477	66 1880	5 3127	3 1144	4 7593	3 3431	.1523
.8478	70 4346	4 2858	2 1464	3 1609	3 4972	.1522
.8479	74 6831	3 2586	1 1779	.5301 5619	3 6511	.1521
.8480	1.0278 9335	.2352 2309	.3590 2089	.5299 9623	.9883 8049	**.1520**
.8481	83 1857	1 2028	.3589 2393	8 3619	3 9587	.1519
.8482	87 4398	.2350 1743	8 2692	6 7609	4 1123	.1518
.8483	91 6957	.2349 1453	7 2986	5 1592	4 2658	.1517
.8484	1.0295 9535	8 1159	6 3274	3 5568	4 4193	.1516
.8485	1.0300 2132	7 0861	5 3556	1 9538	4 5726	.1515
.8486	04 4747	6 0559	4 3834	.5290 3501	4 7258	.1514
.8487	08 7381	5 0252	3 4105	.5288 7457	4 8789	.1513
.8488	13 0034	3 9941	2 4372	7 1406	5 0319	.1512
.8489	17 2706	2 9626	1 4632	5 5349	5 1848	.1511
.8490	1.0321 5396	.2341 9307	.3580 4888	.5283 9285	.9885 3376	**.1510**
.8491	25 8105	.2340 8983	.3579 5138	2 3214	5 4903	.1509
.8492	30 0833	.2339 8655	8 5382	.5280 7136	5 6429	.1508
.8493	34 3580	8 8323	7 5622	.5279 1051	5 7954	.1507
.8494	38 6346	7 7986	6 5855	7 4960	5 9478	.1506
.8495	42 9131	6 7646	5 6083	5 8862	6 1001	.1505
.8496	47 1934	5 7301	4 6306	4 2757	6 2523	.1504
.8497	51 4757	4 6951	3 6523	2 6645	6 4044	.1503
.8498	55 7599	3 6598	2 6735	.5271 0526	6 5563	.1502
.8499	60 0459	2 6240	1 6941	.5269 4401	6 7082	.1501
.8500	1.0364 3339	.2331 5878	.3570 7142	.5267 8269	.9886 8600	**.1500**

$E^{-iL} = .0000,0000,5$ $E^{iL} = .0000,0002$ .0000,0001   .0000,0001   .0000,0001   .0000,0000+

TABLE I

.8500

.1500

p	x	z	$\sqrt{pq}$	$\sqrt{1-p^2}$	$\sqrt{1-q^2}$	q
.8500	1.0364 3339	.2331 5878	.3570 7142	.5267 8269	.9886 8600	**.1500**
.8501	68 6238	.2330 5511	.3569 7337	6 2130	7 0116	.1499
.8502	72 9156	.2329 5140	8 7527	4 5984	7 1632	.1498
.8503	77 2093	8 4765	7 7712	2 9831	7 3147	.1497
.8504	81 5049	7 4386	6 7890	.5261 3671	7 4660	.1496
.8505	85 8024	6 4002	5 8064	.5259 7505	7 6173	.1495
.8506	90 1019	5 3614	4 8231	8 1331	7 7684	.1494
.8507	94 4032	4 3222	3 8394	6 5151	7 9194	.1493
.8508	1.0398 7065	3 2825	2 8550	4 8964	8 0704	.1492
.8509	1.0403 0118	2 2425	1 8702	3 2770	8 2212	.1491
.8510	1.0407 3189	.2321 2019	.3560 8847	.5251 6569	.9888 3720	**.1490**
.8511	11 6280	.2320 1610	.3559 8987	.5250 0361	8 5226	.1489
.8512	15 9390	.2319 1196	8 9122	.5248 4146	8 6731	.1488
.8513	20 2519	8 0778	7 9251	6 7924	8 8235	.1487
.8514	24 5668	7 0356	6 9374	5 1696	8 9739	.1486
.8515	28 8836	5 9929	5 9492	3 5460	9 1241	.1485
.8516	33 2024	4 9498	4 9605	1 9218	9 2742	.1484
.8517	37 5231	3 9063	3 9712	.5240 2968	9 4242	.1483
.8518	41 8458	2 8623	2 9813	.5238 6712	9 5741	.1482
.8519	46 1704	1 8179	1 9909	7 0449	9 7239	.1481
.8520	1.0450 4970	.2310 7730	.3550 9999	.5235 4178	.9889 8736	**.1480**
.8521	54 8255	.2309 7278	.3550 0083	3 7901	.9890 0232	.1479
.8522	59 1560	8 6821	.3549 0162	2 1617	0 1727	.1478
.8523	63 4885	7 6360	8 0235	.5230 5326	0 3221	.1477
.8524	67 8229	6 5894	7 0303	.5228 9028	0 4714	.1476
.8525	72 1593	5 5424	6 0365	7 2722	0 6206	.1475
.8526	76 4976	4 4950	5 0422	5 6410	0 7696	.1474
.8527	80 8380	3 4471	4 0473	4 0091	0 9186	.1473
.8528	85 1803	2 3988	3 0518	2 3765	1 0675	.1472
.8529	89 5245	1 3501	2 0558	.5220 7431	1 2163	.1471
.8530	1.0493 8708	.2300 3009	.3541 0592	.5219 1091	.9891 3649	**.1470**
.8531	1.0498 2190	.2299 2513	.3540 0620	7 4744	1 5135	.1469
.8532	1.0502 5693	8 2012	.3539 0643	5 8390	1 6619	.1468
.8533	06 9215	7 1508	8 0660	4 2028	1 8103	.1467
.8534	11 2757	6 0999	7 0671	2 5660	1 9586	.1466
.8535	15 6319	5 0485	6 0677	.5210 9284	2 1067	.1465
.8536	19 9901	3 9967	5 0678	.5209 2902	2 2547	.1464
.8537	24 3504	2 9445	4 0672	7 6512	2 4027	.1463
.8538	28 7126	1 8919	3 0661	6 0115	2 5505	.1462
.8539	33 0768	.2290 8388	2 0644	4 3711	2 6983	.1461
.8540	1.0537 4430	.2289 7852	.3531 0622	.5202 7301	.9892 8459	**.1460**
.8541	41 8112	8 7313	.3530 0593	.5201 0883	2 9934	.1459
.8542	46 1815	7 6769	.3529 0560	.5199 4457	3 1409	.1458
.8543	50 5537	6 6220	8 0520	7 8025	3 2882	.1457
.8544	54 9280	5 5668	7 0475	6 1586	3 4354	.1456
.8545	59 3043	4 5111	6 0424	4 5139	3 5825	.1455
.8546	63 6826	3 4549	5 0367	2 8686	3 7295	.1454
.8547	68 0630	2 3983	4 0305	.5191 2225	3 8764	.1453
.8548	72 4454	1 3413	3 0237	.5189 5757	4 0232	.1452
.8549	76 8298	.2280 2838	2 0163	7 9282	4 1700	.1451
.8550	1.0581 2162	.2279 2259	.3521 0084	.5186 2800	.9894 3166	**.1450**

$E^{-ii} =$ .000,000,06 $E^{ii} =$ .0000,0003   .0000,0001   .0000,0001   .0000,0001   .0000,0000+

TABLE I

p	x	z	$\sqrt{pq}$	$\sqrt{1-p^2}$	$\sqrt{1-q^2}$	q
.8550	1.0581 2162	.2279 2259	.3521 0084	.5186 2800	.9894 3166	**.1450**
.8551	85 6047	8 1676	.3519 9999	4 6310	4 4630	.1449
.8552	89 9952	7 1088	8 9908	2 9814	4 6094	.1448
.8553	94 3877	6 0496	7 9811	.5181 3310	4 7557	.1447
.8554	1.0598 7823	4 9899	6 9709	.5179 6799	4 9019	.1446
.8555	1.0603 1790	3 9298	5 9600	8 0281	5 0480	.1445
.8556	07 5777	2 8693	4 9486	6 3756	5 1940	.1444
.8557	11 9784	1 8083	3 9367	4 7223	5 3399	.1443
.8558	16 3812	.2270 7469	2 9241	3 0683	5 4856	.1442
.8559	20 7861	.2269 6850	1 9110	.5171 4136	5 6313	.1441
.8560	1.0625 1930	.2268 6227	.3510 8973	.5169 7582	.9895 7769	**.1440**
.8561	29 6020	7 5600	.3509 8830	8 1021	5 9223	.1439
.8562	34 0131	6 4968	8 8682	6 4452	6 0677	.1438
.8563	38 4262	5 4332	7 8528	4 7876	6 2130	.1437
.8564	42 8414	4 3691	6 8368	3 1293	6 3581	.1436
.8565	47 2587	3 3046	5 8202	.5161 4702	6 5032	.1435
.8566	51 6780	2 2397	4 8030	.5159 8105	6 6481	.1434
.8567	56 0995	1 1743	3 7852	8 1500	6 7930	.1433
.8568	60 5230	.2260 1085	2 7669	6 4887	6 9377	.1432
.8569	64 9486	.2259 0422	1 7480	4 8268	7 0823	.1431
.8570	1.0669 3763	.2257 9755	.3500 7285	.5153 1641	.9897 2269	**.1430**
.8571	73 8061	6 9083	.3499 7084	.5151 5007	7 3713	.1429
.8572	78 2380	5 8407	8 6878	.5149 8365	7 5156	.1428
.8573	82 6720	4 7727	7 6665	8 1716	7 6599	.1427
.8574	87 1081	3 7042	6 6447	6 5060	7 8040	.1426
.8575	91 5463	2 6352	5 6223	4 8396	7 9480	.1425
.8576	1.0695 9866	1 5659	4 5993	3 1726	8 0919	.1424
.8577	1.0700 4290	.2250 4960	3 5757	.5141 5047	8 2358	.1423
.8578	04 8735	.2249 4258	2 5515	.5139 8362	8 3795	.1422
.8579	09 3201	8 3551	1 5267	8 1669	8 5231	.1421
.8580	1.0713 7689	.2247 2839	.3490 5014	.5136 4969	.9898 6666	**.1420**
.8581	18 2198	6 2123	.3489 4755	4 8261	8 8100	.1419
.8582	22 6728	5 1403	8 4489	3 1546	8 9533	.1418
.8583	27 1279	4 0678	7 4218	.5131 4823	9 0965	.1417
.8584	31 5852	2 9949	6 3941	.5129 8094	9 2396	.1416
.8585	36 0446	1 9215	5 3658	8 1356	9 3826	.1415
.8586	40 5061	.2240 8476	4 3370	6 4612	9 5254	.1414
.8587	44 9697	.2239 7734	3 3075	4 7859	9 6682	.1413
.8588	49 4355	8 6986	2 2774	3 1100	9 8109	.1412
.8589	53 9035	7 6235	1 2468	.5121 4333	.9899 9535	.1411
.8590	1.0758 3736	.2236 5479	.3480 2155	.5119 7559	.9900 0960	**.1410**
.8591	62 8459	5 4718	.3479 1837	8 0777	0 2383	.1409
.8592	67 3203	4 3953	8 1512	6 3987	0 3806	.1408
.8593	71 7968	3 3183	7 1182	4 7191	0 5228	.1407
.8594	76 2755	2 2409	6 0846	3 0386	0 6648	.1406
.8595	80 7564	1 1631	5 0504	.5111 3575	0 8068	.1405
.8596	85 2395	.2230 0848	4 0155	.5109 6755	0 9486	.1404
.8597	89 7247	.2229 0060	2 9801	7 9929	1 0904	.1403
.8598	94 2121	7 9268	1 9441	6 3094	1 2320	.1402
.8599	1.0798 7017	6 8472	.3470 9075	4 6253	1 3726	.1401
.8600	1.0803 1934	.2225 7671	.3469 8703	.5102 9403	.9901 5150	**.1400**

$E^{-ii}=$ .000,0000p $E^{ii}=$.0000,0003   .0000,0001   .0000,0001   .0000,0001   .0000,0000+

Table I

p	x	z	$\sqrt{pq}$	$\sqrt{1-p^2}$	$\sqrt{1-q^2}$	q
.8600	1.0803 1934	.2225 7671	.3469 8703	.5102 9403	.9901 5150	**.1400**
.8601	07 6873	4 6866	8 8325	.5101 2546	1 6564	.1399
.8602	12 1834	3 6056	7 7941	.5099 5682	1 7976	.1398
.8603	16 6817	2 5241	6 7551	7 8810	1 9387	.1397
.8604	21 1822	1 4422	5 7155	6 1931	2 0798	.1396
.8605	25 6849	.2220 3599	4 6753	4 5044	2 2207	.1395
.8606	30 1898	.2219 2771	3 6345	2 8149	2 3615	.1394
.8607	34 6969	8 1938	2 5931	.5091 1247	2 5023	.1393
.8608	39 2061	7 1102	1 5511	.5089 4338	2 6429	.1392
.8609	43 7176	6 0260	.3460 5085	7 7420	2 7834	.1391
.8610	1.0848 2313	.2214 9414	.3459 4653	.5086 0495	.9902 9238	**.1390**
.8611	52 7472	3 8564	8 4215	4 3563	3 0641	.1389
.8612	57 2653	2 7709	7 3770	2 6623	3 2043	.1388
.8613	61 7856	1 6849	6 3320	.5080 9675	3 3444	.1387
.8614	66 3082	.2210 5985	5 2864	.5079 2720	3 4844	.1386
.8615	70 8330	.2209 5116	4 2401	7 5757	3 6243	.1385
.8616	75 3600	8 4243	3 1933	5 8786	3 7641	.1384
.8617	79 8892	7 3366	2 1459	4 1808	3 9038	.1383
.8618	84 4207	6 2484	1 0978	2 4822	4 0434	.1382
.8619	88 9544	5 1597	.3450 0491	.5070 7829	4 1829	.1381
.8620	1.0893 4903	.2204 0706	.3448 9999	.5069 0828	.9904 3223	**.1380**
.8621	1.0898 0285	2 9810	7 9500	7 3819	4 4616	.1379
.8622	1.0902 5689	1 8910	6 8995	5 6802	4 6007	.1378
.8623	07 1115	.2200 8005	5 8484	3 9778	4 7398	.1377
.8624	11 6565	.2199 7095	4 7967	2 2746	4 8788	.1376
.8625	16 2037	8 6181	3 7443	.5060 5706	5 0177	.1375
.8626	20 7531	7 5263	2 6914	.5058 8659	5 1564	.1374
.8627	25 3048	6 4340	1 6378	7 1604	5 2951	.1373
.8628	29 8588	5 3412	.3440 5837	5 4541	5 4337	.1372
.8629	34 4150	4 2480	.3439 5289	3 7470	5 5721	.1371
.8630	1.0938 9735	.2193 1544	.3438 4735	.5052 0392	.9905 7105	**.1370**
.8631	43 5343	2 0602	7 4175	.5050 3306	5 8487	.1369
.8632	48 0973	.2190 9656	6 3609	.5048 6212	5 9869	.1368
.8633	52 6627	.2189 8706	5 3036	6 9110	6 1249	.1367
.8634	57 2303	8 7751	4 2458	5 2001	6 2629	.1366
.8635	61 8002	7 6792	3 1873	3 4884	6 4007	.1365
.8636	66 3724	6 5828	2 1282	1 7759	6 5384	.1364
.8637	70 9469	5 4859	1 0685	.5040 0626	6 6761	.1363
.8638	75 5237	4 3886	.3430 0082	.5038 3485	6 8136	.1362
.8639	80 1028	3 2908	.3428 9472	6 6337	6 9510	.1361
.8640	1.0984 6842	.2182 1925	.3427 8856	.5034 9181	.9907 0884	**.1360**
.8641	89 2679	.2181 0938	6 8235	3 2017	7 2256	.1359
.8642	93 8539	.2179 9947	5 7606	.5031 4845	7 3627	.1358
.8643	1.0998 4423	8 8951	4 6972	.5029 7665	7 4997	.1357
.8644	1.1003 0329	7 7950	3 6332	8 0477	7 6367	.1356
.8645	07 6259	6 6945	2 5685	6 3282	7 7735	.1355
.8646	12 2212	5 5935	1 5032	4 6078	7 9102	.1354
.8647	16 8188	4 4920	.3420 4373	2 8867	8 0468	.1353
.8648	21 4187	3 3901	.3419 3707	.5021 1648	8 1833	.1352
.8649	26 0210	2 2877	8 3035	.5019 4421	8 3197	.1351
.8650	1.1030 6256	.2171 1849	.3417 2357	.5017 7186	.9908 4560	**.1350**

$E^{-ii}=$ .0000,0000,6 $E^{ii}=$ .0000,0003   .0000,0001   .0000,0001   .0000,0001   .0000,0000+

TABLE I

| | | | | |

p	x	z	$\sqrt{pq}$	$\sqrt{1-p^2}$	$\sqrt{1-q^2}$	q
.8650	1.1030 6256	.2171 1849	.3417 2357	.5017 7186	.9908 4560	**.1350**
.8651	35 2325	.2170 0816	6 1673	5 9943	8 5922	.1349
.8652	39 8418	.2168 9779	5 0982	4 2692	8 7283	.1348
.8653	44 4535	7 8736	4 0286	2 5434	8 8643	.1347
.8654	49 0675	6 7690	2 9582	.5010 8167	9 0002	.1346
.8655	53 6838	5 6638	1 8873	.5009 0892	9 1359	.1345
.8656	58 3025	4 5582	.3410 8157	7 3610	9 2716	.1344
.8657	62 9235	3 4522	.3409 7435	5 6319	9 4072	.1343
.8658	67 5470	2 3456	8 6707	3 9021	9 5427	.1342
.8659	72 1727	1 2387	7 5972	2 1714	9 6780	.1341
.8660	1.1076 8009	.2160 1312	.3406 5232	.5000 4400	.9909 8133	**.1340**
.8661	81 4314	.2159 0233	5 4484	.4998 7077	.9909 9485	.1339
.8662	86 0644	7 9149	4 3731	6 9747	.9910 0836	.1338
.8663	90 6997	6 8061	3 2971	5 2408	0 2185	.1337
.8664	95 3373	5 6968	2 2205	3 5062	0 3534	.1336
.8665	1.1099 9774	4 5870	1 1432	1 7707	0 4881	.1335
.8666	1.1104 6199	3 4768	.3400 0653	.4990 0345	0 6228	.1334
.8667	09 2647	2 3661	.3398 9868	.4988 2974	0 7573	.1333
.8668	13 9120	1 2549	7 9076	6 5595	0 8918	.1332
.8669	18 5616	.2150 1433	6 8278	4 8209	1 0261	.1331
.8670	1.1123 2137	.2149 0312	.3395 7473	.4983 0814	.9911 1604	**.1330**
.8671	27 8682	7 9187	4 6663	.4981 3411	1 2945	.1329
.8672	32 5250	6 8057	3 5845	.4979 6000	1 4286	.1328
.8673	37 1843	5 6922	2 5022	7 8581	1 5625	.1327
.8674	41 8460	4 5782	1 4192	6 1154	1 6963	.1326
.8675	46 5102	3 4638	.3390 3355	4 3718	1 8301	.1325
.8676	51 1767	2 3489	.3389 2512	2 6275	1 9637	.1324
.8677	55 8457	1 2336	8 1663	.4970 8823	2 0972	.1323
.8678	60 5171	.2140 1177	7 0807	.4969 1363	2 2306	.1322
.8679	65 1910	.2139 0015	5 9945	7 3896	2 3639	.1321
.8680	1.1169 8673	.2137 8847	.3384 9077	.4965 6420	.9912 4972	**.1320**
.8681	74 5460	6 7675	3 8202	3 8935	2 6303	.1319
.8682	79 2272	5 6498	2 7320	2 1443	2 7633	.1318
.8683	83 9109	4 5316	1 6432	.4960 3942	2 8962	.1317
.8684	88 5970	3 4130	.3380 5538	.4958 6434	3 0290	.1316
.8685	93 2855	2 2939	.3379 4637	6 8917	3 1617	.1315
.8686	1.1197 9765	1 1744	8 3730	5 1392	3 2943	.1314
.8687	1.1202 6700	.2130 0543	7 2816	3 3858	3 4268	.1313
.8688	07 3660	.2128 9338	6 1896	.4951 6316	3 5592	.1312
.8689	12 0644	7 8129	5 0969	.4949 8767	3 6915	.1311
.8690	1.1216 7653	.2126 6914	.3374 0036	.4948 1209	.9913 8237	**.1310**
.8691	21 4687	5 5695	2 9096	6 3642	3 9558	.1309
.8692	26 1745	4 4471	1 8149	4 6068	4 0878	.1308
.8693	30 8829	3 3243	.3370 7197	2 8485	4 2196	.1307
.8694	35 5937	2 2009	.3369 6237	.4941 0894	4 3514	.1306
.8695	40 3071	.2121 0771	8 5271	.4939 3294	4 4831	.1305
.8696	45 0229	.2119 9529	7 4299	7 5686	4 6147	.1304
.8697	49 7412	8 8281	6 3320	5 8070	4 7461	.1303
.8698	54 4621	7 7029	5 2334	4 0446	4 8775	.1302
.8699	59 1854	6 5772	4 1342	2 2813	5 0088	.1301
.8700	1.1263 9113	.2115 4511	.3363 0343	.4930 5172	.9915 1399	**.1300**

$E^{-ii}=$ $E^{ii}=$.0000,0003   .0000,0001   .0000,0001   .0000,0001   .0000,0000+
.0000,0000,7

TABLE I

.8700

.1300

p	x	z	√pq	√1-p²	√1-q²	q
.8700	1.1263 9113	.2115 4511	.3363 0343	.4930 5172	.9915 1399	**.1300**
.8701	68 6397	4 3245	1 9338	.4928 7523	5 2710	.1299
.8702	73 3706	3 1974	.3360 8326	6 9865	5 4020	.1298
.8703	78 1040	2 0698	.3359 7308	5 2199	5 5328	.1297
.8704	82 8400	.2110 9417	8 6283	3 4524	5 6636	.1296
.8705	87 5785	.2109 8132	7 5251	.4921 6842	5 7942	.1295
.8706	92 3195	8 6842	6 4213	.4919 9150	5 9248	.1294
.8707	1.1297 0630	7 5548	5 3168	8 1451	6 0552	.1293
.8708	1.1301 8091	6 4248	4 2117	6 3743	6 1856	.1292
.8709	06 5578	5 2944	3 1059	4 6026	6 3158	.1291
.8710	1.1311 3090	.2104 1635	.3351 9994	.4912 8301	.9916 4459	**.1290**
.8711	16 0628	3 0321	.3350 8923	.4911 0568	6 5760	.1289
.8712	20 8191	1 9003	.3349 7845	.4909 2826	6 7059	.1288
.8713	25 5780	.2100 7680	8 6760	7 5076	6 8357	.1287
.8714	30 3394	.2099 6352	7 5669	5 7317	6 9655	.1286
.8715	35 1034	8 5019	6 4571	3 9550	7 0951	.1285
.8716	39 8700	7 3682	5 3466	2 1775	7 2246	.1284
.8717	44 6392	6 2339	4 2355	.4900 3991	7 3540	.1283
.8718	49 4109	5 0992	3 1237	.4898 6198	7 4834	.1282
.8719	54 1853	3 9640	2 0112	6 8397	7 6126	.1281
.8720	1.1358 9622	.2092 8284	.3340 8981	.4895 0587	.9917 7417	**.1280**
.8721	63 7417	1 6923	.3339 7843	3 2769	7 8707	.1279
.8722	68 5238	.2090 5556	8 6698	.4891 4943	7 9996	.1278
.8723	73 3086	.2089 4185	7 5546	.4889 7107	8 1284	.1277
.8724	78 0959	8 2810	6 4388	7 9263	8 2571	.1276
.8725	82 8858	7 1429	5 3223	6 1411	8 3857	.1275
.8726	87 6784	6 0044	4 2052	4 3550	8 5142	.1274
.8727	92 4735	4 8654	3 0873	2 5681	8 6426	.1273
.8728	1.1397 2713	3 7259	1 9688	.4880 7803	8 7709	.1272
.8729	1.1402 0718	2 5859	.3330 8496	.4878 9916	8 8991	.1271
.8730	1.1406 8748	.2081 4455	.3329 7297	.4877 2021	.9919 0272	**.1270**
.8731	11 6805	.2080 3046	8 6092	5 4117	9 1552	.1269
.8732	16 4888	.2079 1632	7 4879	3 6204	9 2830	.1268
.8733	21 2997	8 0213	6 3660	1 8283	9 4108	.1267
.8734	26 1133	6 8789	5 2434	.4870 0353	9 5385	.1266
.8735	30 9295	5 7360	4 1202	.4868 2415	9 6661	.1265
.8736	35 7484	4 5927	2 9962	6 4468	9 7935	.1264
.8737	40 5700	3 4489	1 8716	4 6512	.9919 9209	.1263
.8738	45 3942	2 3046	.3320 7463	2 8547	.9920 0482	.1262
.8739	50 2211	1 1598	.3319 6203	.4861 0574	0 1754	.1261
.8740	1.1455 0506	.2070 0146	.3318 4936	.4859 2592	.9920 3024	**.1260**
.8741	59 8828	.2068 8688	7 3663	7 4601	0 4294	.1259
.8742	64 7177	7 7226	6 2382	5 6602	0 5562	.1258
.8743	69 5553	6 5759	5 1095	3 8594	0 6830	.1257
.8744	74 3956	5 4287	3 9801	2 0577	0 8096	.1256
.8745	79 2385	4 2810	2 8500	.4850 2551	0 9362	.1255
.8746	84 0842	3 1328	1 7192	.4848 4517	1 0626	.1254
.8747	88 9325	1 9842	.3310 5877	6 6474	1 1890	.1253
.8748	93 7836	.2060 8350	.3309 4555	4 8422	1 3152	.1252
.8749	1.1498 6373	.2059 6845	8 3227	3 0361	1 4414	.1251
.8750	1.1503 4938	.2058 5353	.3307 1891	.4841 2292	.9921 5674	**.1250**

E⁻ⁱⁱ= .0000,0000,7   Eⁱⁱ=.0000,0003   .0000,0001   .0000,0001   .0000,0001   .0000,0000+

TABLE I

p	x	z	√pq	√(1−p²)	√(1−q²)	q
.8750	1.1503 4938	.2058 5353	.3307 1891	.4841 2292	.9921 5674	.1250
.8751	08 3530	7 3847	6 0549	.4839 4213	1 6934	.1249
.8752	13 2149	6 2336	4 9200	7 6126	1 8192	.1248
.8753	18 0795	5 0821	3 7843	5 8030	1 9449	.1247
.8754	22 9469	3 9300	2 6480	3 9926	2 0705	.1246
.8755	27 8170	2 7775	1 5110	2 1812	2 1961	.1245
.8756	32 6898	1 6245	.3300 3733	.4830 3689	2 3215	.1244
.8757	37 5653	.2050 4709	.3299 2349	.4828 5558	2 4468	.1243
.8758	42 4436	.2049 3169	8 0958	6 7418	2 5720	.1242
.8759	47 3247	8 1625	6 9560	5 9268	2 6972	.1241
.8760	1.1552 2085	.2047 0075	.3295 8155	.4823 1110	.9922 8222	.1240
.8761	57 0951	5 8520	4 6743	.4821 2943	2 9471	.1239
.8762	61 9844	4 6961	3 5325	.4819 4767	3 0719	.1238
.8763	66 8764	3 5396	2 3899	7 6582	3 1966	.1237
.8764	71 7713	2 3827	1 2466	5 8389	3 3212	.1236
.8765	76 6689	1 2253	.3290 1026	4 0186	3 4457	.1235
.8766	81 5693	.2040 0673	.3288 9579	2 1974	3 5701	.1234
.8767	86 4725	.2038 9089	7 8125	.4810 3753	3 6944	.1233
.8768	91 3785	7 7501	6 6664	.4808 5524	3 8186	.1232
.8769	1.1596 2872	6 5907	5 5196	6 7285	3 9427	.1231
.8770	1.1601 1988	.2035 4308	.3284 3721	.4804 9037	.9924 0667	.1230
.8771	06 1132	4 2704	3 2239	3 0781	4 1906	.1229
.8772	11 0303	3 1096	2 0750	.4801 2515	4 3144	.1228
.8773	15 9503	1 9482	.3280 9253	.4799 4240	4 4381	.1227
.8774	20 8731	.2030 7864	.3279 7750	7 5956	4 5617	.1226
.8775	25 7987	.2029 6240	8 6239	5 7664	4 6851	.1225
.8776	30 7271	8 4612	7 4722	3 9362	4 8085	.1224
.8777	35 6584	7 2979	6 3197	2 1051	4 9318	.1223
.8778	40 5925	6 1341	5 1666	.4790 2731	5 0550	.1222
.8779	45 5294	4 9698	4 0127	.4788 4401	5 1780	.1221
.8780	1.1650 4692	.2023 8050	.3272 8581	.4786 6063	.9925 3010	.1220
.8781	55 4118	2 6397	1 7028	4 7716	5 4239	.1219
.8782	60 3573	1 4739	.3270 5467	2 9359	5 5466	.1218
.8783	65 3056	.2020 3076	.3269 3900	.4781 0994	5 6693	.1217
.8784	70 2568	.2019 1408	8 2325	.4779 2619	5 7919	.1216
.8785	75 2108	7 9736	7 0744	7 4235	5 9143	.1215
.8786	80 1677	6 8058	5 9155	5 5842	6 0367	.1214
.8787	85 1275	5 6375	4 7559	3 7439	6 1589	.1213
.8788	90 0901	4 4688	3 5956	1 9028	6 2811	.1212
.8789	1.1695 0557	3 2995	2 4345	.4770 0607	6 4031	.1211
.8790	1.1700 0241	.2012 1298	.3261 2728	.4768 2177	.9926 5251	.1210
.8791	04 9954	.2010 9595	.3260 1103	6 3738	6 6469	.1209
.8792	09 9696	.2009 7888	.3258 9471	4 5289	6 7687	.1208
.8793	14 9467	8 6175	7 7831	2 6832	6 8903	.1207
.8794	19 9267	7 4458	6 6185	.4760 8365	7 0118	.1206
.8795	24 9096	6 2735	5 4531	.4758 9889	7 1333	.1205
.8796	29 8954	5 1008	4 2870	7 1403	7 2546	.1204
.8797	34 8841	3 9276	3 1202	5 2908	7 3758	.1203
.8798	39 8758	2 7538	1 9526	3 4404	7 4970	.1202
.8799	44 8704	1 5796	.3250 7844	.4751 5891	7 6180	.1201
.8800	1.1749 8679	.2000 4048	.3249 6154	.4749 7368	.9927 7389	.1200

$E^{-ii}=$ $E^{ii}=$.0000,0004   .0000,0001   .0000,0001   .0000,0001   .0000,0000+
.0000,0000,7

TABLE I

p	x	z	$\sqrt{pq}$	$\sqrt{1-p^2}$	$\sqrt{1-q^2}$	q
.8800	1.1749 8679	.2000 4048	.3249 6154	.4749 7368	.9927 7389	.1200
.8801	54 8684	.1999 2296	8 4456	7 8836	7 8597	.1199
.8802	59 8718	8 0539	7 2752	6 0295	7 9805	.1198
.8803	64 8781	6 8776	6 1040	4 1744	8 1011	.1197
.8804	69 8874	5 7009	4 9320	2 3184	8 2216	.1196
.8805	74 8997	4 5236	3 7594	.4740 4615	8 3420	.1195
.8806	79 9149	3 3459	2 5860	.4738 6036	8 4623	.1194
.8807	84 9330	2 1677	1 4119	6 7448	8 5825	.1193
.8808	89 9542	.1990 9889	.3240 2370	4 8850	8 7026	.1192
.8809	1.1794 9783	.1989 8097	.3239 0614	3 0243	8 8226	.1191
.8810	1.1800 0054	.1988 6299	.3237 8851	.4731 1626	.9928 9425	.1190
.8811	05 0355	7 4497	6 7080	.4729 3001	9 0623	.1189
.8812	10 0686	6 2689	5 5302	7 4365	9 1820	.1188
.8813	15 1046	5 0877	4 3517	5 5720	9 3016	.1187
.8814	20 1437	3 9059	3 1724	3 7066	9 4211	.1186
.8815	25 1858	2 7236	1 9924	.4721 8402	9 5405	.1185
.8816	30 2308	1 5409	.3230 8117	.4719 9729	9 6598	.1184
.8817	35 2789	.1980 3576	.3229 6302	8 1046	9 7790	.1183
.8818	40 3300	.1979 1738	8 4479	6 2354	.9929 8981	.1182
.8819	45 3841	7 9895	7 2649	4 3652	.9930 0171	.1181
.8820	1.1850 4413	.1976 8047	.3226 0812	.4712 4940	.9930 1360	.1180
.8821	55 5015	5 6194	4 8967	.4710 6219	0 2547	.1179
.8822	60 5647	4 4336	3 7115	.4708 7489	0 3734	.1178
.8823	65 6310	3 2473	2 5256	6 8749	0 4920	.1177
.8824	70 7003	2 0605	1 3389	4 9999	0 6105	.1176
.8825	75 7726	.1970 8732	.3220 1514	3 1240	0 7288	.1175
.8826	80 8480	.1969 6853	.3218 9632	.4701 2471	0 8471	.1174
.8827	85 9265	8 4970	7 7742	.4699 3692	0 9653	.1173
.8828	91 0081	7 3082	6 5845	7 4904	1 0833	.1172
.8829	1.1896 0927	6 1188	5 3941	5 6106	1 2013	.1171
.8830	1.1901 1804	.1964 9289	.3214 2029	.4693 7299	.9931 3191	.1170
.8831	06 2712	3 7386	3 0109	.4691 8481	1 4369	.1169
.8832	11 3651	2 5477	1 8182	.4689 9655	1 5546	.1168
.8833	16 4620	1 3563	.3210 6247	8 0818	1 6721	.1167
.8834	21 5621	.1960 1644	.3209 4305	6 1972	1 7896	.1166
.8835	26 6652	.1958 9720	8 2355	4 3116	1 9069	.1165
.8836	31 7715	7 7791	7 0398	2 4250	2 0242	.1164
.8837	36 8809	6 5856	5 8433	.4680 5375	2 1413	.1163
.8838	41 9934	5 3917	4 6460	.4678 6490	2 2584	.1162
.8839	47 1090	4 1972	3 4480	6 7595	2 3753	.1161
.8840	1.1952 2278	.1953 0023	.3202 2492	.4674 8690	.9932 4921	.1160
.8841	57 3497	1 8068	.3201 0497	2 9775	2 6089	.1159
.8842	62 4747	.1950 6108	.3199 8494	.4671 0851	2 7255	.1158
.8843	67 6029	.1949 4143	8 6483	.4669 1917	2 8420	.1157
.8844	72 7342	8 2173	7 4465	7 2973	2 9585	.1156
.8845	77 8687	7 0197	6 2439	5 4019	3 0748	.1155
.8846	83 0063	5 8217	5 0405	3 5055	3 1910	.1154
.8847	88 1471	4 6231	3 8364	.4661 6082	3 3072	.1153
.8848	93 2911	3 4241	2 6315	.4659 7099	3 4232	.1152
.8849	1.1998 4383	2 2245	1 4259	7 8105	3 5391	.1151
.8850	1.2003 5886	.1941 0244	.3190 2194	.4655 9102	.9933 6549	.1150

$E^{-ii}=$ $E^{ii}=.0000,0004$ $.0000,0001$ $.0000,0001$ $.0000,0001$ $.0000,0000+$
$.0000,00008$

TABLE I

p	x	z	$\sqrt{pq}$	$\sqrt{1-p^2}$	$\sqrt{1-q^2}$	q
.8850	1.2003 5886	.1941 0244	.3190 2194	.4655 9102	.9933 6549	**.1150**
.8851	08 7421	.1939 8238	.3189 0122	4 0089	3 7706	.1149
.8852	13 8988	8 6226	7 8043	2 1066	3 8862	.1148
.8853	19 0587	7 4210	6 5955	.4650 2033	4 0018	.1147
.8854	24 2218	6 2188	5 3860	.4648 2990	4 1172	.1146
.8855	29 3881	5 0161	4 1757	6 3938	4 2325	.1145
.8856	34 5577	3 8129	2 9647	4 4875	4 3477	.1144
.8857	39 7304	2 6092	1 7528	2 5802	4 4628	.1143
.8858	44 9064	1 4050	.3180 5402	.4640 6719	4 5778	.1142
.8859	50 0856	.1930 2002	.3179 3268	.4638 7627	4 6927	.1141
.8860	1.2055 2680	.1928 9950	.3178 1126	.4636 8524	.9934 8075	**.1140**
.8861	60 4537	7 7892	6 8977	4 9411	4 9222	.1139
.8862	65 6426	6 5829	5 6820	3 0288	5 0368	.1138
.8863	70 8347	5 3761	4 4655	.4631 1155	5 1513	.1137
.8864	76 0301	4 1687	3 2482	.4629 2012	5 2657	.1136
.8865	81 2288	2 9609	2 0301	7 2859	5 3800	.1135
.8866	86 4307	1 7525	.3170 8113	5 3696	5 4941	.1134
.8867	91 6360	.1920 5436	.3169 5916	3 4523	5 6082	.1133
.8868	1.2096 8445	.1919 3341	8 3712	.4621 5339	5 7222	.1132
.8869	1.2102 0562	8 1242	7 1500	.4619 6146	5 8361	.1131
.8870	1.2107 2713	.1916 9137	.3165 9280	.4617 6942	.9935 9499	**.1130**
.8871	12 4897	5 7028	4 7052	5 7728	6 0636	.1129
.8872	17 7113	4 4912	3 4816	3 8505	6 1771	.1128
.8873	22 9363	3 2792	2 2573	1 9270	6 2906	.1127
.8874	28 1646	2 0667	.3161 0321	.4610 0026	6 4040	.1126
.8875	33 3962	.1910 8536	.3159 8062	.4608 0771	6 5172	.1125
.8876	38 6311	.1909 6400	8 5794	6 1507	6 6304	.1124
.8877	43 8694	8 4258	7 3519	4 2232	6 7435	.1123
.8878	49 1110	7 2112	6 1236	2 2946	6 8564	.1122
.8879	54 3559	5 9960	4 8945	.4600 3651	6 9693	.1121
.8880	1.2159 6042	.1904 7803	.3153 6645	.4598 4345	.9937 0821	**.1120**
.8881	64 8558	3 5641	2 4338	6 5029	7 1947	.1119
.8882	70 1108	2 3474	.3151 2023	4 5703	7 3073	.1118
.8883	75 3691	.1901 1301	.3149 9700	2 6366	7 4197	.1117
.8884	80 6309	.1899 9123	8 7369	.4590 7019	7 5321	.1116
.8885	85 8960	8 6940	7 5030	.4588 7662	7 6443	.1115
.8886	91 1644	7 4751	6 2683	6 8294	7 7565	.1114
.8887	1.2196 4363	6 2557	5 0328	4 8916	7 8685	.1113
.8888	1.2201 7115	5 0358	3 7964	2 9528	7 9805	.1112
.8889	06 9902	3 8154	2 5593	.4581 0129	8 0923	.1111
.8890	1.2212 2722	.1892 5944	.3141 3214	.4579 0720	.9938 2041	**.1110**
.8891	17 5577	1 3729	.3140 0826	7 1300	8 3157	.1109
.8892	22 8465	.1890 1509	.3138 8431	5 1870	8 4272	.1108
.8893	28 1388	.1888 9284	7 6027	3 2429	8 5387	.1107
.8894	33 4345	7 7053	6 3616	.4571 2978	8 6500	.1106
.8995	38 7337	6 4817	5 1196	.4569 3517	8 7612	.1105
.8896	44 0363	5 2575	3 8768	7 4045	8 8724	.1104
.8897	49 3423	4 0329	2 6332	5 4563	8 9834	.1103
.8898	54 6518	2 8077	1 3888	3 5070	9 0943	.1102
.8899	59 9648	1 5819	.3130 1436	.4561 5566	9 2051	.1101
.8900	1.2265 2812	.1880 3557	.3128 8976	.4559 6052	.9939 3159	**.1100**

$E^{-ii}=$
$.0000,0000\beta$ $E^{ii}=.0000,0004$ .0000,0001   .0000,0001   .0000,0001   .0000,0000+

Table I

p	x	z	$\sqrt{pq}$	$\sqrt{1-p^2}$	$\sqrt{1-q^2}$	q
.8900	1.2265 2812	.1880 3557	.3128 8976	.4559 6052	.9939 3159	.1100
.8901	70 6011	.1879 1289	7 6507	7 6528	9 4265	.1099
.8902	75 9244	7 9015	6 4030	5 6993	9 5370	.1098
.8903	81 2513	6 6737	5 1546	3 7447	9 6474	.1097
.8904	86 5816	5 4453	3 9052	.4551 7891	9 7577	.1096
.8905	91 9154	4 2164	2 6551	.4549 8324	9 8680	.1095
.8906	1.2297 2527	2 9869	1 4042	7 8747	.9939 9781	.1094
.8907	1.2302 5935	1 7569	.3120 1524	5 9159	.9940 0881	.1093
.8908	07 9379	.1870 5264	.3118 8998	3 9560	0 1980	.1092
.8909	13 2857	.1869 2953	7 6464	1 9950	0 3078	.1091
.8910	1.2318 6371	.1868 0637	.3116 3921	.4540 0330	.9940 4175	.1090
.8911	23 9920	6 8316	5 1371	.4538 0700	0 5271	.1089
.8912	29 3504	5 5989	3 8812	6 1058	0 6366	.1088
.8913	34 7124	4 3657	2 6245	4 1406	0 7460	.1087
.8914	40 0779	3 1320	1 3669	2 1743	0 8553	.1086
.8915	45 4470	1 8977	.3110 1085	.4530 2069	0 9645	.1085
.8916	50 8197	.1860 6629	.3108 8493	.4528 2385	1 0736	.1084
.8917	56 1959	.1859 4276	7 5893	6 2690	1 1826	.1083
.8918	61 5757	8 1917	6 3284	4 2984	1 2915	.1082
.8919	66 9590	6 9552	5 0667	2 3267	1 4003	.1081
.8920	1.2372 3460	.1855 7183	.3103 8041	.4520 3540	.9941 5089	.1080
.8921	77 7365	4 4808	2 5407	.4518 3801	1 6175	.1079
.8922	83 1307	3 2427	1 2765	6 4052	1 7260	.1078
.8923	88 5284	2 0041	.3100 0115	4 4292	1 8344	.1077
.8924	93 9298	.1850 7650	.3098 7456	2 4521	1 9427	.1076
.8925	1.2399 3348	.1849 5254	7 4788	.4510 4739	2 0508	.1075
.8926	1.2404 7434	8 2852	6 2112	.4508 4946	2 1589	.1074
.8927	10 1556	7 0444	4 9428	6 5143	2 2669	.1073
.8928	15 5715	5 8031	3 6735	4 5328	2 3748	.1072
.8929	20 9910	4 5613	2 4034	2 5503	2 4825	.1071
.8930	1.2426 4142	.1843 3189	.3091 1325	.4500 5666	.9942 5902	.1070
.8931	31 8410	2 0760	.3089 8607	.4498 5819	2 6978	.1069
.8932	37 2715	.1840 8326	8 5880	6 5960	2 8052	.1068
.8933	42 7057	.1839 5886	7 3145	4 6091	2 9126	.1067
.8934	48 1435	8 3440	6 0402	2 6211	3 0199	.1066
.8935	53 5850	7 0989	4 7650	.4490 6319	3 1270	.1065
.8936	59 0302	5 8533	3 4889	.4488 6417	3 2341	.1064
.8937	64 4791	4 6071	2 2120	6 6503	3 3410	.1063
.8938	69 9317	3 3604	.3080 9343	4 6578	3 4479	.1062
.8939	75 3881	2 1131	.3079 6557	2 6643	3 5546	.1061
.8940	1.2480 8481	.1830 8653	.3078 3762	.4480 6696	.9943 6613	.1060
.8941	86 3119	.1829 6170	7 0959	.4478 6738	3 7678	.1059
.8942	91 7794	8 3681	5 8147	6 6769	3 8743	.1058
.8943	1.2497 2506	7 1186	4 5326	4 6789	3 9806	.1057
.8944	1.2502 7256	5 8686	3 2497	2 6797	4 0869	.1056
.8945	08 2043	4 6181	1 9660	.4470 6795	4 1930	.1055
.8946	13 6868	3 3670	.3070 6814	.4468 6781	4 2991	.1054
.8947	19 1730	2 1153	.3069 3959	6 6756	4 4050	.1053
.8948	24 6630	.1820 8631	8 1095	4 6720	4 5108	.1052
.8949	30 1568	.1819 6104	6 8223	2 6673	4 6166	.1051
.8950	1.2535 6544	.1818 3571	.3065 5342	.4460 6614	.9944 7222	.1050

$E^{-ii}=$ $E^{ii}=.0000,0005$ .0000,0001 .0000,0001 .0000,0001 .0000,0000+
.0000,0000,8

TABLE I

.8950 .1050

p	x	z	$\sqrt{pq}$	$\sqrt{1-p^2}$	$\sqrt{1-q^2}$	q
.8950	1.2535 6544	.1818 3571	.3065 5342	.4460 6614	.9944 7222	.1050
.8951	41 1558	7 1033	4 2453	.4458 6544	4 8278	.1049
.8952	46 6609	5 8489	2 9554	6 6463	4 9332	.1048
.8953	52 1699	4 5939	1 6647	4 6370	5 0385	.1047
.8954	57 6827	3 3384	.3060 3732	2 6266	5 1437	.1046
.8955	63 1993	2 0824	.3059 0807	.4450 6151	5 2489	.1045
.8956	68 7197	.1810 8258	7 7874	.4448 6025	5 3539	.1044
.8957	74 2440	.1809 5687	6 4933	6 5887	5 4588	.1043
.8958	79 7721	8 3110	5 1982	4 5738	5 5636	.1042
.8959	85 3040	7 0527	3 9023	2 5577	5 6684	.1041
.8960	1.2590 8398	.1805 7939	.3052 6054	.4440 5405	.9945 7730	.1040
.8961	1.2596 3795	4 5345	1 3078	.4438 5222	5 8775	.1039
.8962	1.2601 9230	3 2746	.3050 0092	6 5027	5 9819	.1038
.8963	07 4704	2 0141	.3048 7097	4 4820	6 0862	.1037
.8964	13 0217	.1800 7531	7 4094	2 4603	6 1904	.1036
.8965	18 5769	.1799 4915	6 1082	.4430 4373	6 2945	.1035
.8966	24 1360	8 2294	4 8061	.4428 4133	6 3985	.1034
.8967	29 6990	6 9667	3 5031	6 3880	6 5025	.1033
.8968	35 2658	5 7035	2 1992	4 3616	6 6063	.1032
.8969	40 8366	4 4397	.3040 8944	2 3341	6 7100	.1031
.8970	1.2646 4114	.1793 1753	.3039 5888	.4420 3054	.9946 8136	.1030
.8971	51 9901	1 9104	8 2822	.4418 2756	6 9171	.1029
.8972	57 5727	.1790 6449	6 9748	6 2446	7 0205	.1028
.8973	63 1592	.1789 3789	5 6665	4 2124	7 1238	.1027
.8974	68 7497	8 1123	4 3573	2 1791	7 2270	.1026
.8975	74 3442	6 8451	3 0471	.4410 1446	7 3300	.1025
.8976	79 9426	5 5774	1 7361	.4408 0089	7 4330	.1024
.8977	85 5450	4 3091	.3030 4242	6 0721	7 5359	.1023
.8978	91 1514	3 0403	.3029 1114	4 0341	7 6387	.1022
.8979	1.2696 7618	1 7709	7 7977	.4401 9949	7 7414	.1021
.8980	1.2702 3762	.1780 5009	.3026 4831	.4399 9545	.9947 8440	.1020
.8981	07 9946	.1779 2304	5 1676	7 9130	7 9465	.1019
.8982	13 6170	7 9593	3 8512	5 8703	8 0489	.1018
.8983	19 2435	6 6877	2 5339	3 8265	8 1511	.1017
.8984	24 8739	5 4155	.3021 2156	.4391 7814	8 2533	.1016
.8985	30 5084	4 1427	.3019 8965	.4389 7352	8 3554	.1015
.8986	36 1470	2 8694	8 5765	7 6878	8 4574	.1014
.8987	41 7896	1 5955	7 2555	5 6392	8 5592	.1013
.8988	47 4363	.1770 3210	5 9337	3 5894	8 6610	.1012
.8989	53 0870	.1769 0460	4 6109	.4381 5384	8 7627	.1011
.8990	1.2758 7418	.1767 7704	.3013 2872	.4379 4863	.9948 8643	.1010
.8991	64 4007	6 4943	1 9626	7 4329	8 9657	.1009
.8992	70 0637	5 2175	.3010 6371	5 3784	9 0671	.1008
.8993	75 7307	3 9402	.3009 3107	3 3226	9 1684	.1007
.8994	81 4019	2 6624	7 9834	.4371 2657	9 2695	.1006
.8995	87 0772	1 3840	6 6551	.4369 2076	9 3706	.1005
.8996	92 7567	.1760 1050	5 3259	7 1483	9 4715	.1004
.8997	1.2798 4402	.1758 8254	3 9958	5 0877	9 5724	.1003
.8998	1.2804 1279	7 5453	2 6648	3 0260	9 6732	.1002
.8999	09 8197	6 2646	1 3329	.4360 9631	9 7738	.1001
.9000	1.2815 5157	.1754 9833	.3000 0000	.4358 8989	.9949 8744	.1000

$E^{-iL}=$ .0000,0000,9 $E^{iL}=$.0000,0005   .0000,0001   .0000,0001   .0000,0001   .0000,0000+

TABLE I

p	x	z	$\sqrt{pq}$	$\sqrt{1-p^2}$	$\sqrt{1-q^2}$	q
.9000	1.2815 5157	.1754 9833	.3000 0000	.4358 8989	.9949 8744	**.1000**
.9001	21 2158	3 7015	.2998 6662	56 8336	.9949 9748	.0999
.9002	26 9201	2 4191	97 3315	54 7670	.9950 0752	.0998
.9003	32 6286	.1751 1361	95 9958	52 6993	1754	.0997
.9004	38 3413	.1749 8525	94 6592	50 6303	2756	.0996
.9005	44 0581	8 5684	93 3217	48 5601	3756	.0995
.9006	49 7792	7 2837	91 9833	46 4887	4756	.0994
.9007	55 5045	5 9985	90 6439	44 4161	5754	.0993
.9008	61 2340	4 7126	89 3036	42 3422	6752	.0992
.9009	66 9677	3 4262	87 9623	40 2672	7748	.0991
.9010	1.2872 7056	.1742 1392	.2986 6202	.4338 1909	.9950 8743	**.0990**
.9011	78 4478	.1740 8517	85 2770	36 1134	.9950 9738	.0989
.9012	84 1942	.1739 5636	83 9330	34 0346	.9951 0731	.0988
.9013	89 9449	8 2748	82 5880	31 9546	1723	.0987
.9014	1.2895 6999	6 9856	81 2420	29 8734	2715	.0986
.9015	1.2901 4592	5 6957	79 8951	27 7910	3705	.0985
.9016	07 2227	4 4053	78 5473	25 7073	4694	.0984
.9017	12 9905	3 1143	77 1985	23 6224	5683	.0983
.9018	18 7626	1 8227	75 8488	21 5363	6670	.0982
.9019	24 5391	.1730 5305	74 4981	19 4489	7656	.0981
.9020	1.2930 3198	.1729 2378	.2973 1465	.4317 3603	.9951 8641	**.0980**
.9021	36 1048	7 9444	71 7939	15 2704	.9951 9626	.0979
.9022	41 8942	6 6505	70 4404	13 1793	.9952 0609	.0978
.9023	47 6880	5 3561	69 0859	11 0870	1591	.0977
.9024	53 4860	4 0610	67 7304	08 9934	2572	.0976
.9025	59 2885	2 7654	66 3740	06 8985	3552	.0975
.9026	65 0953	1 4691	65 0167	04 8024	4532	.0974
.9027	70 9064	.1720 1723	63 6584	02 7051	5510	.0973
.9028	76 7220	.1718 8750	62 2991	.4300 6065	6487	.0972
.9029	82 5419	7 5770	60 9389	.4298 5066	7463	.0971
.9030	1.2988 3663	.1716 2785	.2959 5777	.4296 4055	.9952 8438	**.0970**
.9031	1.2994 1951	4 9793	58 2155	94 3031	.9952 9412	.0969
.9032	1.3000 0283	3 6796	56 8524	92 1994	.9953 0385	.0968
.9033	05 8659	2 3793	55 4883	90 0945	1357	.0967
.9034	11 7079	.1711 0784	54 1232	87 9883	2328	.0966
.9035	17 5544	.1709 7770	52 7572	85 8809	3298	.0965
.9036	23 4054	8 4749	51 3902	83 7722	4267	.0964
.9037	29 2608	7 1723	50 0222	81 6622	5235	.0963
.9038	35 1207	5 8691	48 6533	79 5509	6202	.0962
.9039	40 9850	4 5653	47 2833	77 4384	7168	.0961
.9040	1.3046 8539	.1703 2609	.2945 9124	.4275 3245	.9953 8133	**.0960**
.9041	52 7272	1 9559	44 5405	73 2094	.9953 9097	.0959
.9042	58 6051	.1700 6503	43 1677	71 0931	.9954 0060	.0958
.9043	64 4874	.1699 3442	41 7938	68 9754	1022	.0957
.9044	70 3743	8 0374	40 4190	66 8565	1983	.0956
.9045	76 2657	6 7301	39 0432	64 7362	2943	.0955
.9046	82 1617	5 4222	37 6664	62 6147	3902	.0954
.9047	88 0622	4 1137	36 2886	60 4919	4860	.0953
.9048	93 9672	2 8046	34 9099	58 3678	5817	.0952
.9049	1.3099 8769	1 4949	33 5301	56 2424	6772	.0951
.9050	1.3105 7911	.1690 1846	.2932 1494	.4254 1157	.9954 7727	**.0950**

$E^{-i \underline{\underline{L}}}_{.0000,0000,9}$ $E^{i \underline{\underline{L}}}.0000,0005$ .0000,0001 .0000,0001 .0000,0002 .0000,0000+

TABLE I

.9050 .0950

p	x	z	$\sqrt{pq}$	$\sqrt{1-p^2}$	$\sqrt{1-q^2}$	q
.9050	1.3105 7911	.1690 1846	.2932 1494	.4254 1157	.9954 7727	.0950
.9051	11 7099	.1688 8737	30 7676	51 9877	8681	.0949
.9052	17 6333	7 5623	29 3849	49 8584	.9954 9634	.0948
.9053	23 5613	6 2502	28 0012	47 7277	.9955 0586	.0947
.9054	29 4940	4 9375	26 6165	45 5958	1536	.0946
.9055	35 4312	3 6243	25 2308	43 4626	2486	.0945
.9056	41 3731	2 3105	23 8440	41 3281	3435	.0944
.9057	47 3196	.1680 9960	22 4563	39 1923	4383	.0943
.9058	53 2708	.1679 6810	21 0676	37 0551	5329	.0942
.9059	59 2267	8 3654	19 6779	34 9166	6275	.0941
.9060	1.3165 1872	.1677 0492	.2918 2872	.4232 7769	.9955 7220	.0940
.9061	71 1524	5 7323	16 8954	30 6358	8163	.0939
.9062	77 1223	4 4149	15 5027	28 4933	.9955 9106	.0938
.9063	83 0969	3 0969	14 1090	26 3496	.9956 0048	.0937
.9064	89 0762	1 7783	12 7142	24 2045	0988	.0936
.9065	1.3195 0602	.1670 4591	11 3184	22 0581	1928	.0935
.9066	1.3201 0489	.1669 1393	09 9216	19 9104	2867	.0934
.9067	07 0424	7 8189	08 5239	17 7614	3804	.0933
.9068	13 0406	6 4979	07 1250	15 6110	4741	.0932
.9069	19 0436	5 1763	05 7252	13 4593	5676	.0931
.9070	1.3225 0514	.1663 8541	.2904 3244	.4211 3062	.9956 6611	.0930
.9071	31 0639	2 5313	02 9225	09 1518	7544	.0929
.9072	37 0812	.1661 2079	01 5196	06 9961	8477	.0928
.9073	43 1034	.1659 8838	.2900 1157	04 8390	.9956 9408	.0927
.9074	49 1303	8 5592	.2898 7107	02 6806	.9957 0339	.0926
.9075	55 1620	7 2340	97 3048	.4200 5208	1268	.0925
.9076	61 1986	5 9082	95 8978	.4198 3597	2197	.0924
.9077	67 2400	4 5818	94 4898	96 1972	3124	.0923
.9078	73 2862	3 2548	93 0807	94 0334	4051	.0922
.9079	79 3373	1 9271	91 6706	91 8682	4976	.0921
.9080	1.3285 3933	.1650 5989	.2890 2595	.4189 7017	.9957 5901	.0920
.9081	91 4541	.1649 2700	88 8473	87 5338	6824	.0919
.9082	1.3297 5199	7 9406	87 4342	85 3645	7747	.0918
.9083	1.3303 5905	6 6105	86 0199	83 1939	8668	.0917
.9084	09 6661	5 2799	84 6047	81 0219	.9957 9588	.0916
.9085	15 7465	3 9486	83 1883	78 8485	.9958 0508	.0915
.9086	21 8319	2 6167	81 7710	76 6738	1426	.0914
.9087	27 9222	.1641 2842	80 3526	74 4977	2343	.0913
.9088	34 0175	.1639 9511	78 9331	72 3202	3260	.0912
.9089	40 1177	8 6174	77 5126	70 1414	4175	.0911
.9090	1.3346 2229	.1637 2831	.2876 0911	.4167 9611	.9958 5089	.0910
.9091	52 3331	5 9482	74 6685	65 7795	6003	.0909
.9092	58 4482	4 6127	73 2449	63 5965	6915	.0908
.9093	64 5684	3 2765	71 8202	61 4121	7826	.0907
.9094	70 6935	1 9397	70 3944	59 2264	8736	.0906
.9095	76 8237	.1630 6024	68 9676	57 0392	.9958 9646	.0905
.9096	82 9589	.1629 2644	67 5397	54 8507	.9959 0554	.0904
.9097	89 0992	7 9258	66 1108	52 6607	1461	.0903
.9098	1.3395 2445	6 5866	64 6808	50 4694	2367	.0902
.9099	1.3401 3949	5 2467	63 2497	48 2766	3272	.0901
.9100	1.3407 5503	.1623 9063	.2861 8176	.4146 0825	.9959 4177	.0900

$E^{-ii}=$ $E^{ii}=.0000,0006$ .0000,0001 .0000,0001 .0000,0002 .0000,0000+
.0000,0001

$E^{-iii}=$ $E^{iii}=.0000,0000,+$
.0000,0000,0+

Table I

.9100 .0900

p	x	z	$\sqrt{pq}$	$\sqrt{1-p^2}$	$\sqrt{1-q^2}$	q
.9100	1.3407 5503	.1623 9063	.2861 8176	.4146 0825	.9959 4177	**.0900**
.9101	13 7108	2 5652	60 3844	43 8869	5080	.0899
.9102	19 8765	.1621 2235	58 9502	41 6900	5982	.0898
.9103	26 0472	.1619 8812	57 5148	39 4916	6883	.0897
.9104	32 2231	8 5383	56 0784	37 2919	7783	.0896
.9105	38 4041	7 1948	54 6410	35 0907	8682	.0895
.9106	44 5902	5 8506	53 2024	32 8881	.9959 9580	.0894
.9107	50 7815	4 5059	51 7628	30 6841	.9960 0477	.0893
.9108	56 9779	3 1605	50 3221	28 4787	1373	.0892
.9109	63 1795	1 8145	48 8803	26 2718	2269	.0891
.9110	1.3469 3863	.1610 4679	.2847 4374	.4124 0635	.9960 3163	**.0890**
.9111	75 5983	.1609 1206	45 9935	21 8538	4056	.0889
.9112	81 8154	7 7727	44 5485	19 6427	4948	.0888
.9113	88 0378	6 4242	43 1024	17 4301	5839	.0887
.9114	1.3494 2654	5 0751	41 6552	15 2162	6729	.0886
.9115	1.3500 4983	3 7254	40 2069	13 0007	7618	.0885
.9116	06 7364	2 3750	38 7575	10 7839	8506	.0884
.9117	12 9797	.1601 0240	37 3070	08 5656	.9960 9393	.0883
.9118	19 2284	.1599 6724	35 8554	06 3458	.9961 0279	.0882
.9119	25 4823	8 3202	34 4028	04 1246	1164	.0881
.9120	1.3531 7415	.1596 9673	.2832 9490	.4101 9020	.9961 2047	**.0880**
.9121	38 0060	5 6138	31 4941	.4099 6779	2930	.0879
.9122	44 2759	4 2597	30 0382	97 4524	3812	.0878
.9123	50 5511	2 9050	28 5811	95 2254	4693	.0877
.9124	56 8316	1 5496	27 1229	92 9969	5573	.0876
.9125	63 1174	.1590 1936	25 6636	90 7670	6452	.0875
.9126	69 4087	.1588 8370	24 2033	88 5357	7330	.0874
.9127	75 7053	7 4797	22 7418	86 3029	8207	.0873
.9128	82 0073	6 1219	21 2791	84 0686	9083	.0872
.9129	88 3147	4 7633	19 8154	81 8328	.9961 9957	.0871
.9130	1.3594 6275	.1583 4042	.2818 3506	.4079 5956	.9962 0831	**.0870**
.9131	1.3600 9457	2 0444	16 8846	77 3569	1704	.0869
.9132	07 2693	.1580 6840	15 4176	75 1167	2576	.0868
.9133	13 5984	.1579 3230	13 9494	72 8750	3447	.0867
.9134	19 9330	7 9613	12 4800	70 6319	4316	.0866
.9135	26 2730	6 5990	11 0096	68 3873	5185	.0865
.9136	32 6185	5 2360	09 5380	66 1412	6053	.0864
.9137	38 9695	3 8724	08 0653	63 8936	6920	.0863
.9138	45 3260	2 5082	06 5915	61 6445	7785	.0862
.9139	51 6880	.1571 1434	05 1166	59 3939	8650	.0861
.9140	1.3658 0556	.1569 7779	.2803 6405	.4057 1419	.9962 9514	**.0860**
.9141	64 4287	8 4118	02 1633	54 8883	.9963 0376	.0859
.9142	70 8074	7 0450	.2800 6849	52 6332	1238	.0858
.9143	77 1916	5 6776	.2799 2054	50 3766	2099	.0857
.9144	83 5814	4 3096	97 7248	48 1186	2958	.0856
.9145	89 9768	2 9409	96 2430	45 8590	3817	.0855
.9146	1.3696 3778	1 5716	94 7601	43 5979	4675	.0854
.9147	1.3702 7844	.1560 2016	93 2760	41 3353	5531	.0853
.9148	09 1967	.1558 8310	91 7908	39 0712	6387	.0852
.9149	15 6146	7 4598	90 3045	36 8055	7242	.0851
.9150	1.3722 0381	.1556 0879	.2788 8170	.4034 5384	.9963 8095	**.0850**

$E^{-ii}=$ .0000,0001 $E^{ii}=$.0000,0007   .0000,0001   .0000,0002   .0000,0002   .0000,0000+

$E^{-iii}=$ .0000,0000,0+ $E^{iii}=$ 0.000,0000,+

TABLE I

p	x	z	$\sqrt{pq}$	$\sqrt{1-p^2}$	$\sqrt{1-q^2}$	q
.9150	1.3722 0381	.1556 0879	.2788 8170	.4034 5384	.9963 8095	**.0850**
.9151	28 4673	4 7154	87 3283	32 2697	8948	.0849
.9152	34 9022	3 3422	85 8385	29 9995	.9963 9799	.0848
.9153	41 3428	1 9684	84 3475	27 7278	.9964 0650	.0847
.9154	47 7891	.1550 5939	82 8554	25 4545	1499	.0846
.9155	54 2411	.1549 2188	81 3621	23 1797	2348	.0845
.9156	60 6988	7 8431	79 8676	20 9034	3195	.0844
.9157	67 1623	6 4667	78 3720	18 6255	4042	.0843
.9158	73 6315	5 0897	76 8752	16 3461	4887	.0842
.9159	80 1065	3 7120	75 3773	14 0651	5732	.0841
.9160	1.3786 5873	.1542 3336	.2773 8782	.4011 7826	.9964 6575	**.0840**
.9161	93 0739	.1540 9546	72 3779	09 4986	7418	.0839
.9162	1.3799 5663	.1539 5750	70 8764	07 2130	8259	.0838
.9163	1.3806 0645	8 1947	69 3738	04 9258	9100	.0837
.9164	12 5685	6 8138	67 8699	02 6371	.9964 9939	.0836
.9165	19 0784	5 4322	66 3649	.4000 3469	.9965 0778	.0835
.9166	25 5942	4 0500	64 8588	.3998 0550	1615	.0834
.9167	32 1158	2 6671	63 3514	95 7616	2452	.0833
.9168	38 6433	.1531 2836	61 8429	93 4667	3287	.0832
.9169	45 1768	.1529 8994	60 3331	91 1701	4121	.0831
.9170	1.3851 7161	.1528 5145	.2758 8222	.3988 8720	.9965 4955	**.0830**
.9171	58 2614	7 1290	57 3101	86 5723	5787	.0829
.9172	64 8126	5 7429	55 7968	84 2711	6618	.0828
.9173	71 3697	4 3561	54 2823	81 9682	7449	.0827
.9174	77 9329	2 9686	52 7666	79 6638	8278	.0826
.9175	84 5020	1 5805	51 2497	77 3578	9106	.0825
.9176	91 0771	.1520 1917	49 7316	75 0502	.9965 9934	.0824
.9177	1.3897 6582	.1518 8023	48 2123	72 7410	.9966 0760	.0823
.9178	1.3904 2454	7 4122	46 6918	70 4302	1585	.0822
.9179	10 8386	6 0214	45 1701	68 1178	2410	.0821
.9180	1.3917 4378	.1514 6300	.2743 6472	.3965 8038	.9966 3233	**.0820**
.9181	24 0431	3 2379	42 1231	63 4882	4055	.0819
.9182	30 6545	1 8452	40 5977	61 1710	4876	.0818
.9183	37 2720	.1510 4518	39 0712	58 8522	5697	.0817
.9184	43 8956	.1509 0577	37 5434	56 5318	6516	.0816
.9185	50 5253	7 6630	36 0144	54 2098	7334	.0815
.9186	57 1611	6 2676	34 4842	51 8861	8151	.0814
.9187	63 8031	4 8716	32 9528	49 5609	8968	.0813
.9188	70 4513	3 4749	31 4201	47 2340	.9966 9783	.0812
.9189	77 1057	2 0775	29 8863	44 9054	.9967 0597	.0811
.9190	1.3983 7662	.1500 6795	.2728 3512	.3942 5753	.9967 1410	**.0810**
.9191	90 4330	.1499 2807	26 8148	40 2435	2222	.0809
.9192	1.3997 1059	7 8814	25 2772	37 9101	3033	.0808
.9193	1.4003 7851	6 4813	23 7384	35 5751	3844	.0807
.9194	10 4706	5 0806	22 1984	33 2384	4653	.0806
.9195	17 1624	3 6792	20 6571	30 9000	5461	.0805
.9196	23 8604	2 2772	19 1146	28 5600	6268	.0804
.9197	30 5647	.1490 8745	17 5708	26 2184	7074	.0803
.9198	37 2753	.1489 4711	16 0258	23 8751	7879	.0802
.9199	43 9923	8 0670	14 4795	21 5302	8683	.0801
.9200	1.4050 7156	.1486 6623	.2712 9320	.3919 1836	.9967 9486	**.0800**

$E^{-ii}=$ .0000,0001 $E^{ii}=$.0000,0008   .0000,0001   .0000,0002   .0000,0002   .0000,0000+

$E^{-iii}=$ .0000,00000,0+ $E^{iii}=$.0000,0000+

TABLE I

p	x	z	$\sqrt{pq}$	$\sqrt{1-p^2}$	$\sqrt{1-q^2}$	q
.9200	1.4050 7156	.1486 6623	.2712 9320	.3919 1836	.9967 9486	.0800
.9201	57 4453	5 2569	11 3832	16 8353	.9968 0288	.0799
.9202	64 1813	3 8508	09 8332	14 4854	1089	.0798
.9203	70 9237	2 4440	08 2819	12 1338	1890	.0797
.9204	77 6725	.1481 0366	06 7294	09 7806	2689	.0796
.9205	84 4278	.1479 6285	05 1756	07 4256	3487	.0795
.9206	91 1894	8 2197	03 6205	05 0690	4284	.0794
.9207	1.4097 9576	6 8102	02 0642	02 7107	5080	.0793
.9208	1.4104 7322	5 4001	.2700 5066	.3900 3508	5875	.0792
.9209	11 5132	3 9893	.2698 9478	.3897 9891	6669	.0791
.9210	1.4118 3008	.1472 5778	.2697 3876	.3895 6258	.9968 7462	.0790
.9211	25 0949	.1471 1656	95 8262	93 2607	8254	.0789
.9212	31 8955	.1469 7528	94 2635	90 8940	9045	.0788
.9213	38 7026	8 3393	92 6996	88 5256	.9968 9834	.0787
.9214	45 5163	6 9250	91 1343	86 1554	.9969 0623	.0786
.9215	52 3366	5 5102	89 5678	83 7836	1411	.0785
.9216	59 1634	4 0946	88 0000	81 4101	2198	.0784
.9217	65 9969	2 6783	86 4309	79 0348	2984	.0783
.9218	72 8370	.1461 2614	84 8605	76 6578	3769	.0782
.9219	79 6837	.1459 8438	83 2888	74 2792	4553	.0781
.9220	1.4186 5371	.1458 4254	.2681 7159	.3871 8988	.9969 5336	.0780
.9221	1.4193 3971	7 0064	80 1416	69 5166	6118	.0779
.9222	1.4200 2639	5 5868	78 5660	67 1328	6899	.0778
.9223	07 1373	4 1664	76 9892	64 7472	7679	.0777
.9224	14 0175	2 7453	75 4110	62 3599	8457	.0776
.9225	20 9043	.1451 3236	73 8315	59 9709	.9969 9235	.0775
.9226	27 7980	.1449 9012	72 2507	57 5801	.9970 0012	.0774
.9227	34 6984	8 4780	70 6686	55 1875	0788	.0773
.9228	41 6056	7 0542	69 0852	52 7933	1563	.0772
.9229	48 5196	5 6297	67 5005	50 3973	2336	.0771
.9230	1.4255 4404	.1444 2045	.2665 9145	.3847 9995	.9970 3109	.0770
.9231	62 3680	2 7786	64 3271	45 6000	3881	.0769
.9232	69 3025	.1441 3520	62 7384	43 1987	4652	.0768
.9233	76 2439	.1439 9248	61 1484	40 7956	5422	.0767
.9234	83 1921	8 4968	59 5571	38 3908	6190	.0766
.9235	90 1473	7 0681	57 9644	35 9842	6958	.0765
.9236	1.4297 1094	5 6388	56 3705	33 5759	7725	.0764
.9237	1.4304 0784	4 2087	54 7751	31 1657	8491	.0763
.9238	11 0543	2 7779	53 1785	28 7538	.9970 9255	.0762
.9239	18 0373	.1431 3465	51 5805	26 3402	.9971 0019	.0761
.9240	1.4325 0272	.1429 9143	.2649 9811	.3823 9247	.9971 0782	.0760
.9241	32 0241	8 4815	48 3804	21 5074	1543	.0759
.9242	39 0281	7 0479	46 7784	19 0884	2304	.0758
.9243	46 0391	5 6137	45 1750	16 6675	3064	.0747
.9244	53 0571	4 1787	43 5703	14 2449	3823	.0756
.9245	60 0823	2 7431	41 9642	11 8204	4580	.0755
.9246	67 1145	.1421 3067	40 3568	09 3942	5337	.0754
.9247	74 1538	.1419 8696	38 7480	06 9661	6092	.0753
.9248	81 2003	8 4319	37 1378	04 5362	6847	.0752
.9249	1.4388 2539	6 9934	35 5263	.3802 1045	7601	.0751
.9250	1.4395 3147	.1415 5542	.2633 9134	.3799 6710	.9971 8353	.0750

$E^{-ii} =$ .0000,0001 $E^{ii} =$.0000,0008   .0000,0001   .0000,0002   .0000,0002   .0000,0000+

$E^{-iii} =$ .0000,0000,0+ $E^{iii} =$.0000,0000,+

TABLE I

.9250   **.0750**

p	x	z	√pq	√1-p²	√1-q²	q
.9250	1.4395 3147	1415 5542	.2633 9134	.3799 6710	.9971 8353	**.0750**
.9251	02 3827	4 1143	32 2992	97 2357	9105	.0749
.9252	09 4578	2 6737	30 6836	94 7985	.9971 9856	.0748
.9253	16 5402	.1411 2324	29 0666	92 3596	.9972 0605	.0747
.9254	23 6299	.1409 7904	27 4482	89 9187	1354	.0746
.9255	30 7268	8 3477	25 8284	87 4761	2101	.0745
.9256	37 8309	6 9043	24 2073	85 0316	2848	.0744
.9257	44 9424	5 4602	22 5848	82 5852	3593	.0743
.9258	52 0611	4 0153	20 9609	80 1370	4338	.0742
.9259	59 1872	2 5697	19 3356	77 6870	5082	.0741
.9260	1.4466 3207	.1401 1235	.2617 7089	.3775 2351	.9972 5824	**.0740**
.9261	73 4615	.1399 6765	16 0808	72 7813	6566	.0739
.9262	80 6097	8 2288	14 4514	70 3257	7306	.0738
.9263	87 7653	6 7804	12 8205	67 8682	8046	.0737
.9264	1.4494 9283	5 3312	11 1882	65 4089	8784	.0736
.9265	1.4502 0988	3 8814	09 5546	62 9476	.9972 9522	.0735
.9266	09 2768	2 4308	07 9195	60 4845	.9973 0258	.0734
.9267	16 4622	.1390 9795	06 2830	58 0196	0994	.0733
.9268	23 6552	.1389 5275	04 6451	55 5527	1728	.0732
.9269	30 8556	8 0748	03 0058	53 0839	2462	.0731
.9270	1.4538 0636	.1386 6213	.2601 3650	.3750 6133	.9973 3194	**.0730**
.9271	45 2792	5 1672	.2599 7229	48 1407	3926	.0729
.9272	52 5023	3 7123	98 0793	45 6663	4656	.0728
.9273	59 7330	2 2567	96 4343	43 1899	5385	.0727
.9274	66 9714	.1380 8003	94 7879	40 7117	6114	.0726
.9275	74 2174	.1379 3433	93 1400	38 2315	6841	.0725
.9276	81 4711	7 8855	91 4907	35 7495	7568	.0724
.9277	88 7324	6 4270	89 8400	33 2655	8293	.0723
.9278	1.4596 0014	4 9677	88 1878	30 7795	9017	.0722
.9279	1.4603 2782	3 5078	86 5342	28 2917	.9973 9741	.0721
.9280	1.4610 5627	.1372 0471	.2584 8791	.3725 8019	.9974 0463	**.0720**
.9281	17 8550	.1370 5857	83 2226	23 3102	1185	.0719
.9282	25 1550	.1369 1235	81 5646	20 8166	1905	.0718
.9283	32 4629	7 6606	79 9052	18 3210	2624	.0717
.9284	39 7785	6 1970	78 2444	15 8235	3343	.0716
.9285	47 1020	4 7327	76 5820	13 3240	4060	.0715
.9286	54 4334	3 2676	74 9183	10 8226	4776	.0714
.9287	61 7727	1 8018	73 2530	08 3192	5492	.0713
.9288	69 1199	.1360 3352	71 5863	05 8138	6206	.0712
.9289	76 4750	.1358 8680	69 9181	03 3065	6919	.0711
.9290	1.4683 8380	.1357 4000	.2568 2484	.3700 7972	.9974 7632	**.0710**
.9291	91 2090	5 9312	66 5773	.3698 2860	8343	.0709
.9292	1.4698 5880	4 4617	64 9047	95 7727	9053	.0708
.9293	1.4705 9750	2 9915	63 2306	93 2575	.9974 9762	.0707
.9294	13 3701	1 5205	61 5550	90 7403	.9975 0471	.0706
.9295	20 7732	.1350 0488	59 8779	88 2211	1178	.0705
.9296	28 1844	.1348 5764	58 1994	85 6999	1884	.0704
.9297	35 6037	7 1032	56 5193	83 1768	2589	.0703
.9298	43 0311	5 6293	54 8378	80 6516	3294	.0702
.9299	50 4666	4 1546	53 1547	78 1244	3997	.0701
.9300	1.4757 9103	.1342 6791	.2551 4702	.3675 5952	.9975 4699	**.0700**

$E^{-ii} = $ .000,0002 $E^{ii} = $.0000,0009   .0000,0001   .0000,0002   .0000,0003   .0000,0000+

$E^{-iii} = $ .0000,0000,0+ $E^{iii} = $.0000,0000,+

TABLE I

p	x	z	√pq	√1-p²	√1-q²	q
.9300	1.4757 9103	.1342 6791	.2551 4702	.3675 5952	.9975 4699	**.0700**
.9301	65 3622	.1341 2030	49 7841	73 0640	5400	.0699
.9302	72 8223	.1339 7261	48 0965	70 5308	6101	.0698
.9303	80 2906	8 2484	46 4075	67 9955	6800	.0697
.9304	87 7672	6 7700	44 7169	65 4582	7498	.0696
.9305	1.4795 2521	5 2909	43 0248	62 9189	8195	.0695
.9306	1.4802 7452	3 8110	41 3311	60 3776	8891	.0694
.9307	10 2467	2 3303	39 6360	57 8342	.9975 9587	.0693
.9308	17 7565	.1330 8489	37 9393	55 2888	.9976 0281	.0692
.9309	25 2747	.1329 3668	36 2411	52 7413	0974	.0691
.9310	1.4832 8013	.1327 8839	.2534 5414	.3650 1918	.9976 1666	**.0690**
.9311	40 3363	6 4002	32 8401	47 6402	2357	.0689
.9312	47 8797	4 9158	31 1373	45 0866	3047	.0688
.9313	55 4316	3 4306	29 4329	42 5309	3736	.0687
.9314	62 9920	1 9447	27 7270	39 9731	4425	.0686
.9315	70 5608	.1320 4580	26 0196	37 4132	5112	.0685
.9316	78 1382	.1318 9706	24 3106	34 8513	5798	.0684
.9317	85 7242	7 4824	22 6000	32 2873	6483	.0683
.9318	1.4893 3187	5 9934	20 8879	29 7212	7167	.0682
.9319	1.4900 9218	4 5037	19 1743	27 1530	7850	.0681
.9320	1.4908 5336	.1313 0133	.2517 4590	.3624 5827	.9976 8532	**.0680**
.9321	16 1540	1 5220	15 7422	22 0104	9213	.0679
.9322	23 7831	.1310 0300	14 0239	19 4359	.9976 9893	.0678
.9323	31 4208	.1308 5373	12 3039	16 8593	.9977 0572	.0677
.9324	39 0673	7 0437	10 5824	14 2806	1250	.0676
.9325	46 7225	5 5495	08 8593	11 6997	1927	.0675
.9326	54 3865	4 0544	07 1346	09 1168	2603	.0674
.9327	62 0593	2 5586	05 4083	06 5317	3278	.0673
.9328	69 7409	.1301 0620	03 6805	03 9445	3953	.0672
.9329	77 4314	.1299 5646	01 9510	.3601 3552	4626	.0671
.9330	1.4985 1307	.1298 0665	.2500 2200	.3598 7637	.9977 5298	**.0670**
.9331	1.4992 8389	6 5676	.2498 4873	96 1700	5969	.0669
.9332	1.5000 5560	5 0679	96 7531	93 5743	6639	.0668
.9333	08 2821	3 5675	95 0172	90 9763	7308	.0667
.9334	16 0172	2 0663	93 2798	88 3762	7976	.0666
.9335	23 7612	.1290 5643	91 5407	85 7740	8643	.0665
.9336	31 5142	.1289 0615	89 8000	83 1695	9308	.0664
.9337	39 2763	7 5580	88 0577	80 5629	.9977 9973	.0663
.9338	47 0475	6 0537	86 3137	77 9542	.9978 0637	.0662
.9339	54 8278	4 5486	84 5682	75 3432	1300	.0661
.9340	1.5062 6172	.1283 0427	.2482 8210	.3572 7300	.9978 1962	**.0660**
.9341	70 4158	1 5361	81 0721	70 1147	2623	.0659
.9342	78 2235	.1280 0286	79 3217	67 4972	3283	.0658
.9343	86 0404	.1278 5204	77 5696	64 8774	3942	.0657
.9344	1.5093 8666	7 0114	75 8158	62 2555	4600	.0656
.9345	1.5101 7020	5 5016	74 0604	59 6313	5257	.0655
.9346	09 5467	3 9911	72 3034	57 0049	5913	.0654
.9347	17 4007	2 4797	70 5447	54 3763	6568	.0653
.9348	25 2641	.1270 9676	68 7843	51 7455	7222	.0652
.9349	33 1368	.1269 4547	67 0223	49 1124	7875	.0651
.9350	1.5141 0189	.1267 9410	.2465 2586	.3546 4771	.9978 8526	**.0650**

$E^{-ii}=$ $E^{ii}=.0000,0011$ ·.0000,0001   .0000,0002   .0000,0003   .0000,0000+
.0000,0001

$E^{-iii}=$ $E^{iii}=.0000,0000,+$
.0000,0000+

TABLE I

.9350   **.0650**

p	x	z	$\sqrt{pq}$	$\sqrt{1-p^2}$	$\sqrt{1-q^2}$	q
.9350	1.5141 0189	.1267 9410	.2465 2586	.3546 4771	.9978 8526	**.0650**
.9351	48 9104	6 4265	63 4933	43 8396	9177	.0649
.9352	56 8114	4 9112	61 7262	41 1998	.9978 9827	.0648
.9353	64 7218	3 3951	59 9575	38 5578	.9979 0476	.0647
.9354	72 6417	1 8782	58 1871	35 9135	1124	.0646
.9355	80 5712	.1260 3606	56 4151	33 2669	1771	.0645
.9356	88 5102	.1258 8421	54 6413	30 6181	2417	.0644
.9357	1.5196 4588	7 3229	52 8659	27 9670	3061	.0643
.9358	1.5204 4170	5 8028	51 0887	25 3136	3705	.0642
.9359	12 3849	4 2820	49 3099	22 6579	4348	.0641
.9360	1.5220 3624	.1252 7604	.2447 5294	.3520 0000	.9979 4990	**.0640**
.9361	28 3496	.1251 2379	45 7471	17 3398	5631	.0639
.9362	36 3466	.1249 7147	43 9632	14 6772	6270	.0638
.9363	44 3533	8 1906	42 1775	12 0124	6909	.0637
.9364	52 3698	6 6658	40 3901	09 3452	7547	.0636
.9365	60 3961	5 1402	38 6010	06 6758	8184	.0635
.9366	68 4323	3 6137	36 8102	04 0040	8820	.0634
.9367	76 4783	2 0865	35 0177	.3501 3299	.9979 9454	.0633
.9368	84 5342	.1240 5584	33 2234	.3498 6535	.9980 0088	.0632
.9369	1.5292 6001	.1239 0296	31 4274	95 9747	0721	.0631
.9370	1.5300 6759	.1237 4999	.2429 6296	.3493 2936	.9980 1353	**.0630**
.9371	08 7617	5 9694	27 8301	90 6101	1983	.0629
.9372	16 8575	4 4382	26 0289	87 9243	2613	.0628
.9373	24 9634	2 9061	24 2259	85 2361	3242	.0627
.9374	33 0794	.1231 3732	22 4211	82 5456	3870	.0626
.9375	41 2054	.1229 8395	20 6146	79 8527	4496	.0625
.9376	49 3417	8 3049	18 8063	77 1575	5122	.0624
.9377	57 4881	6 7696	16 9963	74 4598	5747	.0623
.9378	65 6447	5 2334	15 1845	71 7598	6371	.0622
.9379	73 8115	3 6965	13 3709	69 0574	6993	.0621
.9380	1.5381 9886	.1222 1587	.2411 5555	.3466 3525	.9980 7615	**.0620**
.9381	90 1760	.1220 6201	09 7384	63 6453	8236	.0619
.9382	1.5398 3737	.1219 0806	07 9194	60 9357	8855	.0618
.9383	1.5406 5818	7 5404	06 0987	58 2237	.9980 9474	.0617
.9384	14 8003	5 9993	04 2762	55 5092	.9981 0092	.0616
.9385	23 0292	4 4574	02 4519	52 7923	0708	.0615
.9386	31 2685	2 9147	.2400 6258	50 0730	1324	.0614
.9387	39 5184	.1211 3712	.2398 7978	47 3513	1939	.0613
.9388	47 7788	.1209 8268	96 9681	44 6271	2552	.0612
.9389	56 0497	8 2816	95 1365	41 9005	3165	.0611
.9390	1.5464 3312	.1206 7356	.2393 3032	.3439 1714	.9981 3777	**.0610**
.9391	72 6233	5 1887	91 4680	36 4399	4387	.0609
.9392	80 9261	3 6411	89 6309	33 7059	4997	.0608
.9393	89 2396	2 0926	87 7921	30 9694	5605	.0607
.9394	1.5497 5638	.1200 5432	85 9514	28 2304	6213	.0606
.9395	1.5505 8987	.1198 9930	84 1088	25 4890	6820	.0605
.9396	14 2444	7 4420	82 2645	22 7451	7425	.0604
.9397	22 6010	5 8902	80 4182	19 9987	8030	.0603
.9398	30 9684	4 3375	78 5702	17 2498	8634	.0602
.9399	39 3467	2 7840	76 7202	14 4984	9236	.0601
.9400	1.5547 7359	.1191 2297	.2374 8684	.3411 7444	.9981 9838	**.0600**

$E^{-ii} =$ $E^{ii} = 0000,0013$ .0000,0001   .0000,0003   .0000,0003   .0000,0000+
.0000,0002

$E^{-iii} =$ $E^{iii} = 0000,0000,+$
.0000,0000,+

TABLE I

p	x	z	$\sqrt{pq}$	$\sqrt{1-p^2}$	$\sqrt{1-q^2}$	q
.9400	1.5547 7359	.1191 2297	.2374 8684	.3411 7444	.9981 9838	.0600
.9401	56 1361	.1189 6745	73 0147	08 9880	.9982 0438	.0599
.9402	64 5472	8 1184	71 1592	06 2290	1038	.0598
.9403	72 9694	6 5615	69 3018	03 4675	1636	.0597
.9404	81 4027	5 0038	67 4425	.3400 7035	2234	.0596
.9405	89 8471	3 4453	65 5813	.3397 9369	2831	.0595
.9406	1.5598 3025	1 8859	63 7183	95 1677	3426	.0594
.9407	1.5606 7692	.1180 3256	61 8533	92 3961	4021	.0593
.9408	15 2470	.1178 7645	59 9864	89 6218	4614	.0592
.9409	23 7361	7 2026	58 1177	86 8450	5207	.0591
.9410	1.5632 2365	.1175 6398	.2356 2470	.3384 0656	.9982 5798	.0590
.9411	40 7482	4 0761	54 3744	81 2836	6389	.0589
.9412	49 2712	2 5116	52 4999	78 4991	6978	.0588
.9413	57 8056	.1170 9463	50 6235	75 7119	7567	.0587
.9414	66 3514	.1169 3800	48 7452	72 9222	8154	.0586
.9415	74 9087	7 8130	46 8649	70 1298	8741	.0585
.9416	83 4774	6 2451	44 9827	67 3349	9326	.0584
.9417	1.5692 0577	4 6763	43 0986	64 5373	.9982 9911	.0583
.9418	1.5700 6496	3 1067	41 2125	61 7371	.9983 0494	.0582
.9419	09 2531	.1161 5362	39 3245	58 9342	1077	.0581
.9420	1.5717 8682	.1159 9648	.2337 4345	.3356 1287	.9983 1658	.0580
.9421	26 4950	8 3926	35 5425	53 3206	2239	.0579
.9422	35 1335	6 8195	33 6486	50 5098	2818	.0578
.9423	43 7838	5 2456	31 7528	47 6964	3397	.0577
.9424	52 4459	3 6707	29 8549	44 8803	3974	.0576
.9425	61 1198	2 0951	27 9551	42 0615	4551	.0575
.9426	69 8055	.1150 5185	26 0533	39 2400	5126	.0574
.9427	78 5032	.1148 9411	24 1495	36 4159	5701	.0573
.9428	87 2129	7 3628	22 2437	33 5891	6274	.0572
.9429	1.5795 9345	5 7837	20 3360	30 7595	6846	.0571
.9430	1.5804 6682	.1144 2036	.2318 4262	.3327 9273	.9983 7418	.0570
.9431	13 4139	2 6227	16 5144	25 0923	7988	.0569
.9432	22 1718	.1141 0409	14 6006	22 2547	8558	.0568
.9433	30 9418	.1139 4583	12 6848	19 4143	9126	.0567
.9434	39 7240	7 8748	10 7670	16 5711	.9983 9694	.0566
.9435	48 5185	6 2903	08 8471	13 7252	.9984 0260	.0565
.9436	57 3252	4 7051	06 9252	10 8766	0825	.0564
.9437	66 1442	3 1189	05 0013	08 0252	1390	.0563
.9438	74 9756	.1131 5318	03 0753	05 1711	1953	.0562
.9439	83 8194	.1129 9439	.2301 1473	.3302 3142	2515	.0561
.9440	1.5892 6756	.1128 3551	.2299 2173	.3299 4545	.9984 3077	.0560
.9441	1.5901 5443	6 7654	97 2851	96 5920	3637	.0559
.9442	10 4255	5 1748	95 3510	93 7268	4197	.0558
.9443	19 3193	3 5833	93 4147	90 8587	4755	.0557
.9444	28 2257	1 9909	91 4764	87 9878	5312	.0556
.9445	37 1448	.1120 3976	89 5360	85 1142	5869	.0555
.9446	46 0765	.1118 8035	87 5935	82 2377	6424	.0554
.9447	55 0210	7 2084	85 6489	79 3583	6978	.0553
.9448	63 9783	5 6125	83 7023	76 4762	7532	.0552
.9449	72 9484	4 0156	81 7535	73 5911	8084	.0551
.9450	1.5981 9314	.1112 4179	.2279 8026	.3270 7033	.9984 8635	.0550

$E^{-ii} =$ $E^{ii} = .0000,0015$ .0000,0001   .0000,0003   .0000,0003   .0000,0000+
.0000,0002

$E^{-iii} =$ $E^{iii} = .0000,0000,+$
.0000,0000,0+

TABLE I

p	x	z	$\sqrt{pq}$	$\sqrt{1-p^2}$	$\sqrt{1-q^2}$	q
.9450	1.5981 9314	.1112 4179	.2279 8026	.3270 7033	.9984 8635	**.0550**
.9451	90 9273	.1110 8192	77 8496	67 8126	9186	.0549
.9452	1.5999 9361	.1109 2197	75 8945	64 9190	.9984 9735	.0548
.9453	1.6008 9580	7 6192	73 9373	62 0225	.9985 0283	.0547
.9454	17 9929	6 0179	71 9780	59 1232	0831	.0546
.9455	27 0409	4 4156	70 0165	56 2210	1377	.0545
.9456	36 1020	2 8125	68 0529	53 3158	1922	.0544
.9457	45 1764	.1101 2084	66 0872	50 4078	2467	.0543
.9458	54 2639	.1099 6034	64 1193	47 4969	3010	.0542
.9459	63 3647	7 9976	62 1492	44 5830	3552	.0541
.9460	1.6072 4789	.1096 3908	.2260 1770	.3241 6662	.9985 4094	**.0540**
.9461	81 6064	4 7831	58 2026	38 7465	4634	.0539
.9462	90 7474	3 1744	56 2261	35 8239	5173	.0538
.9463	1.6099 9018	.1091 5649	54 2473	32 8982	5711	.0537
.9464	1.6109 0697	.1089 9545	52 2664	29 9697	6249	.0536
.9465	18 2512	8 3431	50 2833	27 0381	6785	.0535
.9466	27 4463	6 7308	48 2980	24 1036	7320	.0534
.9467	36 6550	5 1176	46 3105	21 1661	7854	.0533
.9468	45 8775	3 5035	44 3208	18 2256	8388	.0532
.9469	55 1137	1 8884	42 3289	15 2821	8920	.0531
.9470	1.6164 3637	.1080 2725	.2240 3348	.3212 3356	.9985 9451	**.0530**
.9471	73 6276	.1078 6556	38 3384	09 3861	.9985 9981	.0529
.9472	82 9053	7 0377	36 3399	06 4335	.9986 0511	.0528
.9473	1.6192 1970	5 4190	34 3391	03 4780	1039	.0527
.9474	1.6201 5027	3 7993	32 3360	.3200 5193	1566	.0526
.9475	10 8225	2 1787	30 3307	.3197 5577	2092	.0525
.9476	20 1564	.1070 5571	28 3231	94 5929	2618	.0524
.9477	29 5044	.1068 9346	26 3133	91 6251	3142	.0523
.9478	38 8666	7 3112	24 3012	88 6543	3665	.0522
.9479	48 2431	5 6869	22 2869	85 6803	4187	.0521
.9480	1.6257 6339	.1064 0616	.2220 2703	.3182 7033	.9986 4708	**.0520**
.9481	67 0390	2 4353	18 2513	79 7231	5229	.0519
.9482	76 4586	.1060 8082	16 2301	76 7398	5748	.0518
.9483	85 8926	.1059 1801	14 2066	73 7535	6266	.0517
.9484	1.6295 3412	7 5510	12 1808	70 7639	6783	.0516
.9485	1.6304 8043	5 9210	10 1527	67 7713	7299	.0515
.9486	14 2820	4 2900	08 1223	64 7755	7815	.0514
.9487	23 7744	2 6581	06 0895	61 7766	8329	.0513
.9488	33 2815	.1051 0253	04 0544	58 7744	8842	.0512
.9489	42 8034	.1049 3915	.2202 0170	55 7692	9354	.0511
.9490	1.6352 3402	.1047 7567	.2199 9773	.3152 7607	.9986 9865	**.0510**
.9491	61 8919	6 1210	97 9352	49 7490	.9987 0375	.0509
.9492	71 4585	4 4843	95 8907	46 7342	0885	.0508
.9493	81 0401	2 8467	93 8439	43 7161	1393	.0507
.9494	1.6390 6368	.1041 2081	91 7947	40 6948	1900	.0506
.9495	1.6400 2486	.1039 5686	89 7431	37 6703	2406	.0505
.9496	09 8755	7 9281	87 6892	34 6426	2911	.0504
.9497	19 5177	6 2866	85 6329	31 6116	3415	.0503
.9498	29 1752	4 6442	83 5741	28 5773	3919	.0502
.9499	38 8481	3 0008	81 5130	25 5398	4421	.0501
.9500	1.6448 5363	.1031 3564	.2179 4495	.3122 4990	.9987 4922	**.0500**

$E^{-ii}=$ .0000,0002 $E^{ii}=$ .0000,0018   .0000,0001   .0000,0003   .0000,0004   .0000,0000+

$E^{-iii}=$ .0000,0000,0+ $E^{iii}=$ .0000,0000,0+

TABLE I

p	x	z	$\sqrt{pq}$	$\sqrt{1-p^2}$	$\sqrt{1-q^2}$	q
.9500	1.6448 5363	.1031 3564	.2179 4495	.3122 4990	.9987 4922	**.0500**
.9501	58 2400	.1029 7111	77 3835	19 4549	5422	.0499
.9502	67 9592	8 0648	75 3151	16 4075	5921	.0498
.9503	77 6940	6 4175	73 2443	13 3569	6419	.0497
.9504	87 4445	4 7692	71 1711	10 3029	6916	.0496
.9505	1.6497 2106	3 1200	69 0954	07 2456	7412	.0495
.9506	1.6506 9925	.1021 4698	67 0173	04 1849	7907	.0494
.9507	16 7903	.1019 8186	64 9367	.3101 1209	8402	.0493
.9508	26 6039	8 1664	62 8537	.3098 0536	8895	.0492
.9509	36 4334	6 5133	60 7682	94 9829	9387	.0491
.9510	1.6546 2790	.1014 8591	.2158 6802	.3091 9088	.9987 9878	**.0490**
.9511	56 1406	3 2040	56 5897	88 8313	.9988 0368	.0489
.9512	66 0184	.1011 5479	54 4967	85 7505	0857	.0488
.9513	75 9123	.1009 8908	52 4012	82 6662	1345	.0487
.9514	85 8225	8 2327	50 3032	79 5785	1832	.0486
.9515	1.6595 7490	6 5736	48 2027	76 4874	2318	.0485
.9516	1.6605 6919	4 9136	46 0997	73 3929	2803	.0484
.9517	15 6513	3 2525	43 9942	70 2949	3287	.0483
.9518	25 6271	.1001 5904	41 8861	67 1935	3770	.0482
.9519	35 6195	.0999 9274	39 7755	64 0886	4253	.0481
.9520	1.6645 6286	.0998 2633	.2137 6623	.3060 9802	.9988 4734	**.0480**
.9521	55 6544	6 5982	35 5465	57 8684	5214	.0479
.9522	65 6969	4 9322	33 4282	54 7530	5693	.0478
.9523	75 7563	3 2651	31 3073	51 6342	6171	.0477
.9524	85 8325	.0991 5970	29 1839	48 5118	6648	.0476
.9525	1.6695 9258	.0989 9279	27 0578	45 3859	7124	.0475
.9526	1.6706 0361	8 2578	24 9292	42 2564	7599	.0474
.9527	16 1634	6 5867	22 7979	39 1234	8073	.0473
.9528	26 3080	4 9146	20 6640	35 9868	8546	.0472
.9529	36 4698	3 2415	18 5276	32 8467	9018	.0471
.9530	1.6746 6489	.0981 5673	.2116 3884	.3029 7030	.9988 9489	**.0470**
.9531	56 8454	.0979 8921	14 2467	26 5556	.9988 9959	.0469
.9532	67 0593	8 2159	12 1023	23 4047	.9989 0428	.0468
.9533	77 2908	6 5387	09 9552	20 2502	0896	.0467
.9534	87 5399	4 8605	07 8055	17 0920	1363	.0466
.9535	1.6797 8066	3 1812	05 6531	13 9302	1829	.0465
.9536	1.6808 0911	.0971 5009	03 4980	10 7647	2294	.0464
.9537	18 3933	.0969 8196	.2101 3403	07 5956	2758	.0463
.9538	28 7135	8 1372	.2099 1798	04 4227	3221	.0462
.9539	39 0516	6 4539	97 0167	.3001 2462	3683	.0461
.9540	1.6849 4077	.0964 7694	.2094 8508	.2998 0660	.9989 4144	**.0460**
.6541	59 7819	3 0840	92 6823	94 8821	4604	.0459
.9542	70 1743	.0961 3975	90 5109	91 6945	5063	.0458
.9543	80 5850	.0959 7099	88 3369	88 5031	5521	.0457
.9544	1.6891 0140	8 0214	86 1601	85 3080	5978	.0456
.9545	1.6901 4614	6 3317	83 9806	82 1092	6434	.0455
.9546	11 9273	4 6411	81 7983	78 9065	6889	.0454
.9547	22 4117	2 9493	79 6132	75 7001	7343	.0453
.9548	32 9147	.0951 2566	77 4253	72 4899	7796	.0452
.9549	43 4365	.0949 5628	75 2347	69 2758	8248	.0451
.9550	1.6953 9771	.0947 8679	.2073 0412	.2966 0580	.9989 8699	**.0450**

$E^{-ii}=$ .0000,0002 $E^{ii}=$.0000,0021   .0000,0001   .0000,0003   .0000,0004   .0000,0000+

$E^{-iii}=$ .0000,0000,0+ $E^{iii}=$.0000,0000,+

TABLE I

p	x	z	$\sqrt{pq}$	$\sqrt{1-p^2}$	$\sqrt{1-q^2}$	q
.9550	1.6953 9771	.0947 8679	.2073 0412	.2966 0580	.9989 8699	.0450
.9551	64 5365	6 1720	70 8450	62 8363	9149	.0449
.9552	75 1149	4 4750	68 6459	59 6108	.9989 9598	.0448
.9553	85 7124	2 7769	66 4440	56 3814	.9990 0046	.0447
.9554	1.6996 3289	.0941 0778	64 2393	53 1482	0492	.0446
.9555	1.7006 9646	.0939 3777	62 0318	49 9110	0938	.0445
.9556	17 6196	7 6765	59 8214	46 6700	1383	.0444
.9557	28 2940	5 9742	57 6081	43 4250	1827	.0443
.9558	38 9878	4 2708	55 3919	40 1762	2270	.0442
.9559	49 7011	2 5664	53 1729	36 9234	2712	.0441
.9560	1.7060 4340	.0930 8608	.2050 9510	.2933 6666	.9990 3153	.0440
.9561	71 1866	.0929 1543	48 7262	30 4059	3593	.0439
.9562	81 9590	7 4466	46 4985	27 1413	4032	.0438
.9563	1.7092 7512	5 7379	44 2678	23 8726	4470	.0437
.9564	1.7103 5634	4 0281	42 0343	20 5999	4907	.0436
.9565	14 3956	2 3172	39 7978	17 3233	5343	.0435
.9566	25 2479	.0920 6052	37 5583	14 0426	5778	.0434
.9567	36 1205	.0918 8921	35 3159	10 7578	6212	.0433
.9568	47 0133	7 1780	33 0706	07 4690	6644	.0432
.9569	57 9265	5 4627	30 8222	04 1761	7076	.0431
.9570	1.7168 8602	.0913 7464	.2028 5709	.2900 8792	.9990 7507	.0430
.9571	79 8144	2 0289	26 3166	.2897 5781	7937	.0429
.9572	1.7190 7893	.0910 3104	24 0593	94 2730	8366	.0428
.9573	1.7201 7850	.0908 5908	21 7990	90 9637	8794	.0427
.9574	12 8015	6 8701	19 5356	87 6503	9221	.0426
.9575	23 8389	5 1482	17 2692	84 3327	.9990 9647	.0425
.9576	34 8973	3 4253	14 9998	81 0109	.9991 0072	.0424
.9577	45 9769	.0901 7012	12 7273	77 6850	0495	.0423
.9578	57 0777	.0899 9761	10 4517	74 3549	0918	.0422
.9579	68 1997	8 2498	08 1731	71 0206	1340	.0421
.9580	1.7279 3432	.0896 5224	.2005 8913	.2867 6820	.9991 1761	.0420
.9581	1.7290 5082	4 7940	03 6065	64 3392	2181	.0419
.9582	1.7301 6948	3 0643	.2001 3186	60 9921	2600	.0418
.9583	12 9030	.0891 3336	.1999 0275	57 6408	3018	.0417
.9584	24 1331	.0889 6018	96 7333	54 2852	3435	.0416
.9585	35 3850	7 8688	94 4360	50 9253	3850	.0415
.9586	46 6590	6 1347	92 1355	47 5611	4265	.0414
.9587	57 9550	4 3995	89 8319	44 1925	4679	.0413
.9588	69 2733	2 6631	87 5251	40 8196	5092	.0412
.9589	80 6138	.0880 9256	85 2151	37 4423	5504	.0411
.9590	1.7391 9767	.0879 1870	.1982 9019	.2834 0607	.9991 5915	.0410
.9591	1.7403 3621	7 4472	80 5855	30 6747	6324	.0409
.9592	14 7701	5 7063	78 2659	27 2842	6733	.0408
.9593	26 2008	3 9642	75 9431	23 8893	7141	.0407
.9594	37 6543	2 2211	73 6170	20 4900	7548	.0406
.9595	49 1308	.0870 4767	71 2877	17 0863	7954	.0405
.9596	60 6303	.0868 7312	68 9551	13 6780	8359	.0404
.9597	72 1529	6 9846	66 6192	10 2653	8763	.0403
.9598	83 6988	5 2368	64 2800	06 8481	9165	.0402
.9599	1.7495 2680	3 4878	61 9376	03 4263	9567	.0401
.9600	1.7506 8607	.0861 7377	.1959 5918	.2800 0000	.9991 9968	.0400

$E^{-ii} =$ $E^{ii} = .0000,0027$ .0000,0002   .0000,0003   .0000,0005   .0000,0000+
.0000,0002

$E^{-iii} =$ $E^{iii} = .0000,0000,+$
.0000,0000,0+

TABLE I

.9600   **.0400**

p	x	z	$\sqrt{pq}$	$\sqrt{1-p^2}$	$\sqrt{1-q^2}$	q
.9600	1.7506 8607	.0861 7377	.1959 5918	.2800 0000	.9991 9968	.0400
.9601	18 4770	.0859 9865	57 2427	.2796 5691	.9992 0368	.0399
.9602	30 1169	8 2340	54 8903	93 1337	0767	.0398
.9603	41 7807	6 4805	52 5345	89 6937	1164	.0397
.9604	53 4683	4 7257	50 1754	86 2491	1561	.0396
.9605	65 1800	2 9698	47 8129	82 7988	1957	.0395
.9606	76 9159	.0851 2127	45 4470	79 3460	2352	.0394
.9607	1.7588 6760	.0849 4544	43 0777	75 8874	2746	.0393
.9608	1.7600 4605	7 6949	40 7050	72 4242	3138	.0392
.9609	12 2694	5 9343	38 3289	68 9563	3530	.0391
.9610	1.7624 1030	.0844 1725	.1935 9494	.2765 4837	.9992 3921	**.0390**
.9611	35 9613	2 4095	33 5664	62 0063	4311	.0389
.9612	47 8445	.0840 6453	31 1800	58 5242	4700	.0388
.9613	59 7526	.0838 8799	28 7900	55 0374	5087	.0387
.9614	71 7858	7 1133	26 3966	51 5457	5474	.0386
.9615	83 6442	5 3455	23 9997	48 0493	5860	.0385
.9616	1.7695 6280	3 5766	21 5993	44 5481	6245	.0384
.9617	1.7707 6373	1 8064	19 1954	41 0420	6629	.0383
.9618	19 6721	.0830 0351	16 7879	37 5310	7011	.0382
.9619	31 7327	.0828 2625	14 3769	34 0152	7393	.0381
.9620	1.7743 8191	.0826 4887	.1911 9623	.2730 4945	.9992 7774	**.0380**
.9621	55 9315	4 7137	09 5442	26 9688	8154	.0379
.9622	68 0700	2 9375	07 1224	23 4383	8532	.0378
.9623	80 2347	.0821 1601	04 6971	19 9028	8910	.0377
.9624	1.7792 4258	.0819 3815	.1902 2681	16 3623	9287	.0376
.9625	1.7804 6434	7 6016	.1899 8355	12 8168	.9992 9663	.0375
.9626	16 8877	5 8205	97 3993	09 2663	.9993 0038	.0374
.9627	29 1587	4 0382	94 9594	05 7108	0411	.0373
.9628	41 4566	2 2547	92 5158	.2702 1503	0784	.0372
.9629	53 7816	.0810 4700	90 0685	.2698 5846	1156	.0371
.9630	1.7866 1337	.0808 6840	.1887 6175	.2695 0139	.9993 1527	**.0370**
.9631	78 5132	6 8967	85 1629	91 4381	1896	.0369
.9632	1.7890 9201	5 1083	82 7044	87 8571	2265	.0368
.9633	1.7903 3546	3 3185	80 2423	84 2710	2633	.0367
.9634	15 8169	.0801 5276	77 7763	80 6798	3000	.0366
.9635	28 3070	.0799 7354	75 3066	77 0833	3365	.0365
.9636	40 8252	7 9419	72 8331	73 4816	3730	.0364
.9637	53 3715	6 1472	70 3558	69 8747	4094	.0363
.9638	65 9462	4 3512	67 8747	66 2626	4457	.0362
.9639	78 5493	2 5540	65 3898	62 6415	4818	.0361
.9640	1.7991 1811	.0790 7555	.1862 9010	.2659 0224	.9993 5179	**.0360**
.9641	1.8003 8416	.0788 9558	60 4083	55 3943	5539	.0359
.9642	16 5311	7 1548	57 9117	51 7609	5897	.0358
.9643	29 2496	5 3525	55 4113	48 1222	6255	.0357
.9644	41 9974	3 5489	52 9069	44 4780	6612	.0356
.9645	54 7746	.0781 7441	50 3986	40 8285	6968	.0355
.9646	67 5813	.0779 9380	47 8864	37 1735	7322	.0354
.9647	80 4177	8 1306	45 3702	33 5131	7676	.0353
.9648	1.8093 2839	6 3219	42 8500	29 8471	8029	.0352
.9649	1.8106 1802	4 5119	40 3258	26 1757	8381	.0351
.9650	1.8119 1067	.0772 7006	.1837 7976	.2622 4988	.9993 8731	**.0350**

$E^{-ii}=$ .0000,0003 $E^{ii}=$.0000,0033   .0000,0002   .0000,0004   .0000,0006   .0000,0000+

$E^{-iii}=$ .0000,0000,0+ $E^{iii}=$.0000,0000,+

TABLE I

.9650   **.0350**

p	x	z	$\sqrt{pq}$	$\sqrt{1-p^2}$	$\sqrt{1-q^2}$	q
.9650	1.8119 1067	.0772 7006	.1837 7976	.2622 4988	.9993 8731	**.0350**
.9651	32 0635	.0770 8881	35 2654	18 8163	9081	.0349
.9652	45 0509	.0769 0742	32 7291	15 1283	9430	.0348
.9653	58 0689	7 2591	30 1888	11 4347	.9993 9777	.0347
.9654	71 1177	5 4426	27 6444	07 7354	.9994 0124	.0346
.9655	84 1976	3 6248	25 0959	04 0305	0470	.0345
.9656	1.8197 3087	.0761 8058	22 5433	.2600 3200	0814	.0344
.9657	1.8210 4511	.0759 9854	19 9865	.2596 6037	1158	.0343
.9658	23 6250	8 1637	17 4257	92 8818	1501	.0342
.9659	36 8307	6 3407	14 8606	89 1541	1843	.0341
.9660	1.8250 0682	.0754 5163	.1812 2914	.2585 4207	.9994 2183	**.0340**
.9661	63 3378	2 6906	09 7179	81 6814	2523	.0339
.9662	76 6396	.0750 8636	07 1403	77 9364	2862	.0338
.9663	1.8289 9738	.0749 0353	04 5584	74 1855	3199	.0337
.9664	1.8303 3407	7 2056	.1801 9723	70 4288	3536	.0336
.9665	16 7403	5 3746	.1799 3818	66 6661	3872	.0335
.9666	30 1729	3 5423	96 7871	62 8976	4206	.0334
.9667	43 6386	.0741 7086	94 1881	59 1231	4540	.0333
.9668	57 1377	.0739 8736	91 5848	55 3426	4873	.0332
.9669	70 6704	8 0372	88 9771	51 5562	5204	.0331
.9670	1.8384 2367	.0736 1994	.1786 3650	.2547 7637	.9994 5535	**.0330**
.9671	1.8397 8370	4 3603	83 7486	43 9652	5856	.0329
.9672	1.8411 4713	2 5199	81 1277	40 1606	6194	.0328
.9673	25 1400	.0730 6780	78 5025	36 3499	6521	.0327
.9674	38 8433	.0728 8348	75 8727	32 5331	6848	.0326
.9675	52 5812	6 9903	73 2386	28 7101	7174	.0325
.9676	66 3540	5 1443	70 5999	24 8810	7498	.0324
.9677	80 1620	3 2970	67 9567	21 0456	7822	.0323
.9678	1.8494 0052	.0721 4483	65 3090	17 2040	8145	.0322
.9679	1.8507 8840	.0719 5982	62 6568	13 3561	8466	.0321
.9680	1.8521 7986	.0717 7467	.1760 0000	.2509 5019	.9994 8787	**.0320**
.9681	35 7491	5 8938	57 3386	05 6414	9107	.0319
.9682	49 7358	4 0396	54 6726	.2501 7746	9425	.0318
.9683	63 7588	2 1839	52 0020	.2497 9013	.9994 9743	.0317
.9684	77 8185	.0710 3268	49 3267	94 0217	.9995 0060	.0316
.9685	1.8591 9149	.0708 4683	46 6468	90 1355	0375	.0315
.9686	1.8606 0485	6 6084	43 9622	86 2429	0690	.0314
.9687	20 2192	4 7471	41 2728	82 3439	1003	.0313
.9688	34 4275	2 8844	38 5787	78 4382	1316	.0312
.9689	48 6735	.0701 0202	35 8799	74 5260	1628	.0311
.9690	1.8662 9574	.0699 1546	.1733 1763	.2470 6072	.9995 1938	**.0310**
.9691	77 2795	7 2876	30 4679	66 6818	2248	.0309
.9692	1.8691 6400	5 4192	27 7546	62 7497	2557	.0308
.9693	1.8706 0392	3 5493	25 0365	58 8109	2864	.0307
.9694	20 4773	.0691 6780	22 3136	54 8654	3171	.0306
.9695	34 9545	.0689 8052	19 5857	50 9131	3477	.0305
.9696	49 4711	7 9310	16 8529	46 9540	3781	.0304
.9697	64 0273	6 0553	14 1152	42 9881	4085	.0303
.9698	78 6234	4 1782	11 3725	39 0154	4388	.0302
.9699	1.8793 2596	2 2996	08 6249	35 0357	4689	.0301
.9700	1.8807 9361	.0680 4195	.1705 8722	.2431 0492	.9995 4990	**.0300**

$E^{-ii}=$ .0000,0003 $E^{ii}=$ .0000,0044   .0000,0002   .0000,0006   .0000,0008   .0000,0000+

$E^{-iii}=$ .0000,0000+ $E^{iii}=$ .0000,0000,+

TABLE I

p	x	z	$\sqrt{pq}$	$\sqrt{1-p^2}$	$\sqrt{1-q^2}$	q
.9700	1.8807 9361	.0680 4195	.1705 8722	.2431 0492	.9995 4990	.0300
.9701	1.8822 6533	78 5380	03 1145	27 0556	5290	.0299
.9702	1.8837 4113	76 6550	.1700 3517	23 0551	5588	.0298
.9703	1.8852 2105	74 7705	.1697 5839	19 0475	5886	.0297
.9704	1.8867 0511	72 8845	94 8109	15 0329	6182	.0296
.9705	1.8881 9334	70 9971	92 0328	11 0112	6478	.0295
.9706	1.8896 8576	69 1082	89 2495	06 9823	6773	.0294
.9707	1.8911 8240	67 2177	86 4611	.2402 9463	7066	.0293
.9708	1.8926 8329	65 3258	83 6674	.2398 9031	7359	.0292
.9709	1.8941 8845	63 4323	80 8685	94 8526	7651	.0291
.9710	1.8956 9792	.0661 5374	.1678 0644	.2390 7948	.9995 7941	.0290
.9711	1.8972 1172	59 6410	75 2549	86 7298	8231	.0289
.9712	1.8987 2988	57 7430	72 4401	82 6573	8519	.0288
.9713	1.9002 5243	55 8435	69 6200	78 5775	8807	.0287
.9714	1.9017 7940	53 9425	66 7945	74 4903	9094	.0286
.9715	1.9033 1082	52 0399	63 9636	70 3955	9379	.0285
.9716	1.9048 4671	50 1358	61 1273	66 2933	9664	.0284
.9717	1.9063 8711	48 2302	58 2856	62 1835	.9995 9947	.0283
.9718	1.9079 3204	46 3231	55 4383	58 0662	.9996 0230	.0282
.9719	1.9094 8155	44 4144	52 5855	53 9412	0512	.0281
.9720	1.9110 3565	.0642 5041	.1649 7273	.2349 8085	.9996 0792	.0280
.9721	1.9125 9438	40 5923	46 8634	45 6681	1072	.0279
.9722	1.9141 5777	38 6789	43 9939	41 5200	1351	.0278
.9723	1.9157 2585	36 7640	41 1188	37 3641	1628	.0277
.9724	1.9172 9866	34 8475	38 2381	33 2004	1905	.0276
.9725	1.9188 7623	32 9294	35 3516	29 0288	2180	.0275
.9726	1.9204 5858	31 0097	32 4595	24 8492	2455	.0274
.9727	1.9220 4576	29 0885	29 5616	20 6618	2729	.0273
.9728	1.9236 3779	27 1656	26 6579	16 4663	3001	.0272
.9729	1.9252 3472	25 2412	23 7484	12 2627	3273	.0271
.9730	1.9268 3657	.0623 3151	.1620 8331	.2308 0511	.9996 3543	.0270
.9731	1.9284 4338	21 3875	17 9119	.2303 8314	3813	.0269
.9732	1.9300 5518	19 4583	14 9848	.2299 6034	4082	.0268
.9733	1.9316 7202	17 5274	12 0518	95 3673	4349	.0267
.9734	1.9332 9392	15 5949	09 1128	91 1229	4616	.0266
.9735	1.9349 2093	13 6608	06 1678	86 8701	4881	.0265
.9736	1.9365 5307	11 7251	03 2168	82 6090	5146	.0264
.9737	1.9381 9038	09 7877	.1600 2597	78 3395	5410	.0263
.9738	1.9398 3290	07 8487	.1597 2965	74 0616	5672	.0262
.9739	1.9414 8068	05 9080	94 3271	69 7751	5934	.0261
.9740	1.9431 3375	.0603 9657	.1591 3516	.2265 4801	.9996 6194	.0260
.9741	1.9447 9214	02 0218	88 3699	61 1765	6454	.0259
.9742	1.9464 5590	.0600 0761	85 3820	56 8642	6712	.0258
.9743	1.9481 2506	.0598 1288	82 3878	52 5432	6970	.0257
.9744	1.9497 9968	96 1799	79 3872	48 2135	7227	.0256
.9745	1.9514 7977	94 2292	76 3803	43 8750	7482	.0255
.9746	1.9531 6540	92 2769	73 3671	39 5276	7737	.0254
.9747	1.9548 5658	90 3229	70 3474	35 1714	7990	.0253
.9748	1.9565 5338	88 3672	67 3213	30 8061	8243	.0252
.9749	1.9582 5583	86 4098	64 2887	26 4319	8495	.0251
.9750	1.9599 6398	.0584 4507	.1561 2495	.2222 0486	.9996 8745	.0250

$E^{-ii}=$ .0000,0004 $E^{ii}=$ .0000,0060   .0000,0002   .0000,0007   .0000,0010   .0000,0000+

$E^{-iii}=$ .0000,0000,0+ $E^{iii}=$ .0000,0000,0+

TABLE I

.9750   .0250

p	x	z	$\sqrt{pq}$	$\sqrt{1-p^2}$	$\sqrt{1-q^2}$	q
.9750	1.9599 6398	.0584 4507	.1561 2495	.2222 0486	.9996 8745	**.0250**
.9751	1.9616 7787	82 4899	58 2038	17 6562	8995	.0249
.9752	1.9633 9753	80 5273	55 1514	13 2546	9243	.0248
.9753	1.9651 2302	78 5631	52 0925	08 8438	9491	.0247
.9754	1.9668 5439	76 5971	49 0268	.2204 4237	9737	.0246
.9755	1.9685 9167	74 6294	45 9544	.2199 9943	.9996 9983	.0245
.9756	1.9703 3491	72 6599	42 8752	95 5555	.9997 0228	.0244
.9757	1.9720 8416	70 6887	39 7893	91 1073	0471	.0243
.9758	1.9738 3946	68 7157	36 6965	86 6495	0714	.0242
.9759	1.9756 0087	66 7410	33 5968	82 1822	0955	.0241
.9760	1.9773 6843	.0564 7645	.1530 4901	.2177 7052	.9997 1196	**.0240**
.9761	1.9791 4219	62 7863	27 3765	73 2186	1435	.0239
.9762	1.9809 2219	60 8062	24 2559	68 7222	1674	.0238
.9763	1.9827 0850	58 8244	20 1282	64 2160	1912	.0237
.9764	1.9845 0115	56 8408	17 9934	59 7000	2148	.0236
.9765	1.9863 0021	54 8554	14 8515	55 1740	2384	.0235
.9766	1.9881 0571	52 8682	11 7024	50 6380	2618	.0234
.9767	1.9899 1772	50 8792	08 5460	· 46 0920	2852	.0233
.9768	1.9917 3629	48 8884	05 3823	41 5359	3084	.0232
.9769	1.9935 6147	46 8957	.1502 2114	36 9696	3316	.0231
.9770	1.9953 9331	.0544 9013	.1499 0330	.2132 3930	.9997 3547	**.0230**
.9771	1.9972 3187	42 9049	95 8473	27 8061	3776	.0229
.9772	1.9990 7721	40 9068	92 6540	23 2089	4005	.0228
.9773	2.0009 2939	38 9068	89 4533	18 6012	4232	.0227
.9774	2.0027 8845	36 9049	86 2449	13 9830	4459	.0226
.9775	2.0046 5446	34 9012	83 0290	09 3542	4684	.0225
.9776	2.0065 2748	32 8956	79 8054	04 7147	4909	.0224
.9777	2.0084 0756	30 8882	76 5741	.2100 0645	5132	.0223
.9778	2.0102 9477	28 8788	73 3350	.2095 4035	5355	.0222
.9779	2.0121 8917	26 8676	70 0881	90 7317	5577	.0221
.9780	2.0140 9081	.0524 8544	.1466 8333	.2086 0489	.9997 5797	**.0220**
.9781	2.0159 9977	22 8394	63 5706	81 3551	6017	.0219
.9782	2.0179 1610	20 8224	60 3000	76 6502	6235	.0218
.9783	2.0198 3987	18 8035	57 0213	71 9341	6453	.0217
.9784	2.0217 7115	16 7827	53 7345	67 2068	6669	.0216
.9785	2.0237 0999	14 7600	50 4396	62 4682	6885	.0215
.9786	2.0256 5648	12 7353	47 1365	57 7182	7099	.0214
.9787	2.0276 1066	10 7087	43 8251	52 9566	7313	.0213
.9788	2.0295 7262	08 6801	40 5055	48 1836	7525	.0212
.9789	2.0315 4243	06 6495	37 1774	43 3989	7737	.0211
.9790	2.0335 2015	.0504 6170	.1433 8410	.2038 6025	.9997 7948	**.0210**
.9791	2.0355 0586	02 5825	30 4961	33 7942	8157	.0209
.9792	2.0374 9962	.0500 5460	27 1426	28 9741	8366	.0208
.9793	2.0395 0152	.0498 5075	23 7805	24 1420	8573	.0207
.9794	2.0415 1162	96 4670	20 4098	19 2979	8780	.0206
.9795	2.0435 3001	94 4245	17 0303	14 4416	8985	.0205
.9796	2.0455 5676	92 3799	13 6421	09 5731	9190	.0204
.9797	2.0475 9194	90 3334	10 2450	.2004 6922	9393	.0203
.9798	2.0496 3564	88 2847	06 8390	.1999 7990	9596	.0202
.9799	2.0516 8794	86 2341	03 4240	94 8932	9797	.0201
.9800	2.0537 4891	.0484 1814	.1400 0000	.1989 9749	.9997 9998	**.0200**

E⁻ⁱⁱ=
.0000,0005

Eⁱⁱ=.0000,0088   .0000,0002   .0000,0010   .0000,0013   .0000,0000+

E⁻ⁱⁱ =
.0000,0000,0+

Eⁱⁱ =.0000,0000,+

TABLE I

.9800   .0200

p	x	z	$\sqrt{pq}$	$\sqrt{1-p^2}$	$\sqrt{1-q^2}$	q
.9800	2.0537 4891	.0484 1814	.1400 0000	.1989 9749	.9997 9998	**.0200**
.9801	2.0558 1865	82 1266	.1396 5669	85 0438	.9998 0198	.0199
.9802	2.0578 9723	80 0697	93 1245	80 1000	0396	.0198
.9803	2.0599 8474	78 0108	89 6730	75 1433	0594	.0197
.9804	2.0620 8126	75 9497	86 2121	70 1736	0790	.0196
.9805	2.0641 8689	73 8866	82 7418	65 1908	0986	.0195
.9806	2.0663 0171	71 8214	79 2621	60 1949	1180	.0194
.9807	2.0684 2581	69 7540	75 7729	55 1857	1374	.0193
.9808	2.0705 5929	67 6845	72 2740	50 1631	1566	.0192
.9809	2.0727 0223	65 6129	68 7655	45 1270	1758	.0191
.9810	2.0748 5473	.0463 5391	.1365 2472	.1940 0773	.9998 1948	**.0190**
.9811	2.0770 1689	61 4632	61 7191	35 0140	2138	.0189
.9812	2.0791 8881	59 3851	58 1811	29 9368	2316	.0188
.9813	2.0813 7057	57 3048	54 6332	24 8457	2514	.0187
.9814	2.0835 6229	55 2223	51 0751	19 7406	2701	.0186
.9815	2.0857 6407	53 1377	47 5070	14 6214	2886	.0185
.9816	2.0879 7600	51 0508	43 9286	09 4879	3071	.0184
.9817	2.0901 9819	48 9617	40 3399	.1904 3400	3254	.0183
.9818	2.0924 3076	46 8704	36 7408	.1899 1777	3437	.0182
.9819	2.0946 7380	44 7768	33 1313	94 0008	3618	.0181
.9820	2.0969 2743	.0442 6810	.1329 5112	.1888 8091	.9998 3799	**.0180**
.9821	2.0991 9176	40 5830	25 8805	83 6027	3978	.0179
.9822	2.1014 6691	38 4827	22 2390	78 3812	4157	.0178
.9823	2.1037 5299	36 3800	18 5867	73 1447	4334	.0177
.9824	2.1060 5012	34 2751	14 9236	67 8929	4511	.0176
.9825	2.1083 5841	32 1679	11 2494	62 6258	4686	.0175
.9826	2.1106 7799	30 0584	07 5641	57 3433	4861	.0174
.9827	2.1130 0898	27 9466	03 8677	52 0451	5034	.0173
.9828	2.1153 5151	25 8342	.1300 1600	46 7312	5207	.0172
.9829	2.1177 0570	23 7159	.1296 4409	41 4014	5378	.0171
.9830	2.1200 7169	.0421 5970	.1292 7103	.1836 0556	.9998 5549	**.0170**
.9831	2.1224 4961	19 4757	88 9682	30 6936	5718	.0169
.9832	2.1248 3959	17 3521	85 2144	25 3153	5887	.0168
.9833	2.1272 4177	15 2260	81 4488	19 9206	6055	.0167
.9834	2.1296 5629	13 0976	77 6713	14 5093	6221	.0166
.9835	2.1320 8329	10 9667	73 8819	09 0813	6387	.0165
.9836	2.1345 2291	08 8334	70 0803	.1803 6363	6551	.0164
.9837	2.1369 7531	06 6977	66 2666	.1798 1744	6715	.0163
.9838	2.1394 4063	04 5595	62 4405	92 6952	6877	.0162
.9839	2.1419 1901	02 4188	58 6020	87 1986	7039	.0161
.9840	2.1444 1062	.0400 2756	.1254 7510	.1781 6846	.9998 7199	**.0160**
.9841	2.1469 1562	.0398 1300	50 8873	76 1529	7359	.0159
.9842	2.1494 3416	95 9818	47 0108	70 6033	7517	.0158
.9843	2.1519 6641	93 8311	43 1215	65 0357	7675	.0157
.9844	2.1545 1253	91 6779	39 2191	59 4499	7831	.0156
.9845	2.1570 7270	89 5221	35 3036	53 8458	7987	.0155
.9846	2.1596 4709	87 3637	31 3748	48 2231	8141	.0154
.9847	2.1622 3587	85 2028	27 4327	42 5817	8295	.0153
.9848	2.1648 3922	83 0392	23 4770	36 9214	8447	.0152
.9849	2.1674 5733	80 8731	19 5077	31 2420	8599	.0151
.9850	2.1700 9038	.0378 7043	.1215 5246	.1725 5434	.9998 8749	**.0150**

$E^{-ii}=$ $E^{ii}=.0000,014$ .0000,0003   .0000,0014   .0000,0020   .0000,0000+
.0000,0007

$E^{-iii}=$ $E^{ii}=.0000,0001$
.0000,0000,0+

TABLE I

.9850   **.0150**

p	x	z	$\sqrt{pq}$	$\sqrt{1-p^2}$	$\sqrt{1-q^2}$	q
.9850	2.1700 9038	.0378 7043	.1215 5246	.1725 5434	.9998 8749	**.0150**
.9851	2.1727 3856	76 5329	11 5275	19 8253	8899	.0149
.9852	2.1754 0207	74 3588	07 5165	14 0875	9047	.0148
.9853	2.1780 8110	72 1821	.1203 4912	08 3299	9195	.0147
.9854	2.1807 7585	70 0027	.1199 4515	.1702 5522	9341	.0146
.9855	2.1834 8653	67 8205	95 3974	.1696 7543	9487	.0145
.9856	2.1862 1335	65 6357	91 3287	90 9358	9631	.0144
.9857	2.1889 5653	63 4481	87 2451	85 0967	9775	.0143
.9858	2.1917 1628	61 2578	83 1467	79 2367	.9998 9917	.0142
.9859	2.1944 9282	59 0647	79 0331	73 3556	.9999 0059	.0141
.9860	2.1972 8638	.0356 8688	.1174 9043	.1667 4531	.9999 0200	**.0140**
.9861	2.2000 9720	54 6701	70 7600	61 5291	0339	.0139
.9862	2.2029 2550	52 4686	66 6002	55 5833	0478	.0138
.9863	2.2057 7154	50 2642	62 4246	49 6154	0615	.0137
.9864	2.2086 3556	48 0570	58 2331	43 6253	0752	.0136
.9865	2.2115 1781	45 8469	54 0256	37 6126	0887	.0135
.9866	2.2144 1855	43 6340	49 8017	31 5772	1022	.0134
.9867	2.2173 3805	41 4181	45 5614	25 5187	1155	.0133
.9868	2.2202 7656	39 1993	41 3045	19 4369	1288	.0132
.9869	2.2232 3438	36 9775	37 0308	13 3316	1419	.0131
.9870	2.2262 1177	.0334 7528	.1132 7400	.1607 2025	.9999 1550	**.0130**
.9871	2.2292 0903	32 5251	28 4321	.1601 0493	1679	.0129
.9872	2.2322 2645	30 2944	24 1068	.1594 8718	1808	.0128
.9873	2.2352 6433	28 0607	19 7638	88 6696	1935	.0127
.9874	2.2383 2298	25 8239	15 4031	82 4424	2062	.0126
.9875	2.2414 0272	23 5840	11 0243	76 1900	2187	.0125
.9876	2.2445 0387	21 3410	06 6273	69 9121	2312	.0124
.9877	2.2476 2676	19 0950	.1102 2119	63 6083	2435	.0123
.9878	2.2507 7171	16 8458	.1097 7778	57 2784	2558	.0122
.9879	2.2539 3909	14 5934	93 3247	50 9220	2679	.0121
.9880	2.2571 2924	.0312 3379	.1088 8526	.1544 5388	.9999 2800	**.0120**
.9881	2.2603 4254	10 0792	84 3611	38 1284	2919	.0119
.9882	2.2635 7934	07 8172	79 8500	31 6906	3038	.0118
.9883	2.2668 4003	05 5520	75 3190	25 2249	3155	.0117
.9884	2.2701 2500	03 2835	70 7679	18 7310	3272	.0116
.9885	2.2734 3465	.0301 0117	66 1965	12 2086	3387	.0115
.9886	2.2767 6940	.0298 7366	61 6044	.1505 6573	3502	.0114
.9887	2.2801 2965	96 4582	56 9915	.1499 0767	3615	.0113
.9888	2.2835 1586	94 1764	52 3574	92 4664	3728	.0112
.9889	2.2869 2845	91 8912	47 7018	85 8260	3839	.0111
.9890	2.2903 6788	.0289 6025	.1043 0244	.1479 1552	.9999 3950	**.0110**
.9891	2.2938 3462	87 3104	38 3251	72 4534	4059	.0109
.9892	2.2973 2915	85 0148	33 6034	65 7203	4168	.0108
.9893	2.3008 5197	82 7157	28 8591	58 9554	4275	.0107
.9894	2.3044 0357	80 4131	24 0918	52 1584	4382	.0106
.9895	2.3079 8448	78 1069	19 3012	45 3287	4487	.0105
.9896	2.3115 9523	75 7971	14 4871	38 4658	4592	.0104
.9897	2.3152 3637	73 4837	09 6489	31 5694	4695	.0103
.9898	2.3189 0847	71 1667	.1004 7865	24 6389	4798	.0102
.9899	2.3226 1210	68 8459	.0999 8995	17 6738	4899	.0101
.9900	2.3263 4787	.0266 5214	.0994 9874	.1410 6736	.9999 5000	**.0100**

E^{-ii}= E^{ii}=.0000,027   .0000,0004   .0000,0023   .0000,0032   .0000,0000+
.0000,0009

E^{-iii}= E^{iii}=.0000,0002
.0000,0000,0+

TABLE I

p	x	z	√pq	√1−p²	√1−q²	q
.9900	2.3263 4787	.0266 5214	.0994 9874	.1410 6736	.9999 5000	**.0100**
.9901	2.3301 1640	64 1932	90 0500	.1403 6378	5099	.0099
.9902	2.3339 1831	61 8612	85 0868	.1396 5658	5198	.0098
.9903	2.3377 5425	59 5253	80 0975	89 4571	5295	.0097
.9904	2.3416 2491	57 1857	75 0815	82 3111	5392	.0096
.9905	2.3455 3097	54 8421	70 0387	75 1273	5487	.0095
.9906	2.3494 7315	52 4946	64 9684	67 9050	5582	.0094
.9907	2.3534 5218	50 1431	59 8703	60 6436	5675	.0093
.9908	2.3574 6883	47 7877	54 7439	53 3425	5768	.0092
.9909	2.3615 2387	45 4282	49 5889	46 0011	5859	.0091
.9910	2.3656 1813	.0243 0646	.0944 4046	.1338 6187	.9999 5950	**.0090**
.9911	2.3697 5242	40 6969	39 1906	31 1946	6039	.0089
.9912	2.3739 2762	38 3251	33 9465	23 7281	6128	.0088
.9913	2.3781 4462	35 9491	28 6716	16 2184	6215	.0087
.9914	2.3824 0434	33 5688	23 3656	08 6650	6302	.0086
.9915	2.3867 0773	31 1842	18 0278	.1301 0669	6387	.0085
.9916	2.3910 5579	28 7954	12 6577	.1293 4234	6472	.0084
.9917	2.3954 4952	26 4021	07 2547	85 7336	6555	.0083
.9918	2.3998 8998	24 0044	.0901 8182	77 9969	6638	.0082
.9919	2.4043 7828	21 6023	.0896 3476	70 2122	6719	.0081
.9920	2.4089 1555	.0219 1957	.0890 8423	.1262 3787	.9999 6800	**.0080**
.9921	2.4135 0295	16 7845	85 3016	54 4955	6879	.0079
.9922	2.4181 4171	14 3686	79 7250	46 5617	6958	.0078
.9923	2.4228 3310	11 9482	74 1115	38 5762	7035	.0077
.9924	2.4275 7843	09 5230	68 4607	30 5381	7112	.0076
.9925	2.4323 7906	07 0930	62 7717	22 4463	7187	.0075
.9926	2.4372 3641	04 6582	57 0438	14 2998	7262	.0074
.9927	2.4421 5195	.0202 2185	51 2761	.1206 0974	7335	.0073
.9928	2.4471 2722	.0199 7739	45 4679	.1197 8381	7408	.0072
.9929	2.4521 6381	97 3242	39 6184	89 5205	7479	.0071
.9930	2.4572 6339	.0194 8695	.0833 7266	.1181 1435	.9999 7550	**.0070**
.9931	2.4624 2768	92 4097	27 7916	72 7058	7619	.0069
.9932	2.4676 5849	89 9446	21 8126	64 2062	7688	.0068
.9933	2.4729 5771	87 4743	15 7886	55 6431	7755	.0067
.9934	2.4783 2729	84 9987	09 7185	47 0153	7822	.0066
.9935	2.4837 6929	82 5177	.0803 6013	38 3211	7887	.0065
.9936	2.4892 8586	80 0311	.0797 4359	29 5592	7952	.0064
.9937	2.4948 7925	77 5391	91 2212	20 7279	8015	.0063
.9938	2.5005 5179	75 0413	84 9560	11 8255	8078	.0062
.9939	2.5063 0596	72 5379	78 6394	.1102 8504	8139	.0061
.9940	2.5121 4433	.0170 0287	.0772 2694	.1093 8007	.9999 8200	**.0060**
.9941	2.5180 6960	67 5136	65 8453	84 6746	8259	.0059
.9942	2.5240 8463	64 9925	59 3655	75 4701	8318	.0058
.9943	2.5301 9238	62 4654	52 8287	66 1883	8375	.0057
.9944	2.5363 9601	59 9321	46 2332	56 8179	8432	.0056
.9945	2.5426 9882	57 3926	39 5776	47 3657	8487	.0055
.9946	2.5491 0428	54 8467	32 8602	37 8266	8542	.0054
.9947	2.5556 1608	52 2943	26 0792	28 1979	8595	.0053
.9948	2.5622 3808	49 7354	19 2329	18 4773	8648	.0052
.9949	2.5689 7437	47 1698	12 3195	.1008 6620	8699	.0051
.9950	2.5758 2930	.0144 5974	.0705 3368	.0998 7492	.9999 8750	**.0050**

$E^{-ii}=$ .0000,0015 $E^{ii}=$.0000,073   .0000,0006   .0000,0050   .0000,0069   .0000,0000+

$E^{-iii}=$ .0000,00001 $E^{iii}=$.0000,0009   .0000,0001   .0000,0001

TABLE I

p	x	z	$\sqrt{pq}$	$\sqrt{1-p^2}$	$\sqrt{1-q^2}$	q
.9950	2.5758 2930	.0144 5974	.0705 3368	.0998 7492	.9999 8750	.0050
.9951	2.5828 0745	42 0181	.0698 2829	.0988 7361	8799	.0049
.9952	2.5899 1368	39 4318	91 1556	.0978 6194	8848	.0048
.9953	2.5971 5316	36 8383	83 9525	.0968 3961	8895	.0047
.9954	2.6045 3136	34 2374	76 6713	.0958 0626	8942	.0046
.9955	2.6120 5414	31 6291	69 3093	.0947 6154	8987	.0045
.9956	2.6197 2771	29 0133	61 8640	.0937 0507	9032	.0044
.9957	2.6275 5871	26 3896	54 3325	.0926 3644	9075	.0043
.9958	2.6355 5424	23 7581	46 7117	.0915 5523	9118	.0042
.9959	2.6437 2189	21 1185	38 9984	.0904 6099	9159	.0041
.9960	2.6520 6981	.0118 4706	.0631 1894	.0893 5323	.9999 9200	.0040
.9961	2.6606 0674	15 8143	23 2808	.0882 3146	9239	.0039
.9962	2.6693 4209	13 1493	15 2690	.0870 9512	9278	.0038
.9963	2.6782 8601	10 4755	.0607 1499	.0859 4364	9315	.0037
.9964	2.6874 4945	07 7927	.0598 9190	.0847 7641	9352	.0036
.9965	2.6968 4426	05 1005	90 5718	.0835 9276	9387	.0035
.9966	2.7064 8331	.0102 3989	82 1031	.0823 9199	9422	.0034
.9967	2.7163 8058	.0099 68748	73 5076	.0811 7333	9455	.0033
.9968	2.7265 5132	96 96604	64 7796	.0799 3597	9488	.0032
.9969	2.7370 1217	94 23428	55 9128	.0786 7903	9519	.0031
.9970	2.7477 8139	.0091 49191	.0546 9004	.0774 0155	.9999 9550	.0030
.9971	2.7588 7903	88 73862	37 7351	.0761 0250	9579	.0029
.9972	2.7703 2723	85 97405	28 4089	.0747 8075	9608	.0028
.9973	2.7821 5045	83 19784	18 9133	.0734 3507	9635	0027
.9974	2.7943 7587	80 40961	.0509 2386	.0720 6414	9662	.0026
.9975	2.8070 3377	77 60893	.0499 3746	.0706 6647	9687	.0025
.9976	2.8201 5806	74 79541	89 3097	.0692 4045	9712	.0024
.9977	2.8337 8687	71 96851	79 0313	.0677 8429	9735	.0023
.9978	2.8479 6329	69 12768	68 5253	.0662 9600	9758	.0022
.9979	2.8627 3626	66 27237	57 7761	.0647 7337	9779	.0021
.9980	2.8781 6174	.0063 40193	.0446 7662	.0632 1392	.9999 9800	.0020
.9981	2.8943 0405	60 51576	35 4756	.0616 1485	9819	.0019
.9982	2.9112 3773	57 61306	23 8821	.0599 7299	9838	.0018
.9983	2.9290 4975	54 69299	.0411 9599	.0582 8473	9855	.0017
.9984	2.9478 4255	51 75463	.0399 6799	.0565 4592	9872	.0016
.9985	2.9677 3793	48 79694	87 0078	.0547 5171	9887	.0015
.9986	2.9888 8227	45 81874	73 9037	.0528 9650	9902	.0014
.9987	3.0114 5376	42 81870	60 3207	.0509 7362	9915	.0013
.9988	3.0356 7237	39 79529	46 2023	.0489 7499	9928	.0012
.9989	3.0618 1415	36 74672	31 4800	.0468 9126	9939	.0011
.9990	3.0902 3231	.0033 67090	.0316 0696	.0447 1018	.9999 9950	.0010
.9991	3.1213 8915	30 56534	.0299 8650	.0424 1686	9959	.0009
.9992	3.1559 0676	27 42700	82 7296	.0399 9200	9968	.0008
.9993	3.1946 5105	24 25212	64 4825	.0374 1003	9975	.0007
.9994	3.2388 8012	21 03588	44 8755	.0346 3582	9982	.0006
.9995	3.2905 2673	17 77190	.0223 5509	.0316 1882	9987	.0005
.9996	3.3527 9478	14 45132	.0199 9600	.0282 8144	9992	.0004
.9997	3.4316 1440	11 06086	73 1791	.0244 9306	9995	.0003
.9998	3.5400 8380	07 57842	.0141 4072	.0199 9900	9998	.0002
.9999	3.7190 1649	.0003 95848	.0099 9950	.0141 4178	.9999 9999	.0001

For Interpolation Errors and for continuation of this Table
see page 36

TABLE II. THREE-POINT INTERPOLATION COEFFICIENTS

p	c_{-1} $-$	c_0 $+$	c_1 $+$	p	c_{-1} $-$	c_0 $+$	c_1 $+$
.0000	.00000	1.00000	.00000	.0050	.00249	.99998	.00251
.0001	.00005	1.00000	.00005	.0051	.00254	.99998	.00256
.0002	.00010	1.00000	.00010	.0052	.00259	.99998	.00261
.0003	.00015	1.00000	.00015	.0053	.00264	.99998	.00266
.0004	.00020	1.00000	.00020	.0054	.00269	.99998	.00271
.0005	.00025	1.00000	.00025	.0055	.00273	.99996	.00277
.0006	.00030	1.00000	.00030	.0056	.00278	.99996	.00282
.0007	.00035	1.00000	.00035	.0057	.00283	.99996	.00287
.0008	.00040	1.00000	.00040	.0058	.00288	.99996	.00292
.0009	.00045	1.00000	.00045	.0059	.00293	.99996	.00297
.0010	.00050	1.00000	.00050	.0060	.00298	.99996	.00302
.0011	.00055	1.00000	.00055	.0061	.00303	.99996	.00307
.0012	.00060	1.00000	.00060	.0062	.00308	.99996	.00312
.0013	.00065	1.00000	.00065	.0063	.00313	.99996	.00317
.0014	.00070	1.00000	.00070	.0064	.00318	.99996	.00322
.0015	.00075	1.00000	.00075	.0065	.00323	.99996	.00327
.0016	.00080	1.00000	.00080	.0066	.00328	.99996	.00332
.0017	.00085	1.00000	.00085	.0067	.00333	.99996	.00337
.0018	.00090	1.00000	.00090	.0068	.00338	.99996	.00342
.0019	.00095	1.00000	.00095	.0069	.00343	.99996	.00347
.0020	.00100	1.00000	.00100	.0070	.00348	.99996	.00352
.0021	.00105	1.00000	.00105	.0071	.00352	.99994	.00358
.0022	.00110	1.00000	.00110	.0072	.00357	.99994	.00363
.0023	.00115	1.00000	.00115	.0073	.00362	.99994	.00368
.0024	.00120	1.00000	.00120	.0074	.00367	.99994	.00373
.0025	.00125	1.00000	.00125	.0075	.00372	.99994	.00378
.0026	.00130	1.00000	.00130	.0076	.00377	.99994	.00383
.0027	.00135	1.00000	.00135	.0077	.00382	.99994	.00388
.0028	.00140	1.00000	.00140	.0078	.00387	.99994	.00393
.0029	.00145	1.00000	.00145	.0079	.00392	.99994	.00398
.0030	.00150	1.00000	.00150	.0080	.00397	.99994	.00403
.0031	.00155	1.00000	.00155	.0081	.00402	.99994	.00408
.0032	.00159	.99998	.00161	.0082	.00407	.99994	.00413
.0033	.00164	.99998	.00166	.0083	.00412	.99994	.00418
.0034	.00169	.99998	.00171	.0084	.00416	.99992	.00424
.0035	.00174	.99998	.00176	.0085	.00421	.99992	.00429
.0036	.00179	.99998	.00181	.0086	.00426	.99992	.00434
.0037	.00184	.99998	.00186	.0087	.00431	.99992	.00439
.0038	.00189	.99998	.00191	.0088	.00436	.99992	.00444
.0039	.00194	.99998	.00196	.0089	.00441	.99992	.00449
.0040	.00199	.99998	.00201	.0090	.00446	.99992	.00454
.0041	.00204	.99998	.00206	.0091	.00451	.99992	.00459
.0042	.00209	.99998	.00211	.0092	.00456	.99992	.00464
.0043	.00214	.99998	.00216	.0093	.00461	.99992	.00469
.0044	.00219	.99998	.00221	.0094	.00466	.99992	.00474
.0045	.00224	.99998	.00226	.0095	.00470	.99990	.00480
.0046	.00229	.99998	.00231	.0096	.00475	.99990	.00485
.0047	.00234	.99998	.00236	.0097	.00480	.99990	.00490
.0048	.00239	.99998	.00241	.0098	.00485	.99990	.00495
.0049	.00244	.99998	.00246	.0099	.00490	.99990	.00500
.0050	.00249	.99998	.00251	.0100	.00495	.99990	.00505

TABLE II. THREE-POINT INTERPOLATION COEFFICIENTS

p	c_{-1} −	c_0 +	c_1 +	p	c_{-1} −	c_0 +	c_1 +
.0100	.00495	.99990	.00505	.0150	.00739	.99978	.00761
.0101	.00500	.99990	.00510	.0151	.00744	.99978	.00766
.0102	.00505	.99990	.00515	.0152	.00748	.99976	.00772
.0103	.00510	.99990	.00520	.0153	.00753	.99976	.00777
.0104	.00515	.99990	.00525	.0154	.00758	.99976	.00782
.0105	.00519	.99988	.00531	.0155	.00763	.99976	.00787
.0106	.00524	.99988	.00536	.0156	.00768	.99976	.00792
.0107	.00529	.99988	.00541	.0157	.00773	.99976	.00797
.0108	.00534	.99988	.00546	.0158	.00778	.99976	.00802
.0109	.00539	.99988	.00551	.0159	.00782	.99974	.00808
.0110	.00544	.99988	.00556	.0160	.00787	.99974	.00813
.0111	.00549	.99988	.00561	.0161	.00792	.99974	.00818
.0112	.00554	.99988	.00566	.0162	.00797	.99974	.00823
.0113	.00559	.99988	.00571	.0163	.00802	.99974	.00828
.0114	.00564	.99988	.00576	.0164	.00807	.99974	.00833
.0115	.00568	.99986	.00582	.0165	.00811	.99972	.00839
.0116	.00573	.99986	.00587	.0166	.00816	.99972	.00844
.0117	.00578	.99986	.00592	.0167	.00821	.99972	.00849
.0118	.00583	.99986	.00597	.0168	.00826	.99972	.00854
.0119	.00588	.99986	.00602	.0169	.00831	.99972	.00859
.0120	.00593	.99986	.00607	.0170	.00836	.99972	.00864
.0121	.00598	.99986	.00612	.0171	.00840	.99970	.00870
.0122	.00603	.99986	.00617	.0172	.00845	.99970	.00875
.0123	.00607	.99984	.00623	.0173	.00850	.99970	.00880
.0124	.00612	.99984	.00628	.0174	.00855	.99970	.00885
.0125	.00617	.99984	.00633	.0175	.00860	.99970	.00890
.0126	.00622	.99984	.00638	.0176	.00865	.99970	.00895
.0127	.00627	.99984	.00643	.0177	.00869	.99968	.00901
.0128	.00632	.99984	.00648	.0178	.00874	.99968	.00906
.0129	.00637	.99984	.00653	.0179	.00879	.99968	.00911
.0130	.00642	.99984	.00658	.0180	.00884	.99968	.00916
.0131	.00646	.99982	.00664	.0181	.00889	.99968	.00921
.0132	.00651	.99982	.00669	.0182	.00893	.99966	.00927
.0133	.00656	.99982	.00674	.0183	.00898	.99966	.00932
.0134	.00661	.99982	.00679	.0184	.00903	.99966	.00937
.0135	.00666	.99982	.00684	.0185	.00908	.99966	.00942
.0136	.00671	.99982	.00689	.0186	.00913	.99966	.00947
.0137	.00676	.99982	.00694	.0187	.00918	.99966	.00952
.0138	.00680	.99980	.00700	.0188	.00922	.99964	.00958
.0139	.00685	.99980	.00705	.0189	.00927	.99964	.00963
.0140	.00690	.99980	.00710	.0190	.00932	.99964	.00968
.0141	.00695	.99980	.00715	.0191	.00937	.99964	.00973
.0142	.00700	.99980	.00720	.0192	.00942	.99964	.00978
.0143	.00705	.99980	.00725	.0193	.00946	.99962	.00984
.0144	.00710	.99980	.00730	.0194	.00951	.99962	.00989
.0145	.00714	.99978	.00736	.0195	.00956	.99962	.00994
.0146	.00719	.99978	.00741	.0196	.00961	.99962	.00999
.0147	.00724	.99978	.00746	.0197	.00966	.99962	.01004
.0148	.00729	.99978	.00751	.0198	.00970	.99960	.01010
.0149	.00734	.99978	.00756	.0199	.00975	.99960	.01015
.0150	.00739	.99978	.00761	.0200	.00980	.99960	.01020

TABLE II. THREE-POINT INTERPOLATION COEFFICIENTS

p	c_{-1} $-$	c_0 $+$	c_1 $+$	p	c_{-1} $-$	c_0 $+$	c_1 $+$
.0200	.00980	.99960	.01020	.0250	.01219	.99938	.01281
.0201	.00985	.99960	.01025	.0251	.01223	.99936	.01287
.0202	.00990	.99960	.01030	.0252	.01228	.99936	.01292
.0203	.00994	.99958	.01036	.0253	.01233	.99936	.01297
.0204	.00999	.99958	.01041	.0254	.01238	.99936	.01302
.0205	.01004	.99958	.01046	.0255	.01242	.99934	.01308
.0206	.01009	.99958	.01051	.0256	.01247	.99934	.01313
.0207	.01014	.99958	.01056	.0257	.01252	.99934	.01318
.0208	.01018	.99956	.01062	.0258	.01257	.99934	.01323
.0209	.01023	.99956	.01067	.0259	.01261	.99932	.01329
.0210	.01028	.99956	.01072	.0260	.01266	.99932	.01334
.0211	.01033	.99956	.01077	.0261	.01271	.99932	.01339
.0212	.01038	.99956	.01082	.0262	.01276	.99932	.01344
.0213	.01042	.99954	.01088	.0263	.01280	.99930	.01350
.0214	.01047	.99954	.01093	.0264	.01285	.99930	.01355
.0215	.01052	.99954	.01098	.0265	.01290	.99930	.01360
.0216	.01057	.99954	.01103	.0266	.01295	.99930	.01365
.0217	.01061	.99952	.01109	.0267	.01299	.99928	.01371
.0218	.01066	.99952	.01114	.0268	.01304	.99928	.01376
.0219	.01071	.99952	.01119	.0269	.01309	.99928	.01381
.0220	.01076	.99952	.01124	.0270	.01314	.99928	.01386
.0221	.01081	.99952	.01129	.0271	.01318	.99926	.01392
.0222	.01085	.99950	.01135	.0272	.01323	.99926	.01397
.0223	.01090	.99950	.01140	.0273	.01328	.99926	.01402
.0224	.01095	.99950	.01145	.0274	.01332	.99924	.01408
.0225	.01100	.99950	.01150	.0275	.01337	.99924	.01413
.0226	.01104	.99948	.01156	.0276	.01342	.99924	.01418
.0227	.01109	.99948	.01161	.0277	.01347	.99924	.01423
.0228	.01114	.99948	.01166	.0278	.01351	.99922	.01429
.0229	.01119	.99948	.01171	.0279	.01356	.99922	.01434
.0230	.01124	.99948	.01176	.0280	.01361	.99922	.01439
.0231	.01128	.99946	.01182	.0281	.01366	.99922	.01444
.0232	.01133	.99946	.01187	.0282	.01370	.99920	.01450
.0233	.01138	.99946	.01192	.0283	.01375	.99920	.01455
.0234	.01143	.99946	.01197	.0284	.01380	.99920	.01460
.0235	.01147	.99944	.01203	.0285	.01384	.99918	.01466
.0236	.01152	.99944	.01208	.0286	.01389	.99918	.01471
.0237	.01157	.99944	.01213	.0287	.01394	.99918	.01476
.0238	.01162	.99944	.01218	.0288	.01399	.99918	.01481
.0239	.01166	.99942	.01224	.0289	.01403	.99916	.01487
.0240	.01171	.99942	.01229	.0290	.01408	.99916	.01492
.0241	.01176	.99942	.01234	.0291	.01413	.99916	.01497
.0242	.01181	.99942	.01239	.0292	.01417	.99914	.01503
.0243	.01185	.99940	.01245	.0293	.01422	.99914	.01508
.0244	.01190	.99940	.01250	.0294	.01427	.99914	.01513
.0245	.01195	.99940	.01255	.0295	.01431	.99912	.01519
.0246	.01200	.99940	.01260	.0296	.01436	.99912	.01524
.0247	.01204	.99938	.01266	.0297	.01441	.99912	.01529
.0248	.01209	.99938	.01271	.0298	.01446	.99912	.01534
.0249	.01214	.99938	.01276	.0299	.01450	.99910	.01540
.0250	.01219	.99938	.01281	.0300	.01455	.99910	.01545

TABLE II. THREE-POINT INTERPOLATION COEFFICIENTS

p	c_{-1} $-$	c_0 $+$	c_1 $+$	p	c_{-1} $-$	c_0 $+$	c_1 $+$
.0300	.01455	.99910	.01545	.0350	.01689	.99878	.01811
.0301	.01460	.99910	.01550	.0351	.01693	.99876	.01817
.0302	.01464	.99908	.01556	.0352	.01698	.99876	.01822
.0303	.01469	.99908	.01561	.0353	.01703	.99876	.01827
.0304	.01474	.99908	.01566	.0354	.01707	.99874	.01833
.0305	.01478	.99906	.01572	.0355	.01712	.99874	.01838
.0306	.01483	.99906	.01577	.0356	.01717	.99874	.01843
.0307	.01488	.99906	.01582	.0357	.01721	.99872	.01849
.0308	.01493	.99906	.01587	.0358	.01726	.99872	.01854
.0309	.01497	.99904	.01593	.0359	.01731	.99872	.01859
.0310	.01502	.99904	.01598	.0360	.01735	.99870	.01865
.0311	.01507	.99904	.01603	.0361	.01740	.99870	.01870
.0312	.01511	.99902	.01609	.0362	.01744	.99868	.01876
.0313	.01516	.99902	.01614	.0363	.01749	.99868	.01881
.0314	.01521	.99902	.01619	.0364	.01754	.99868	.01886
.0315	.01525	.99900	.01625	.0365	.01758	.99866	.01892
.0316	.01530	.99900	.01630	.0366	.01763	.99866	.01897
.0317	.01535	.99900	.01635	.0367	.01768	.99866	.01902
.0318	.01539	.99898	.01641	.0368	.01772	.99864	.01908
.0319	.01544	.99898	.01646	.0369	.01777	.99864	.01913
.0320	.01549	.99898	.01651	.0370	.01782	.99864	.01918
.0321	.01553	.99896	.01657	.0371	.01786	.99862	.01924
.0322	.01558	.99896	.01662	.0372	.01791	.99862	.01929
.0323	.01563	.99896	.01667	.0373	.01795	.99860	.01935
.0324	.01568	.99896	.01672	.0374	.01800	.99860	.01940
.0325	.01572	.99894	.01678	.0375	.01805	.99860	.01945
.0326	.01577	.99894	.01683	.0376	.01809	.99858	.01951
.0327	.01582	.99894	.01688	.0377	.01814	.99858	.01956
.0328	.01586	.99892	.01694	.0378	.01819	.99858	.01961
.0329	.01591	.99892	.01699	.0379	.01823	.99856	.01967
.0330	.01596	.99892	.01704	.0380	.01828	.99856	.01972
.0331	.01600	.99890	.01710	.0381	.01832	.99854	.01978
.0332	.01605	.99890	.01715	.0382	.01837	.99854	.01983
.0333	.01610	.99890	.01720	.0383	.01842	.99854	.01988
.0334	.01614	.99888	.01726	.0384	.01846	.99852	.01994
.0335	.01619	.99888	.01731	.0385	.01851	.99852	.01999
.0336	.01624	.99888	.01736	.0386	.01856	.99852	.02004
.0337	.01628	.99886	.01742	.0387	.01860	.99850	.02010
.0338	.01633	.99886	.01747	.0388	.01865	.99850	.02015
.0339	.01638	.99886	.01752	.0389	.01869	.99848	.02021
.0340	.01642	.99884	.01758	.0390	.01874	.99848	.02026
.0341	.01647	.99884	.01763	.0391	.01879	.99848	.02031
.0342	.01652	.99884	.01768	.0392	.01883	.99846	.02037
.0343	.01656	.99882	.01774	.0393	.01888	.99846	.02042
.0344	.01661	.99882	.01779	.0394	.01892	.99844	.02048
.0345	.01665	.99880	.01785	.0395	.01897	.99844	.02053
.0346	.01670	.99880	.01790	.0396	.01902	.99844	.02058
.0347	.01675	.99880	.01795	.0397	.01906	.99842	.02064
.0348	.01679	.99878	.01801	.0398	.01911	.99842	.02069
.0349	.01684	.99878	.01806	.0399	.01915	.99840	.02075
.0350	.01689	.99878	.01811	.0400	.01920	.99840	.02080

TABLE II. THREE-POINT INTERPOLATION COEFFICIENTS

p	c_{-1} −	c_0 +	c_1 +	p	c_{-1} −	c_0 +	c_1 +
.0400	.01920	.99840	.02080	.0450	.02149	.99798	.02351
.0401	.01925	.99840	.02085	.0451	.02153	.99796	.02357
.0402	.01929	.99838	.02091	.0452	.02158	.99796	.02362
.0403	.01934	.99838	.02096	.0453	.02162	.99794	.02368
.0404	.01938	.99836	.02102	.0454	.02167	.99794	.02373
.0405	.01943	.99836	.02107	.0455	.02171	.99792	.02379
.0406	.01948	.99836	.02112	.0456	.02176	.99792	.02384
.0407	.01952	.99834	.02118	.0457	.02181	.99792	.02389
.0408	.01957	.99834	.02123	.0458	.02185	.99790	.02395
.0409	.01961	.99832	.02129	.0459	.02190	.99790	.02400
.0410	.01966	.99832	.02134	.0460	.02194	.99788	.02406
.0411	.01971	.99832	.02139	.0461	.02199	.99788	.02411
.0412	.01975	.99830	.02145	.0462	.02203	.99786	.02417
.0413	.01980	.99830	.02150	.0463	.02208	.99786	.02422
.0414	.01984	.99828	.02156	.0464	.02212	.99784	.02428
.0415	.01989	.99828	.02161	.0465	.02217	.99784	.02433
.0416	.01993	.99826	.02167	.0466	.02221	.99782	.02439
.0417	.01998	.99826	.02172	.0467	.02226	.99782	.02444
.0418	.02003	.99826	.02177	.0468	.02230	.99780	.02450
.0419	.02007	.99824	.02183	.0469	.02235	.99780	.02455
.0420	.02012	.99824	.02188	.0470	.02240	.99780	.02460
.0421	.02016	.99822	.02194	.0471	.02244	.99778	.02466
.0422	.02021	.99822	.02199	.0472	.02249	.99778	.02471
.0423	.02026	.99822	.02204	.0473	.02253	.99776	.02477
.0424	.02030	.99820	.02210	.0474	.02258	.99776	.02482
.0425	.02035	.99820	.02215	.0475	.02262	.99774	.02488
.0426	.02039	.99818	.02221	.0476	.02267	.99774	.02493
.0427	.02044	.99818	.02226	.0477	.02271	.99772	.02499
.0428	.02048	.99816	.02232	.0478	.02276	.99772	.02504
.0429	.02053	.99816	.02237	.0479	.02280	.99770	.02510
.0430	.02058	.99816	.02242	.0480	.02285	.99770	.02515
.0431	.02062	.99814	.02248	.0481	.02289	.99768	.02521
.0432	.02067	.99814	.02253	.0482	.02294	.99768	.02526
.0433	.02071	.99812	.02259	.0483	.02298	.99766	.02532
.0434	.02076	.99812	.02264	.0484	.02303	.99766	.02537
.0435	.02080	.99810	.02270	.0485	.02307	.99764	.02543
.0436	.02085	.99810	.02275	.0486	.02312	.99764	.02548
.0437	.02090	.99810	.02280	.0487	.02316	.99762	.02554
.0438	.02094	.99808	.02286	.0488	.02321	.99762	.02559
.0439	.02099	.99808	.02291	.0489	.02325	.99760	.02565
.0440	.02103	.99806	.02297	.0490	.02330	.99760	.02570
.0441	.02108	.99806	.02302	.0491	.02334	.99758	.02576
.0442	.02112	.99804	.02308	.0492	.02339	.99758	.02581
.0443	.02117	.99804	.02313	.0493	.02343	.99756	.02587
.0444	.02121	.99802	.02319	.0494	.02348	.99756	.02592
.0445	.02126	.99802	.02324	.0495	.02352	.99754	.02598
.0446	.02131	.99802	.02329	.0496	.02357	.99754	.02603
.0447	.02135	.99800	.02335	.0497	.02361	.99752	.02609
.0448	.02140	.99800	.02340	.0498	.02366	.99752	.02614
.0449	.02144	.99798	.02346	.0499	.02370	.99750	.02620
.0450	.02149	.99798	.02351	.0500	.02375	.99750	.02625

TABLE II. THREE-POINT INTERPOLATION COEFFICIENTS

p	c_{-1} $-$	c_0 $+$	c_1 $+$	p	c_{-1} $-$	c_0 $+$	c_1 $+$
.0500	.02375	.99750	.02625	.0550	.02599	.99698	.02901
.0501	.02379	.99748	.02631	.0551	.02603	.99696	.02907
.0502	.02384	.99748	.02636	.0552	.02608	.99696	.02912
.0503	.02388	.99746	.02642	.0553	.02612	.99694	.02918
.0504	.02393	.99746	.02647	.0554	.02617	.99694	.02923
.0505	.02397	.99744	.02653	.0555	.02621	.99692	.02929
.0506	.02402	.99744	.02658	.0556	.02625	.99690	.02935
.0507	.02406	.99742	.02664	.0557	.02630	.99690	.02940
.0508	.02411	.99742	.02669	.0558	.02634	.99688	.02946
.0509	.02415	.99740	.02675	.0559	.02639	.99688	.02951
.0510	.02420	.99740	.02680	.0560	.02643	.99686	.02957
.0511	.02424	.99738	.02686	.0561	.02648	.99686	.02962
.0512	.02429	.99738	.02691	.0562	.02652	.99684	.02968
.0513	.02433	.99736	.02697	.0563	.02657	.99684	.02973
.0514	.02438	.99736	.02702	.0564	.02661	.99682	.02979
.0515	.02442	.99734	.02708	.0565	.02665	.99680	.02985
.0516	.02447	.99734	.02713	.0566	.02670	.99680	.02990
.0517	.02451	.99732	.02719	.0567	.02674	.99678	.02996
.0518	.02456	.99732	.02724	.0568	.02679	.99678	.03001
.0519	.02460	.99730	.02730	.0569	.02683	.99676	.03007
.0520	.02465	.99730	.02735	.0570	.02688	.99676	.03012
.0521	.02469	.99728	.02741	.0571	.02692	.99674	.03018
.0522	.02474	.99728	.02746	.0572	.02696	.99672	.03024
.0523	.02478	.99726	.02752	.0573	.02701	.99672	.03029
.0524	.02483	.99726	.02757	.0574	.02705	.99670	.03035
.0525	.02487	.99724	.02763	.0575	.02710	.99670	.03040
.0526	.02492	.99724	.02768	.0576	.02714	.99668	.03046
.0527	.02496	.99722	.02774	.0577	.02719	.99668	.03051
.0528	.02501	.99722	.02779	.0578	.02723	.99666	.03057
.0529	.02505	.99720	.02785	.0579	.02727	.99664	.03063
.0530	.02510	.99720	.02790	.0580	.02732	.99664	.03068
.0531	.02514	.99718	.02796	.0581	.02736	.99662	.03074
.0532	.02518	.99716	.02802	.0582	.02741	.99662	.03079
.0533	.02523	.99716	.02807	.0583	.02745	.99660	.03085
.0534	.02527	.99714	.02813	.0584	.02749	.99658	.03091
.0535	.02532	.99714	.02818	.0585	.02754	.99658	.03096
.0536	.02536	.99712	.02824	.0586	.02758	.99656	.03102
.0537	.02541	.99712	.02829	.0587	.02763	.99656	.03107
.0538	.02545	.99710	.02835	.0588	.02767	.99654	.03113
.0539	.02550	.99710	.02840	.0589	.02772	.99654	.03118
.0540	.02554	.99708	.02846	.0590	.02776	.99652	.03124
.0541	.02559	.99708	.02851	.0591	.02780	.99650	.03130
.0542	.02563	.99706	.02857	.0592	.02785	.99650	.03135
.0543	.02568	.99706	.02862	.0593	.02789	.99648	.03141
.0544	.02572	.99704	.02868	.0594	.02794	.99648	.03146
.0545	.02576	.99702	.02874	.0595	.02798	.99646	.03152
.0546	.02581	.99702	.02879	.0596	.02802	.99644	.03158
.0547	.02585	.99700	.02885	.0597	.02807	.99644	.03163
.0548	.02590	.99700	.02890	.0598	.02811	.99642	.03169
.0549	.02594	.99698	.02896	.0599	.02816	.99642	.03174
.0550	.02599	.99698	.02901	.0600	.02820	.99640	.03180

TABLE II. THREE-POINT INTERPOLATION COEFFICIENTS

p	c_{-1} −	c_0 +	c_1 +	p	c_{-1} −	c_0 +	c_1 +
.0600	.02820	.99640	.03180	.0650	.03039	.99578	.03461
.0601	.02824	.99638	.03186	.0651	.03043	.99576	.03467
.0602	.02829	.99638	.03191	.0652	.03047	.99574	.03473
.0603	.02833	.99636	.03197	.0653	.03052	.99574	.03478
.0604	.02838	.99636	.03202	.0654	.03056	.99572	.03484
.0605	.02842	.99634	.03208	.0655	.03060	.99570	.03490
.0606	.02846	.99632	.03214	.0656	.03065	.99570	.03495
.0607	.02851	.99632	.03219	.0657	.03069	.99568	.03501
.0608	.02855	.99630	.03225	.0658	.03074	.99568	.03506
.0609	.02860	.99630	.03230	.0659	.03078	.99566	.03512
.0610	.02864	.99628	.03236	.0660	.03082	.99564	.03518
.0611	.02868	.99626	.03242	.0661	.03087	.99564	.03523
.0612	.02873	.99626	.03247	.0662	.03091	.99562	.03529
.0613	.02877	.99624	.03253	.0663	.03095	.99560	.03535
.0614	.02882	.99624	.03258	.0664	.03100	.99560	.03540
.0615	.02886	.99622	.03264	.0665	.03104	.99558	.03546
.0616	.02890	.99620	.03270	.0666	.03108	.99556	.03552
.0617	.02895	.99620	.03275	.0667	.03113	.99556	.03557
.0618	.02899	.99618	.03281	.0668	.03117	.99554	.03563
.0619	.02903	.99616	.03287	.0669	.03121	.99552	.03569
.0620	.02908	.99616	.03292	.0670	.03126	.99552	.03574
.0621	.02912	.99614	.03298	.0671	.03130	.99550	.03580
.0622	.02917	.99614	.03303	.0672	.03134	.99548	.03586
.0623	.02921	.99612	.03309	.0673	.03139	.99548	.03591
.0624	.02925	.99610	.03315	.0674	.03143	.99546	.03597
.0625	.02930	.99610	.03320	.0675	.03147	.99544	.03603
.0626	.02934	.99608	.03326	.0676	.03152	.99544	.03608
.0627	.02938	.99606	.03332	.0677	.03156	.99542	.03614
.0628	.02943	.99606	.03337	.0678	.03160	.99540	.03620
.0629	.02947	.99604	.03343	.0679	.03164	.99538	.03626
.0630	.02952	.99604	.03348	.0680	.03169	.99538	.03631
.0631	.02956	.99602	.03354	.0681	.03173	.99536	.03637
.0632	.02960	.99600	.03360	.0682	.03177	.99534	.03643
.0633	.02965	.99600	.03365	.0683	.03182	.99534	.03648
.0634	.02969	.99598	.03371	.0684	.03186	.99532	.03654
.0635	.02973	.99596	.03377	.0685	.03190	.99530	.03660
.0636	.02978	.99596	.03382	.0686	.03195	.99530	.03665
.0637	.02982	.99594	.03388	.0687	.03199	.99528	.03671
.0638	.02986	.99592	.03394	.0688	.03203	.99526	.03677
.0639	.02991	.99592	.03399	.0689	.03208	.99526	.03682
.0640	.02995	.99590	.03405	.0690	.03212	.99524	.03688
.0641	.03000	.99590	.03410	.0691	.03216	.99522	.03694
.0642	.03004	.99588	.03416	.0692	.03221	.99522	.03699
.0643	.03008	.99586	.03422	.0693	.03225	.99520	.03705
.0644	.03013	.99586	.03427	.0694	.03229	.99518	.03711
.0645	.03017	.99584	.03433	.0695	.03233	.99516	.03717
.0646	.03021	.99582	.03439	.0696	.03238	.99516	.03722
.0647	.03026	.99582	.03444	.0697	.03242	.99514	.03728
.0648	.03030	.99580	.03450	.0698	.03246	.99512	.03734
.0649	.03034	.99578	.03456	.0699	.03251	.99512	.03739
.0650	.03039	.99578	.03461	.0700	.03255	.99510	.03745

TABLE II. THREE-POINT INTERPOLATION COEFFICIENTS

p	c_{-1} −	c_0 +	c_1 +	p	c_{-1} −	c_0 +	c_1 +
.0700	.03255	.99510	.03745	.0750	.03469	.99438	.04031
.0701	.03259	.99508	.03751	.0751	.03473	.99436	.04037
.0702	.03264	.99508	.03756	.0752	.03477	.99434	.04043
.0703	.03268	.99506	.03762	.0753	.03481	.99432	.04049
.0704	.03272	.99504	.03768	.0754	.03486	.99432	.04054
.0705	.03276	.99502	.03774	.0755	.03490	.99430	.04060
.0706	.03281	.99502	.03779	.0756	.03494	.99428	.04066
.0707	.03285	.99500	.03785	.0757	.03498	.99426	.04072
.0708	.03289	.99498	.03791	.0758	.03503	.99426	.04077
.0709	.03294	.99498	.03796	.0759	.03507	.99424	.04083
.0710	.03298	.99496	.03802	.0760	.03511	.99422	.04089
.0711	.03302	.99494	.03808	.0761	.03515	.99420	.04095
.0712	.03307	.99494	.03813	.0762	.03520	.99420	.04100
.0713	.03311	.99492	.03819	.0763	.03524	.99418	.04106
.0714	.03315	.99490	.03825	.0764	.03528	.99416	.04112
.0715	.03319	.99488	.03831	.0765	.03532	.99414	.04118
.0716	.03324	.99488	.03836	.0766	.03537	.99414	.04123
.0717	.03328	.99486	.03842	.0767	.03541	.99412	.04129
.0718	.03332	.99484	.03848	.0768	.03545	.99410	.04135
.0719	.03337	.99484	.03853	.0769	.03549	.99408	.04141
.0720	.03341	.99482	.03859	.0770	.03554	.99408	.04146
.0721	.03345	.99480	.03865	.0771	.03558	.99406	.04152
.0722	.03349	.99478	.03871	.0772	.03562	.99404	.04158
.0723	.03354	.99478	.03876	.0773	.03566	.99402	.04164
.0724	.03358	.99476	.03882	.0774	.03570	.99400	.04170
.0725	.03362	.99474	.03888	.0775	.03575	.99400	.04175
.0726	.03366	.99472	.03894	.0776	.03579	.99398	.04181
.0727	.03371	.99472	.03899	.0777	.03583	.99396	.04187
.0728	.03375	.99470	.03905	.0778	.03587	.99394	.04193
.0729	.03379	.99468	.03911	.0779	.03592	.99394	.04198
.0730	.03384	.99468	.03916	.0780	.03596	.99392	.04204
.0731	.03388	.99466	.03922	.0781	.03600	.99390	.04210
.0732	.03392	.99464	.03928	.0782	.03604	.99388	.04216
.0733	.03396	.99462	.03934	.0783	.03608	.99386	.04222
.0734	.03401	.99462	.03939	.0784	.03613	.99386	.04227
.0735	.03405	.99460	.03945	.0785	.03617	.99384	.04233
.0736	.03409	.99458	.03951	.0786	.03621	.99382	.04239
.0737	.03413	.99456	.03957	.0787	.03625	.99380	.04245
.0738	.03418	.99456	.03962	.0788	.03630	.99380	.04250
.0739	.03422	.99454	.03968	.0789	.03634	.99378	.04256
.0740	.03426	.99452	.03974	.0790	.03638	.99376	.04262
.0741	.03430	.99450	.03980	.0791	.03642	.99374	.04268
.0742	.03435	.99450	.03985	.0792	.03646	.99372	.04274
.0743	.03439	.99448	.03991	.0793	.03651	.99372	.04279
.0744	.03443	.99446	.03997	.0794	.03655	.99370	.04285
.0745	.03447	.99444	.04003	.0795	.03659	.99368	.04291
.0746	.03452	.99444	.04008	.0796	.03663	.99366	.04297
.0747	.03456	.99442	.04014	.0797	.03667	.99364	.04303
.0748	.03460	.99440	.04020	.0798	.03672	.99364	.04308
.0749	.03464	.99438	.04026	.0799	.03676	.99362	.04314
.0750	.03469	.99438	.04031	.0800	.03680	.99360	.04320

TABLE II. THREE-POINT INTERPOLATION COEFFICIENTS

p	c_{-1} −	c_0 +	c_1 +	p	c_{-1} −	c_0 +	c_1 +
.0800	.03680	.99360	.04320	.0850	.03889	.99278	.04611
.0801	.03684	.99358	.04326	.0851	.03893	.99276	.04617
.0802	.03688	.99356	.04332	.0852	.03897	.99274	.04623
.0803	.03693	.99356	.04337	.0853	.03901	.99272	.04629
.0804	.03697	.99354	.04343	.0854	.03905	.99270	.04635
.0805	.03701	.99352	.04349	.0855	.03909	.99268	.04641
.0806	.03705	.99350	.04355	.0856	.03914	.99268	.04646
.0807	.03709	.99348	.04361	.0857	.03918	.99266	.04652
.0808	.03714	.99348	.04366	.0858	.03922	.99264	.04658
.0809	.03718	.99346	.04372	.0859	.03926	.99262	.04664
.0810	.03722	.99344	.04378	.0860	.03930	.99260	.04670
.0811	.03726	.99342	.04384	.0861	.03934	.99258	.04676
.0812	.03730	.99340	.04390	.0862	.03938	.99256	.04682
.0813	.03735	.99340	.04395	.0863	.03943	.99256	.04687
.0814	.03739	.99338	.04401	.0864	.03947	.99254	.04693
.0815	.03743	.99336	.04407	.0865	.03951	.99252	.04699
.0816	.03747	.99334	.04413	.0866	.03955	.99250	.04705
.0817	.03751	.99332	.04419	.0867	.03959	.99248	.04711
.0818	.03755	.99330	.04425	.0868	.03963	.99246	.04717
.0819	.03760	.99330	.04430	.0869	.03967	.99244	.04723
.0820	.03764	.99328	.04436	.0870	.03972	.99244	.04728
.0821	.03768	.99326	.04442	.0871	.03976	.99242	.04734
.0822	.03772	.99324	.04448	.0872	.03980	.99240	.04740
.0823	.03776	.99322	.04454	.0873	.03984	.99238	.04746
.0824	.03781	.99322	.04459	.0874	.03988	.99236	.04752
.0825	.03785	.99320	.04465	.0875	.03992	.99234	.04758
.0826	.03789	.99318	.04471	.0876	.03996	.99232	.04764
.0827	.03793	.99316	.04477	.0877	.04000	.99230	.04770
.0828	.03797	.99314	.04483	.0878	.04005	.99230	.04775
.0829	.03801	.99312	.04489	.0879	.04009	.99228	.04781
.0830	.03806	.99312	.04494	.0880	.04013	.99226	.04787
.0831	.03810	.99310	.04500	.0881	.04017	.99224	.04793
.0832	.03814	.99308	.04506	.0882	.04021	.99222	.04799
.0833	.03818	.99306	.04512	.0883	.04025	.99220	.04805
.0834	.03822	.99304	.04518	.0884	.04029	.99218	.04811
.0835	.03826	.99302	.04524	.0885	.04033	.99216	.04817
.0836	.03831	.99302	.04529	.0886	.04038	.99216	.04822
.0837	.03835	.99300	.04535	.0887	.04042	.99214	.04828
.0838	.03839	.99298	.04541	.0888	.04046	.99212	.04834
.0839	.03843	.99296	.04547	.0889	.04050	.99210	.04840
.0840	.03847	.99294	.04553	.0890	.04054	.99208	.04846
.0841	.03851	.99292	.04559	.0891	.04058	.99206	.04852
.0842	.03856	.99292	.04564	.0892	.04062	.99204	.04858
.0843	.03860	.99290	.04570	.0893	.04066	.99202	.04864
.0844	.03864	.99288	.04576	.0894	.04070	.99200	.04870
.0845	.03868	.99286	.04582	.0895	.04074	.99198	.04876
.0846	.03872	.99284	.04588	.0896	.04079	.99198	.04881
.0847	.03876	.99282	.04594	.0897	.04083	.99196	.04887
.0848	.03880	.99280	.04600	.0898	.04087	.99194	.04893
.0849	.03885	.99280	.04605	.0899	.04091	.99192	.04899
.0850	.03889	.99278	.04611	.0900	.04095	.99190	.04905

TABLE II. THREE-POINT INTERPOLATION COEFFICIENTS

p	c_{-1} $-$	c_0 $+$	c_1 $+$	p	c_{-1} $-$	c_0 $+$	c_1 $+$
.0900	.04095	.99190	.04905	.0950	.04299	.99098	.05201
.0901	.04099	.99188	.04911	.0951	.04303	.99096	.05207
.0902	.04103	.99186	.04917	.0952	.04307	.99094	.05213
.0903	.04107	.99184	.04923	.0953	.04311	.99092	.05219
.0904	.04111	.99182	.04929	.0954	.04315	.99090	.05225
.0905	.04115	.99180	.04935	.0955	.04319	.99088	.05231
.0906	.04120	.99180	.04940	.0956	.04323	.99086	.05237
.0907	.04124	.99178	.04946	.0957	.04327	.99084	.05243
.0908	.04128	.99176	.04952	.0958	.04331	.99082	.05249
.0909	.04132	.99174	.04958	.0959	.04335	.99080	.05255
.0910	.04136	.99172	.04964	.0960	.04339	.99078	.05261
.0911	.04140	.99170	.04970	.0961	.04343	.99076	.05267
.0912	.04144	.99168	.04976	.0962	.04347	.99074	.05273
.0913	.04148	.99166	.04982	.0963	.04351	.99072	.05279
.0914	.04152	.99164	.04988	.0964	.04355	.99070	.05285
.0915	.04156	.99162	.04994	.0965	.04359	.99068	.05291
.0916	.04160	.99160	.05000	.0966	.04363	.99066	.05297
.0917	.04165	.99160	.05005	.0967	.04367	.99064	.05303
.0918	.04169	.99158	.05011	.0968	.04371	.99062	.05309
.0919	.04173	.99156	.05017	.0969	.04376	.99062	.05314
.0920	.04177	.99154	.05023	.0970	.04380	.99060	.05320
.0921	.04181	.99152	.05029	.0971	.04384	.99058	.05326
.0922	.04185	.99150	.05035	.0972	.04388	.99056	.05332
.0923	.04189	.99148	.05041	.0973	.04392	.99054	.05338
.0924	.04193	.99146	.05047	.0974	.04396	.99052	.05344
.0925	.04197	.99144	.05053	.0975	.04400	.99050	.05350
.0926	.04201	.99142	.05059	.0976	.04404	.99048	.05356
.0927	.04205	.99140	.05065	.0977	.04408	.99046	.05362
.0928	.04209	.99138	.05071	.0978	.04412	.99044	.05368
.0929	.04213	.99136	.05077	.0979	.04416	.99042	.05374
.0930	.04218	.99136	.05082	.0980	.04420	.99040	.05380
.0931	.04222	.99134	.05088	.0981	.04424	.99038	.05386
.0932	.04226	.99132	.05094	.0982	.04428	.99036	.05392
.0933	.04230	.99130	.05100	.0983	.04432	.99034	.05398
.0934	.04234	.99128	.05106	.0984	.04436	.99032	.05404
.0935	.04238	.99126	.05112	.0985	.04440	.99030	.05410
.0936	.04242	.99124	.05118	.0986	.04444	.99028	.05416
.0937	.04246	.99122	.05124	.0987	.04448	.99026	.05422
.0938	.04250	.99120	.05130	.0988	.04452	.99024	.05428
.0939	.04254	.99118	.05136	.0989	.04456	.99022	.05434
.0940	.04258	.99116	.05142	.0990	.04460	.99020	.05440
.0941	.04262	.99114	.05148	.0991	.04464	.99018	.05446
.0942	.04266	.99112	.05154	.0992	.04468	.99016	.05452
.0943	.04270	.99110	.05160	.0993	.04472	.99014	.05458
.0944	.04274	.99108	.05166	.0994	.04476	.99012	.05464
.0945	.04278	.99106	.05172	.0995	.04480	.99010	.05470
.0946	.04283	.99106	.05177	.0996	.04484	.99008	.05476
.0947	.04287	.99104	.05183	.0997	.04488	.99006	.05482
.0948	.04291	.99102	.05189	.0998	.04492	.99004	.05488
.0949	.04295	.99100	.05195	.0999	.04496	.99002	.05494
.0950	.04299	.99098	.05201	.1000	.04500	.99000	.05500

TABLE II. THREE-POINT INTERPOLATION COEFFICIENTS

p	c_{-1} −	c_0 +	c_1 +	p	c_{-1} −	c_0 +	c_1 +
.1000	.04500	.99000	.05500	.1050	.04699	.98898	.05801
.1001	.04504	.98998	.05506	.1051	.04703	.98896	.05807
.1002	.04508	.98996	.05512	.1052	.04707	.98894	.05813
.1003	.04512	.98994	.05518	.1053	.04711	.98892	.05819
.1004	.04516	.98992	.05524	.1054	.04715	.98890	.05825
.1005	.04520	.98990	.05530	.1055	.04718	.98886	.05832
.1006	.04524	.98988	.05536	.1056	.04722	.98884	.05838
.1007	.04528	.98986	.05542	.1057	.04726	.98882	.05844
.1008	.04532	.98984	.05548	.1058	.04730	.98880	.05850
.1009	.04536	.98982	.05554	.1059	.04734	.98878	.05856
.1010	.04540	.98980	.05560	.1060	.04738	.98876	.05862
.1011	.04544	.98978	.05566	.1061	.04742	.98874	.05868
.1012	.04548	.98976	.05572	.1062	.04746	.98872	.05874
.1013	.04552	.98974	.05578	.1063	.04750	.98870	.05880
.1014	.04556	.98972	.05584	.1064	.04754	.98868	.05886
.1015	.04560	.98970	.05590	.1065	.04758	.98866	.05892
.1016	.04564	.98968	.05596	.1066	.04762	.98864	.05898
.1017	.04568	.98966	.05602	.1067	.04766	.98862	.05904
.1018	.04572	.98964	.05608	.1068	.04770	.98860	.05910
.1019	.04576	.98962	.05614	.1069	.04774	.98858	.05916
.1020	.04580	.98960	.05620	.1070	.04778	.98856	.05922
.1021	.04584	.98958	.05626	.1071	.04781	.98852	.05929
.1022	.04588	.98956	.05632	.1072	.04785	.98850	.05935
.1023	.04592	.98954	.05638	.1073	.04789	.98848	.05941
.1024	.04596	.98952	.05644	.1074	.04793	.98846	.05947
.1025	.04600	.98950	.05650	.1075	.04797	.98844	.05953
.1026	.04604	.98948	.05656	.1076	.04801	.98842	.05959
.1027	.04608	.98946	.05662	.1077	.04805	.98840	.05965
.1028	.04612	.98944	.05668	.1078	.04809	.98838	.05971
.1029	.04616	.98942	.05674	.1079	.04813	.98836	.05977
.1030	.04620	.98940	.05680	.1080	.04817	.98834	.05983
.1031	.04624	.98938	.05686	.1081	.04821	.98832	.05989
.1032	.04627	.98934	.05693	.1082	.04825	.98830	.05995
.1033	.04631	.98932	.05699	.1083	.04829	.98828	.06001
.1034	.04635	.98930	.05705	.1084	.04832	.98824	.06008
.1035	.04639	.98928	.05711	.1085	.04836	.98822	.06014
.1036	.04643	.98926	.05717	.1086	.04840	.98820	.06020
.1037	.04647	.98924	.05723	.1087	.04844	.98818	.06026
.1038	.04651	.98922	.05729	.1088	.04848	.98816	.06032
.1039	.04655	.98920	.05735	.1089	.04852	.98814	.06038
.1040	.04659	.98918	.05741	.1090	.04856	.98812	.06044
.1041	.04663	.98916	.05747	.1091	.04860	.98810	.06050
.1042	.04667	.98914	.05753	.1092	.04864	.98808	.06056
.1043	.04671	.98912	.05759	.1093	.04868	.98806	.06062
.1044	.04675	.98910	.05765	.1094	.04872	.98804	.06068
.1045	.04679	.98908	.05771	.1095	.04875	.98800	.06075
.1046	.04683	.98906	.05777	.1096	.04879	.98798	.06081
.1047	.04687	.98904	.05783	.1097	.04883	.98796	.06087
.1048	.04691	.98902	.05789	.1098	.04887	.98794	.06093
.1049	.04695	.98900	.05795	.1099	.04891	.98792	.06099
.1050	.04699	.98898	.05801	.1100	.04895	.98790	.06105

TABLE II. THREE-POINT INTERPOLATION COEFFICIENTS

p	c_{-1} −	c_0 +	c_1 +	p	c_{-1} −	c_0 +	c_1 +
.1100	.04895	.98790	.06105	.1150	.05089	.98678	.06411
.1101	.04899	.98788	.06111	.1151	.05093	.98676	.06417
.1102	.04903	.98786	.06117	.1152	.05096	.98672	.06424
.1103	.04907	.98784	.06123	.1153	.05100	.98670	.06430
.1104	.04911	.98782	.06129	.1154	.05104	.98668	.06436
.1105	.04914	.98778	.06136	.1155	.05108	.98666	.06442
.1106	.04918	.98776	.06142	.1156	.05112	.98664	.06448
.1107	.04922	.98774	.06148	.1157	.05116	.98662	.06454
.1108	.04926	.98772	.06154	.1158	.05120	.98660	.06460
.1109	.04930	.98770	.06160	.1159	.05123	.98656	.06467
.1110	.04934	.98768	.06166	.1160	.05127	.98654	.06473
.1111	.04938	.98766	.06172	.1161	.05131	.98652	.06479
.1112	.04942	.98764	.06178	.1162	.05135	.98650	.06485
.1113	.04946	.98762	.06184	.1163	.05139	.98648	.06491
.1114	.04950	.98760	.06190	.1164	.05143	.98646	.06497
.1115	.04953	.98756	.06197	.1165	.05146	.98642	.06504
.1116	.04957	.98754	.06203	.1166	.05150	.98640	.06510
.1117	.04961	.98752	.06209	.1167	.05154	.98638	.06516
.1118	.04965	.98750	.06215	.1168	.05158	.98636	.06522
.1119	.04969	.98748	.06221	.1169	.05162	.98634	.06528
.1120	.04973	.98746	.06227	.1170	.05166	.98632	.06534
.1121	.04977	.98744	.06233	.1171	.05169	.98628	.06541
.1122	.04981	.98742	.06239	.1172	.05173	.98626	.06547
.1123	.04984	.98738	.06246	.1173	.05177	.98624	.06553
.1124	.04988	.98736	.06252	.1174	.05181	.98622	.06559
.1125	.04992	.98734	.06258	.1175	.05185	.98620	.06565
.1126	.04996	.98732	.06264	.1176	.05189	.98618	.06571
.1127	.05000	.98730	.06270	.1177	.05192	.98614	.06578
.1128	.05004	.98728	.06276	.1178	.05196	.98612	.06584
.1129	.05008	.98726	.06282	.1179	.05200	.98610	.06590
.1130	.05012	.98724	.06288	.1180	.05204	.98608	.06596
.1131	.05015	.98720	.06295	.1181	.05208	.98606	.06602
.1132	.05019	.98718	.06301	.1182	.05211	.98602	.06609
.1133	.05023	.98716	.06307	.1183	.05215	.98600	.06615
.1134	.05027	.98714	.06313	.1184	.05219	.98598	.06621
.1135	.05031	.98712	.06319	.1185	.05223	.98596	.06627
.1136	.05035	.98710	.06325	.1186	.05227	.98594	.06633
.1137	.05039	.98708	.06331	.1187	.05231	.98592	.06639
.1138	.05042	.98704	.06338	.1188	.05234	.98588	.06646
.1139	.05046	.98702	.06344	.1189	.05238	.98586	.06652
.1140	.05050	.98700	.06350	.1190	.05242	.98584	.06658
.1141	.05054	.98698	.06356	.1191	.05246	.98582	.06664
.1142	.05058	.98696	.06362	.1192	.05250	.98580	.06670
.1143	.05062	.98694	.06368	.1193	.05253	.98576	.06677
.1144	.05066	.98692	.06374	.1194	.05257	.98574	.06683
.1145	.05069	.98688	.06381	.1195	.05261	.98572	.06689
.1146	.05073	.98686	.06387	.1196	.05265	.98570	.06695
.1147	.05077	.98684	.06393	.1197	.05269	.98568	.06701
.1148	.05081	.98682	.06399	.1198	.05272	.98564	.06708
.1149	.05085	.98680	.06405	.1199	.05276	.98562	.06714
.1150	.05089	.98678	.06411	.1200	.05280	.98560	.06720

TABLE II. THREE-POINT INTERPOLATION COEFFICIENTS

p	c_{-1} $-$	c_0 $+$	c_1 $+$	p	c_{-1} $-$	c_0 $+$	c_1 $+$
.1200	.05280	.98560	.06720	.1250	.05469	.98438	.07031
.1201	.05284	.98558	.06726	.1251	.05472	.98434	.07038
.1202	.05288	.98556	.06732	.1252	.05476	.98432	.07044
.1203	.05291	.98552	.06739	.1253	.05480	.98430	.07050
.1204	.05295	.98550	.06745	.1254	.05484	.98428	.07056
.1205	.05299	.98548	.06751	.1255	.05487	.98424	.07063
.1206	.05303	.98546	.06757	.1256	.05491	.98422	.07069
.1207	.05307	.98544	.06763	.1257	.05495	.98420	.07075
.1208	.05310	.98540	.06770	.1258	.05499	.98418	.07081
.1209	.05314	.98538	.06776	.1259	.05502	.98414	.07088
.1210	.05318	.98536	.06782	.1260	.05506	.98412	.07094
.1211	.05322	.98534	.06788	.1261	.05510	.98410	.07100
.1212	.05326	.98532	.06794	.1262	.05514	.98408	.07106
.1213	.05329	.98528	.06801	.1263	.05517	.98404	.07113
.1214	.05333	.98526	.06807	.1264	.05521	.98402	.07119
.1215	.05337	.98524	.06813	.1265	.05525	.98400	.07125
.1216	.05341	.98522	.06819	.1266	.05529	.98398	.07131
.1217	.05344	.98518	.06826	.1267	.05532	.98394	.07138
.1218	.05348	.98516	.06832	.1268	.05536	.98392	.07144
.1219	.05352	.98514	.06838	.1269	.05540	.98390	.07150
.1220	.05356	.98512	.06844	.1270	.05544	.98388	.07156
.1221	.05360	.98510	.06850	.1271	.05547	.98384	.07163
.1222	.05363	.98506	.06857	.1272	.05551	.98382	.07169
.1223	.05367	.98504	.06863	.1273	.05555	.98380	.07175
.1224	.05371	.98502	.06869	.1274	.05558	.98376	.07182
.1225	.05375	.98500	.06875	.1275	.05562	.98374	.07188
.1226	.05378	.98496	.06882	.1276	.05566	.98372	.07194
.1227	.05382	.98494	.06888	.1277	.05570	.98370	.07200
.1228	.05386	.98492	.06894	.1278	.05573	.98366	.07207
.1229	.05390	.98490	.06900	.1279	.05577	.98364	.07213
.1230	.05394	.98488	.06906	.1280	.05581	.98362	.07219
.1231	.05397	.98484	.06913	.1281	.05585	.98360	.07225
.1232	.05401	.98482	.06919	.1282	.05588	.98356	.07232
.1233	.05405	.98480	.06925	.1283	.05592	.98354	.07238
.1234	.05409	.98478	.06931	.1284	.05596	.98352	.07244
.1235	.05412	.98474	.06938	.1285	.05599	.98348	.07251
.1236	.05416	.98472	.06944	.1286	.05603	.98346	.07257
.1237	.05420	.98470	.06950	.1287	.05607	.98344	.07263
.1238	.05424	.98468	.06956	.1288	.05611	.98342	.07269
.1239	.05427	.98464	.06963	.1289	.05614	.98338	.07276
.1240	.05431	.98462	.06969	.1290	.05618	.98336	.07282
.1241	.05435	.98460	.06975	.1291	.05622	.98334	.07288
.1242	.05439	.98458	.06981	.1292	.05625	.98330	.07295
.1243	.05442	.98454	.06988	.1293	.05629	.98328	.07301
.1244	.05446	.98452	.06994	.1294	.05633	.98326	.07307
.1245	.05450	.98450	.07000	.1295	.05636	.98322	.07314
.1246	.05454	.98448	.07006	.1296	.05640	.98320	.07320
.1247	.05457	.98444	.07013	.1297	.05644	.98318	.07326
.1248	.05461	.98442	.07019	.1298	.05648	.98316	.07332
.1249	.05465	.98440	.07025	.1299	.05651	.98312	.07339
.1250	.05469	.98438	.07031	.1300	.05655	.98310	.07345

TABLE II. THREE-POINT INTERPOLATION COEFFICIENTS

p	c_{-1} $-$	c_0 $+$	c_1 $+$	p	c_{-1} $-$	c_0 $+$	c_1 $+$
.1300	.05655	.98310	.07345	.1350	.05839	.98178	.07661
.1301	.05659	.98308	.07351	.1351	.05842	.98174	.07668
.1302	.05662	.98304	.07358	.1352	.05846	.98172	.07674
.1303	.05666	.98302	.07364	.1353	.05850	.98170	.07680
.1304	.05670	.98300	.07370	.1354	.05853	.98166	.07687
.1305	.05673	.98296	.07377	.1355	.05857	.98164	.07693
.1306	.05677	.98294	.07383	.1356	.05861	.98162	.07699
.1307	.05681	.98292	.07389	.1357	.05864	.98158	.07706
.1308	.05685	.98290	.07395	.1358	.05868	.98156	.07712
.1309	.05688	.98286	.07402	.1359	.05872	.98154	.07718
.1310	.05692	.98284	.07408	.1360	.05875	.98150	.07725
.1311	.05696	.98282	.07414	.1361	.05879	.98148	.07731
.1312	.05699	.98278	.07421	.1362	.05882	.98144	.07738
.1313	.05703	.98276	.07427	.1363	.05886	.98142	.07744
.1314	.05707	.98274	.07433	.1364	.05890	.98140	.07750
.1315	.05710	.98270	.07440	.1365	.05893	.98136	.07757
.1316	.05714	.98268	.07446	.1366	.05897	.98134	.07763
.1317	.05718	.98266	.07452	.1367	.05901	.98132	.07769
.1318	.05721	.98262	.07459	.1368	.05904	.98128	.07776
.1319	.05725	.98260	.07465	.1369	.05908	.98126	.07782
.1320	.05729	.98258	.07471	.1370	.05912	.98124	.07788
.1321	.05732	.98254	.07478	.1371	.05915	.98120	.07795
.1322	.05736	.98252	.07484	.1372	.05919	.98118	.07801
.1323	.05740	.98250	.07490	.1373	.05922	.98114	.07808
.1324	.05744	.98248	.07496	.1374	.05926	.98112	.07814
.1325	.05747	.98244	.07503	.1375	.05930	.98110	.07820
.1326	.05751	.98242	.07509	.1376	.05933	.98106	.07827
.1327	.05755	.98240	.07515	.1377	.05937	.98104	.07833
.1328	.05758	.98236	.07522	.1378	.05941	.98102	.07839
.1329	.05762	.98234	.07528	.1379	.05944	.98098	.07846
.1330	.05766	.98232	.07534	.1380	.05948	.98096	.07852
.1331	.05769	.98228	.07541	.1381	.05951	.98092	.07859
.1332	.05773	.98226	.07547	.1382	.05955	.98090	.07865
.1333	.05777	.98224	.07553	.1383	.05959	.98088	.07871
.1334	.05780	.98220	.07560	.1384	.05962	.98084	.07878
.1335	.05784	.98218	.07566	.1385	.05966	.98082	.07884
.1336	.05788	.98216	.07572	.1386	.05970	.98080	.07890
.1337	.05791	.98212	.07579	.1387	.05973	.98076	.07897
.1338	.05795	.98210	.07585	.1388	.05977	.98074	.07903
.1339	.05799	.98208	.07591	.1389	.05980	.98070	.07910
.1340	.05802	.98204	.07598	.1390	.05984	.98068	.07916
.1341	.05806	.98202	.07604	.1391	.05988	.98066	.07922
.1342	.05810	.98200	.07610	.1392	.05991	.98062	.07929
.1343	.05813	.98196	.07617	.1393	.05995	.98060	.07935
.1344	.05817	.98194	.07623	.1394	.05998	.98056	.07942
.1345	.05820	.98190	.07630	.1395	.06002	.98054	.07948
.1346	.05824	.98188	.07636	.1396	.06006	.98052	.07954
.1347	.05828	.98186	.07642	.1397	.06009	.98048	.07961
.1348	.05831	.98182	.07649	.1398	.06013	.98046	.07967
.1349	.05835	.98180	.07655	.1399	.06016	.98042	.07974
.1350	.05839	.98178	.07661	.1400	.06020	.98040	.07980

TABLE II. THREE-POINT INTERPOLATION COEFFICIENTS

p	c_{-1} −	c_0 +	c_1 +	p	c_{-1} −	c_0 +	c_1 +
.1400	.06020	.98040	.07980	.1450	.06199	.97898	.08301
.1401	.06024	.98038	.07986	.1451	.06202	.97894	.08308
.1402	.06027	.98034	.07993	.1452	.06206	.97892	.08314
.1403	.06031	.98032	.07999	.1453	.06209	.97888	.08321
.1404	.06034	.98028	.08006	.1454	.06213	.97886	.08327
.1405	.06038	.98026	.08012	.1455	.06216	.97882	.08334
.1406	.06042	.98024	.08018	.1456	.06220	.97880	.08340
.1407	.06045	.98020	.08025	.1457	.06224	.97878	.08346
.1408	.06049	.98018	.08031	.1458	.06227	.97874	.08353
.1409	.06052	.98014	.08038	.1459	.06231	.97872	.08359
.1410	.06056	.98012	.08044	.1460	.06234	.97868	.08366
.1411	.06060	.98010	.08050	.1461	.06238	.97866	.08372
.1412	.06063	.98006	.08057	.1462	.06241	.97862	.08379
.1413	.06067	.98004	.08063	.1463	.06245	.97860	.08385
.1414	.06070	.98000	.08070	.1464	.06248	.97856	.08392
.1415	.06074	.97998	.08076	.1465	.06252	.97854	.08398
.1416	.06077	.97994	.08083	.1466	.06255	.97850	.08405
.1417	.06081	.97992	.08089	.1467	.06259	.97848	.08411
.1418	.06085	.97990	.08095	.1468	.06262	.97844	.08418
.1419	.06088	.97986	.08102	.1469	.06266	.97842	.08424
.1420	.06092	.97984	.08108	.1470	.06270	.97840	.08430
.1421	.06095	.97980	.08115	.1471	.06273	.97836	.08437
.1422	.06099	.97978	.08121	.1472	.06277	.97834	.08443
.1423	.06103	.97976	.08127	.1473	.06280	.97830	.08450
.1424	.06106	.97972	.08134	.1474	.06284	.97828	.08456
.1425	.06110	.97970	.08140	.1475	.06287	.97824	.08463
.1426	.06113	.97966	.08147	.1476	.06291	.97822	.08469
.1427	.06117	.97964	.08153	.1477	.06294	.97818	.08476
.1428	.06120	.97960	.08160	.1478	.06298	.97816	.08482
.1429	.06124	.97958	.08166	.1479	.06301	.97812	.08489
.1430	.06128	.97956	.08172	.1480	.06305	.97810	.08495
.1431	.06131	.97952	.08179	.1481	.06308	.97806	.08502
.1432	.06135	.97950	.08185	.1482	.06312	.97804	.08508
.1433	.06138	.97946	.08192	.1483	.06315	.97800	.08515
.1434	.06142	.97944	.08198	.1484	.06319	.97798	.08521
.1435	.06145	.97940	.08205	.1485	.06322	.97794	.08528
.1436	.06149	.97938	.08211	.1486	.06326	.97792	.08534
.1437	.06153	.97936	.08217	.1487	.06329	.97788	.08541
.1438	.06156	.97932	.08224	.1488	.06333	.97786	.08547
.1439	.06160	.97930	.08230	.1489	.06336	.97782	.08554
.1440	.06163	.97926	.08237	.1490	.06340	.97780	.08560
.1441	.06167	.97924	.08243	.1491	.06343	.97776	.08567
.1442	.06170	.97920	.08250	.1492	.06347	.97774	.08573
.1443	.06174	.97918	.08256	.1493	.06350	.97770	.08580
.1444	.06177	.97914	.08263	.1494	.06354	.97768	.08586
.1445	.06181	.97912	.08269	.1495	.06357	.97764	.08593
.1446	.06185	.97910	.08275	.1496	.06361	.97762	.08599
.1447	.06188	.97906	.08282	.1497	.06364	.97758	.08606
.1448	.06192	.97904	.08288	.1498	.06368	.97756	.08612
.1449	.06195	.97900	.08295	.1499	.06371	.97752	.08619
.1450	.06199	.97898	.08301	.1500	.06375	.97750	.08625

TABLE II. THREE-POINT INTERPOLATION COEFFICIENTS

p	c_{-1} $-$	c_0 $+$	c_1 $+$	p	c_{-1} $-$	c_0 $+$	c_1 $+$
.1500	.06375	.97750	.08625	.1550	.06549	.97598	.08951
.1501	.06378	.97746	.08632	.1551	.06552	.97594	.08958
.1502	.06382	.97744	.08638	.1552	.06556	.97592	.08964
.1503	.06385	.97740	.08645	.1553	.06559	.97588	.08971
.1504	.06389	.97738	.08651	.1554	.06563	.97586	.08977
.1505	.06392	.97734	.08658	.1555	.06566	.97582	.08984
.1506	.06396	.97732	.08664	.1556	.06569	.97578	.08991
.1507	.06399	.97728	.08671	.1557	.06573	.97576	.08997
.1508	.06403	.97726	.08677	.1558	.06576	.97572	.09004
.1509	.06406	.97722	.08684	.1559	.06580	.97570	.09010
.1510	.06410	.97720	.08690	.1560	.06583	.97566	.09017
.1511	.06413	.97716	.08697	.1561	.06587	.97564	.09023
.1512	.06417	.97714	.08703	.1562	.06590	.97560	.09030
.1513	.06420	.97710	.08710	.1563	.06594	.97558	.09036
.1514	.06424	.97708	.08716	.1564	.06597	.97554	.09043
.1515	.06427	.97704	.08723	.1565	.06600	.97550	.09050
.1516	.06431	.97702	.08729	.1566	.06604	.97548	.09056
.1517	.06434	.97698	.08736	.1567	.06607	.97544	.09063
.1518	.06438	.97696	.08742	.1568	.06611	.97542	.09069
.1519	.06441	.97692	.08749	.1569	.06614	.97538	.09076
.1520	.06445	.97690	.08755	.1570	.06618	.97536	.09082
.1521	.06448	.97686	.08762	.1571	.06621	.97532	.09089
.1522	.06452	.97684	.08768	.1572	.06624	.97528	.09096
.1523	.06455	.97680	.08775	.1573	.06628	.97526	.09102
.1524	.06459	.97678	.08781	.1574	.06631	.97522	.09109
.1525	.06462	.97674	.08788	.1575	.06635	.97520	.09115
.1526	.06466	.97672	.08794	.1576	.06638	.97516	.09122
.1527	.06469	.97668	.08801	.1577	.06642	.97514	.09128
.1528	.06473	.97666	.08807	.1578	.06645	.97510	.09135
.1529	.06476	.97662	.08814	.1579	.06648	.97506	.09142
.1530	.06480	.97660	.08820	.1580	.06652	.97504	.09148
.1531	.06483	.97656	.08827	.1581	.06655	.97500	.09155
.1532	.06486	.97652	.08834	.1582	.06659	.97498	.09161
.1533	.06490	.97650	.08840	.1583	.06662	.97494	.09168
.1534	.06493	.97646	.08847	.1584	.06665	.97490	.09175
.1535	.06497	.97644	.08853	.1585	.06669	.97488	.09181
.1536	.06500	.97640	.08860	.1586	.06672	.97484	.09188
.1537	.06504	.97638	.08866	.1587	.06676	.97482	.09194
.1538	.06507	.97634	.08873	.1588	.06679	.97478	.09201
.1539	.06511	.97632	.08879	.1589	.06683	.97476	.09207
.1540	.06514	.97628	.08886	.1590	.06686	.97472	.09214
.1541	.06518	.97626	.08892	.1591	.06689	.97468	.09221
.1542	.06521	.97622	.08899	.1592	.06693	.97466	.09227
.1543	.06525	.97620	.08905	.1593	.06696	.97462	.09234
.1544	.06528	.97616	.08912	.1594	.06700	.97460	.09240
.1545	.06531	.97612	.08919	.1595	.06703	.97456	.09247
.1546	.06535	.97610	.08925	.1596	.06706	.97452	.09254
.1547	.06538	.97606	.08932	.1597	.06710	.97450	.09260
.1548	.06542	.97604	.08938	.1598	.06713	.97446	.09267
.1549	.06545	.97600	.08945	.1599	.06717	.97444	.09273
.1550	.06549	.97598	.08951	.1600	.06720	.97440	.09280

TABLE II. THREE-POINT INTERPOLATION COEFFICIENTS

p	c_{-1} −	c_0 +	c_1 +	p	c_{-1} −	c_0 +	c_1 +
.1600	.06720	.97440	.09280	.1650	.06889	.97278	.09611
.1601	.06723	.97436	.09287	.1651	.06892	.97274	.09618
.1602	.06727	.97434	.09293	.1652	.06895	.97270	.09625
.1603	.06730	.97430	.09300	.1653	.06899	.97268	.09631
.1604	.06734	.97428	.09306	.1654	.06902	.97264	.09638
.1605	.06737	.97424	.09313	.1655	.06905	.97260	.09645
.1606	.06740	.97420	.09320	.1656	.06909	.97258	.09651
.1607	.06744	.97418	.09326	.1657	.06912	.97254	.09658
.1608	.06747	.97414	.09333	.1658	.06916	.97252	.09664
.1609	.06751	.97412	.09339	.1659	.06919	.97248	.09671
.1610	.06754	.97408	.09346	.1660	.06922	.97244	.09678
.1611	.06757	.97404	.09353	.1661	.06926	.97242	.09684
.1612	.06761	.97402	.09359	.1662	.06929	.97238	.09691
.1613	.06764	.97398	.09366	.1663	.06932	.97234	.09698
.1614	.06768	.97396	.09372	.1664	.06936	.97232	.09704
.1615	.06771	.97392	.09379	.1665	.06939	.97228	.09711
.1616	.06774	.97388	.09386	.1666	.06942	.97224	.09718
.1617	.06778	.97386	.09392	.1667	.06946	.97222	.09724
.1618	.06781	.97382	.09399	.1668	.06949	.97218	.09731
.1619	.06784	.97378	.09406	.1669	.06952	.97214	.09738
.1620	.06788	.97376	.09412	.1670	.06956	.97212	.09744
.1621	.06791	.97372	.09419	.1671	.06959	.97208	.09751
.1622	.06795	.97370	.09425	.1672	.06962	.97204	.09758
.1623	.06798	.97366	.09432	.1673	.06966	.97202	.09764
.1624	.06801	.97362	.09439	.1674	.06969	.97198	.09771
.1625	.06805	.97360	.09445	.1675	.06972	.97194	.09778
.1626	.06808	.97356	.09452	.1676	.06976	.97192	.09784
.1627	.06811	.97352	.09459	.1677	.06979	.97188	.09791
.1628	.06815	.97350	.09465	.1678	.06982	.97184	.09798
.1629	.06818	.97346	.09472	.1679	.06985	.97180	.09805
.1630	.06822	.97344	.09478	.1680	.06989	.97178	.09811
.1631	.06825	.97340	.09485	.1681	.06992	.97174	.09818
.1632	.06828	.97336	.09492	.1682	.06995	.97170	.09825
.1633	.06832	.97334	.09498	.1683	.06999	.97168	.09831
.1634	.06835	.97330	.09505	.1684	.07002	.97164	.09838
.1635	.06838	.97326	.09512	.1685	.07005	.97160	.09845
.1636	.06842	.97324	.09518	.1686	.07009	.97158	.09851
.1637	.06845	.97320	.09525	.1687	.07012	.97154	.09858
.1638	.06848	.97316	.09532	.1688	.07015	.97150	.09865
.1639	.06852	.97314	.09538	.1689	.07019	.97148	.09871
.1640	.06855	.97310	.09545	.1690	.07022	.97144	.09878
.1641	.06859	.97308	.09551	.1691	.07025	.97140	.09885
.1642	.06862	.97304	.09558	.1692	.07029	.97138	.09891
.1643	.06865	.97300	.09565	.1693	.07032	.97134	.09898
.1644	.06869	.97298	.09571	.1694	.07035	.97130	.09905
.1645	.06872	.97294	.09578	.1695	.07038	.97126	.09912
.1646	.06875	.97290	.09585	.1696	.07042	.97124	.09918
.1647	.06879	.97288	.09591	.1697	.07045	.97120	.09925
.1648	.06882	.97284	.09598	.1698	.07048	.97116	.09932
.1649	.06885	.97280	.09605	.1699	.07052	.97114	.09938
.1650	.06889	.97278	.09611	.1700	.07055	.97110	.09945

TABLE II. THREE-POINT INTERPOLATION COEFFICIENTS

p	c_{-1} $-$	c_0 $+$	c_1 $+$	p	c_{-1} $-$	c_0 $+$	c_1 $+$
.1700	.07055	.97110	.09945	.1750	.07219	.96938	.10281
.1701	.07058	.97106	.09952	.1751	.07222	.96934	.10288
.1702	.07062	.97104	.09958	.1752	.07225	.96930	.10295
.1703	.07065	.97100	.09965	.1753	.07228	.96926	.10302
.1704	.07068	.97096	.09972	.1754	.07232	.96924	.10308
.1705	.07071	.97092	.09979	.1755	.07235	.96920	.10315
.1706	.07075	.97090	.09985	.1756	.07238	.96916	.10322
.1707	.07078	.97086	.09992	.1757	.07241	.96912	.10329
.1708	.07081	.97082	.09999	.1758	.07245	.96910	.10335
.1709	.07085	.97080	.10005	.1759	.07248	.96906	.10342
.1710	.07088	.97076	.10012	.1760	.07251	.96902	.10349
.1711	.07091	.97072	.10019	.1761	.07254	.96898	.10356
.1712	.07095	.97070	.10025	.1762	.07258	.96896	.10362
.1713	.07098	.97066	.10032	.1763	.07261	.96892	.10369
.1714	.07101	.97062	.10039	.1764	.07264	.96888	.10376
.1715	.07104	.97058	.10046	.1765	.07267	.96884	.10383
.1716	.07108	.97056	.10052	.1766	.07271	.96882	.10389
.1717	.07111	.97052	.10059	.1767	.07274	.96878	.10396
.1718	.07114	.97048	.10066	.1768	.07277	.96874	.10403
.1719	.07118	.97046	.10072	.1769	.07280	.96870	.10410
.1720	.07121	.97042	.10079	.1770	.07284	.96868	.10416
.1721	.07124	.97038	.10086	.1771	.07287	.96864	.10423
.1722	.07127	.97034	.10093	.1772	.07290	.96860	.10430
.1723	.07131	.97032	.10099	.1773	.07293	.96856	.10437
.1724	.07134	.97028	.10106	.1774	.07296	.96852	.10444
.1725	.07137	.97024	.10113	.1775	.07300	.96850	.10450
.1726	.07140	.97020	.10120	.1776	.07303	.96846	.10457
.1727	.07144	.97018	.10126	.1777	.07306	.96842	.10464
.1728	.07147	.97014	.10133	.1778	.07309	.96838	.10471
.1729	.07150	.97010	.10140	.1779	.07313	.96836	.10477
.1730	.07154	.97008	.10146	.1780	.07316	.96832	.10484
.1731	.07157	.97004	.10153	.1781	.07319	.96828	.10491
.1732	.07160	.97000	.10160	.1782	.07322	.96824	.10498
.1733	.07163	.96996	.10167	.1783	.07325	.96820	.10505
.1734	.07167	.96994	.10173	.1784	.07329	.96818	.10511
.1735	.07170	.96990	.10180	.1785	.07332	.96814	.10518
.1736	.07173	.96986	.10187	.1786	.07335	.96810	.10525
.1737	.07176	.96982	.10194	.1787	.07338	.96806	.10532
.1738	.07180	.96980	.10200	.1788	.07342	.96804	.10538
.1739	.07183	.96976	.10207	.1789	.07345	.96800	.10545
.1740	.07186	.96972	.10214	.1790	.07348	.96796	.10552
.1741	.07189	.96968	.10221	.1791	.07351	.96792	.10559
.1742	.07193	.96966	.10227	.1792	.07354	.96788	.10566
.1743	.07196	.96962	.10234	.1793	.07358	.96786	.10572
.1744	.07199	.96958	.10241	.1794	.07361	.96782	.10579
.1745	.07202	.96954	.10248	.1795	.07364	.96778	.10586
.1746	.07206	.96952	.10254	.1796	.07367	.96774	.10593
.1747	.07209	.96948	.10261	.1797	.07370	.96770	.10600
.1748	.07212	.96944	.10268	.1798	.07374	.96768	.10606
.1749	.07215	.96940	.10275	.1799	.07377	.96764	.10613
.1750	.07219	.96938	.10281	.1800	.07380	.96760	.10620

TABLE II. THREE-POINT INTERPOLATION COEFFICIENTS

p	c_{-1} $-$	c_0 $+$	c_1 $+$	p	c_{-1} $-$	c_0 $+$	c_1 $+$
.1800	.07380	.96760	.10620	.1850	.07539	.96578	.10961
.1801	.07383	.96756	.10627	.1851	.07542	.96574	.10968
.1802	.07386	.96752	.10634	.1852	.07545	.96570	.10975
.1803	.07390	.96750	.10640	.1853	.07548	.96566	.10982
.1804	.07393	.96746	.10647	.1854	.07551	.96562	.10989
.1805	.07396	.96742	.10654	.1855	.07554	.96558	.10996
.1806	.07399	.96738	.10661	.1856	.07558	.96556	.11002
.1807	.07402	.96734	.10668	.1857	.07561	.96552	.11009
.1808	.07406	.96732	.10674	.1858	.07564	.96548	.11016
.1809	.07409	.96728	.10681	.1859	.07567	.96544	.11023
.1810	.07412	.96724	.10688	.1860	.07570	.96540	.11030
.1811	.07415	.96720	.10695	.1861	.07573	.96536	.11037
.1812	.07418	.96716	.10702	.1862	.07576	.96532	.11044
.1813	.07422	.96714	.10708	.1863	.07580	.96530	.11050
.1814	.07425	.96710	.10715	.1864	.07583	.96526	.11057
.1815	.07428	.96706	.10722	.1865	.07586	.96522	.11064
.1816	.07431	.96702	.10729	.1866	.07589	.96518	.11071
.1817	.07434	.96698	.10736	.1867	.07592	.96514	.11078
.1818	.07437	.96694	.10743	.1868	.07595	.96510	.11085
.1819	.07441	.96692	.10749	.1869	.07598	.96506	.11092
.1820	.07444	.96688	.10756	.1870	.07602	.96504	.11098
.1821	.07447	.96684	.10763	.1871	.07605	.96500	.11105
.1822	.07450	.96680	.10770	.1872	.07608	.96496	.11112
.1823	.07453	.96676	.10777	.1873	.07611	.96492	.11119
.1824	.07457	.96674	.10783	.1874	.07614	.96488	.11126
.1825	.07460	.96670	.10790	.1875	.07617	.96484	.11133
.1826	.07463	.96666	.10797	.1876	.07620	.96480	.11140
.1827	.07466	.96662	.10804	.1877	.07623	.96476	.11147
.1828	.07469	.96658	.10811	.1878	.07627	.96474	.11153
.1829	.07472	.96654	.10818	.1879	.07630	.96470	.11160
.1830	.07476	.96652	.10824	.1880	.07633	.96466	.11167
.1831	.07479	.96648	.10831	.1881	.07636	.96462	.11174
.1832	.07482	.96644	.10838	.1882	.07639	.96458	.11181
.1833	.07485	.96640	.10845	.1883	.07642	.96454	.11188
.1834	.07488	.96636	.10852	.1884	.07645	.96450	.11195
.1835	.07491	.96632	.10859	.1885	.07648	.96446	.11202
.1836	.07495	.96630	.10865	.1886	.07652	.96444	.11208
.1837	.07498	.96626	.10872	.1887	.07655	.96440	.11215
.1838	.07501	.96622	.10879	.1888	.07658	.96436	.11222
.1839	.07504	.96618	.10886	.1889	.07661	.96432	.11229
.1840	.07507	.96614	.10893	.1890	.07664	.96428	.11236
.1841	.07510	.96610	.10900	.1891	.07667	.96424	.11243
.1842	.07514	.96608	.10906	.1892	.07670	.96420	.11250
.1843	.07517	.96604	.10913	.1893	.07673	.96416	.11257
.1844	.07520	.96600	.10920	.1894	.07676	.96412	.11264
.1845	.07523	.96596	.10927	.1895	.07679	.96408	.11271
.1846	.07526	.96592	.10934	.1896	.07683	.96406	.11277
.1847	.07529	.96588	.10941	.1897	.07686	.96402	.11284
.1848	.07532	.96584	.10948	.1898	.07689	.96398	.11291
.1849	.07536	.96582	.10954	.1899	.07692	.96394	.11298
.1850	.07539	.96578	.10961	.1900	.07695	.96390	.11305

TABLE II. THREE-POINT INTERPOLATION COEFFICIENTS

p	c_{-1} $-$	c_0 $+$	c_1 $+$	p	c_{-1} $-$	c_0 $+$	c_1 $+$
.1900	.07695	.96390	.11305	.1950	.07849	.96198	.11651
.1901	.07698	.96386	.11312	.1951	.07852	.96194	.11658
.1902	.07701	.96382	.11319	.1952	.07855	.96190	.11665
.1903	.07704	.96378	.11326	.1953	.07858	.96186	.11672
.1904	.07707	.96374	.11333	.1954	.07861	.96182	.11679
.1905	.07710	.96370	.11340	.1955	.07864	.96178	.11686
.1906	.07714	.96368	.11346	.1956	.07867	.96174	.11693
.1907	.07717	.96364	.11353	.1957	.07870	.96170	.11700
.1908	.07720	.96360	.11360	.1958	.07873	.96166	.11707
.1909	.07723	.96356	.11367	.1959	.07876	.96162	.11714
.1910	.07726	.96352	.11374	.1960	.07879	.96158	.11721
.1911	.07729	.96348	.11381	.1961	.07882	.96154	.11728
.1912	.07732	.96344	.11388	.1962	.07885	.96150	.11735
.1913	.07735	.96340	.11395	.1963	.07888	.96146	.11742
.1914	.07738	.96336	.11402	.1964	.07891	.96142	.11749
.1915	.07741	.96332	.11409	.1965	.07894	.96138	.11756
.1916	.07744	.96328	.11416	.1966	.07897	.96134	.11763
.1917	.07748	.96326	.11422	.1967	.07900	.96130	.11770
.1918	.07751	.96322	.11429	.1968	.07903	.96126	.11777
.1919	.07754	.96318	.11436	.1969	.07907	.96124	.11783
.1920	.07757	.96314	.11443	.1970	.07910	.96120	.11790
.1921	.07760	.96310	.11450	.1971	.07913	.96116	.11797
.1922	.07763	.96306	.11457	.1972	.07916	.96112	.11804
.1923	.07766	.96302	.11464	.1973	.07919	.96108	.11811
.1924	.07769	.96298	.11471	.1974	.07922	.96104	.11818
.1925	.07772	.96294	.11478	.1975	.07925	.96100	.11825
.1926	.07775	.96290	.11485	.1976	.07928	.96096	.11832
.1927	.07778	.96286	.11492	.1977	.07931	.96092	.11839
.1928	.07781	.96282	.11499	.1978	.07934	.96088	.11846
.1929	.07784	.96278	.11506	.1979	.07937	.96084	.11853
.1930	.07788	.96276	.11512	.1980	.07940	.96080	.11860
.1931	.07791	.96272	.11519	.1981	.07943	.96076	.11867
.1932	.07794	.96268	.11526	.1982	.07946	.96072	.11874
.1933	.07797	.96264	.11533	.1983	.07949	.96068	.11881
.1934	.07800	.96260	.11540	.1984	.07952	.96064	.11888
.1935	.07803	.96256	.11547	.1985	.07955	.96060	.11895
.1936	.07806	.96252	.11554	.1986	.07958	.96056	.11902
.1937	.07809	.96248	.11561	.1987	.07961	.96052	.11909
.1938	.07812	.96244	.11568	.1988	.07964	.96048	.11916
.1939	.07815	.96240	.11575	.1989	.07967	.96044	.11923
.1940	.07818	.96236	.11582	.1990	.07970	.96040	.11930
.1941	.07821	.96232	.11589	.1991	.07973	.96036	.11937
.1942	.07824	.96228	.11596	.1992	.07976	.96032	.11944
.1943	.07827	.96224	.11603	.1993	.07979	.96028	.11951
.1944	.07830	.96220	.11610	.1994	.07982	.96024	.11958
.1945	.07833	.96216	.11617	.1995	.07985	.96020	.11965
.1946	.07837	.96214	.11623	.1996	.07988	.96016	.11972
.1947	.07840	.96210	.11630	.1997	.07991	.96012	.11979
.1948	.07843	.96206	.11637	.1998	.07994	.96008	.11986
.1949	.07846	.96202	.11644	.1999	.07997	.96004	.11993
.1950	.07849	.96198	.11651	.2000	.08000	.96000	.12000

TABLE II. THREE-POINT INTERPOLATION COEFFICIENTS

p	c_{-1} −	c_0 +	c_1 +	p	c_{-1} −	c_0 +	c_1 +
.2000	.08000	.96000	.12000	.2050	.08149	.95798	.12351
.2001	.08003	.95996	.12007	.2051	.08152	.95794	.12358
.2002	.08006	.95992	.12014	.2052	.08155	.95790	.12365
.2003	.08009	.95988	.12021	.2053	.08158	.95786	.12372
.2004	.08012	.95984	.12028	.2054	.08161	.95782	.12379
.2005	.08015	.95980	.12035	.2055	.08163	.95776	.12387
.2006	.08018	.95976	.12042	.2056	.08166	.95772	.12394
.2007	.08021	.95972	.12049	.2057	.08169	.95768	.12401
.2008	.08024	.95968	.12056	.2058	.08172	.95764	.12408
.2009	.08027	.95964	.12063	.2059	.08175	.95760	.12415
.2010	.08030	.95960	.12070	.2060	.08178	.95756	.12422
.2011	.08033	.95956	.12077	.2061	.08181	.95752	.12429
.2012	.08036	.95952	.12084	.2062	.08184	.95748	.12436
.2013	.08039	.95948	.12091	.2063	.08187	.95744	.12443
.2014	.08042	.95944	.12098	.2064	.08190	.95740	.12450
.2015	.08045	.95940	.12105	.2065	.08193	.95736	.12457
.2016	.08048	.95936	.12112	.2066	.08196	.95732	.12464
.2017	.08051	.95932	.12119	.2067	.08199	.95728	.12471
.2018	.08054	.95928	.12126	.2068	.08202	.95724	.12478
.2019	.08057	.95924	.12133	.2069	.08205	.95720	.12485
.2020	.08060	.95920	.12140	.2070	.08208	.95716	.12492
.2021	.08063	.95916	.12147	.2071	.08210	.95710	.12500
.2022	.08066	.95912	.12154	.2072	.08213	.95706	.12507
.2023	.08069	.95908	.12161	.2073	.08216	.95702	.12514
.2024	.08072	.95904	.12168	.2074	.08219	.95698	.12521
.2025	.08075	.95900	.12175	.2075	.08222	.95694	.12528
.2026	.08078	.95896	.12182	.2076	.08225	.95690	.12535
.2027	.08081	.95892	.12189	.2077	.08228	.95686	.12542
.2028	.08084	.95888	.12196	.2078	.08231	.95682	.12549
.2029	.08087	.95884	.12203	.2079	.08234	.95678	.12556
.2030	.08090	.95880	.12210	.2080	.08237	.95674	.12563
.2031	.08093	.95876	.12217	.2081	.08240	.95670	.12570
.2032	.08095	.95870	.12225	.2082	.08243	.95666	.12577
.2033	.08098	.95866	.12232	.2083	.08246	.95662	.12584
.2034	.08101	.95862	.12239	.2084	.08248	.95656	.12592
.2035	.08104	.95858	.12246	.2085	.08251	.95652	.12599
.2036	.08107	.95854	.12253	.2086	.08254	.95648	.12606
.2037	.08110	.95850	.12260	.2087	.08257	.95644	.12613
.2038	.08113	.95846	.12267	.2088	.08260	.95640	.12620
.2039	.08116	.95842	.12274	.2089	.08263	.95636	.12627
.2040	.08119	.95838	.12281	.2090	.08266	.95632	.12634
.2041	.08122	.95834	.12288	.2091	.08269	.95628	.12641
.2042	.08125	.95830	.12295	.2092	.08272	.95624	.12648
.2043	.08128	.95826	.12302	.2093	.08275	.95620	.12655
.2044	.08131	.95822	.12309	.2094	.08278	.95616	.12662
.2045	.08134	.95818	.12316	.2095	.08280	.95610	.12670
.2046	.08137	.95814	.12323	.2096	.08283	.95606	.12677
.2047	.08140	.95810	.12330	.2097	.08286	.95602	.12684
.2048	.08143	.95806	.12337	.2098	.08289	.95598	.12691
.2049	.08146	.95802	.12344	.2099	.08292	.95594	.12698
.2050	.08149	.95798	.12351	.2100	.08295	.95590	.12705

TABLE II. THREE-POINT INTERPOLATION COEFFICIENTS

p	c_{-1} $-$	c_0 $+$	c_1 $+$	p	c_{-1} $-$	c_0 $+$	c_1 $+$
.2100	.08295	.95590	.12705	.2150	.08439	.95378	.13061
.2101	.08298	.95586	.12712	.2151	.08442	.95374	.13068
.2102	.08301	.95582	.12719	.2152	.08444	.95368	.13076
.2103	.08304	.95578	.12726	.2153	.08447	.95364	.13083
.2104	.08307	.95574	.12733	.2154	.08450	.95360	.13090
.2105	.08309	.95568	.12741	.2155	.08453	.95356	.13097
.2106	.08312	.95564	.12748	.2156	.08456	.95352	.13104
.2107	.08315	.95560	.12755	.2157	.08459	.95348	.13111
.2108	.08318	.95556	.12762	.2158	.08462	.95344	.13118
.2109	.08321	.95552	.12769	.2159	.08464	.95338	.13126
.2110	.08324	.95548	.12776	.2160	.08467	.95334	.13133
.2111	.08327	.95544	.12783	.2161	.08470	.95330	.13140
.2112	.08330	.95540	.12790	.2162	.08473	.95326	.13147
.2113	.08333	.95536	.12797	.2163	.08476	.95322	.13154
.2114	.08336	.95532	.12804	.2164	.08479	.95318	.13161
.2115	.08338	.95526	.12812	.2165	.08481	.95312	.13169
.2116	.08341	.95522	.12819	.2166	.08484	.95308	.13176
.2117	.08344	.95518	.12826	.2167	.08487	.95304	.13183
.2118	.08347	.95514	.12833	.2168	.08490	.95300	.13190
.2119	.08350	.95510	.12840	.2169	.08493	.95296	.13197
.2120	.08353	.95506	.12847	.2170	.08496	.95292	.13204
.2121	.08356	.95502	.12854	.2171	.08498	.95286	.13212
.2122	.08359	.95498	.12861	.2172	.08501	.95282	.13219
.2123	.08361	.95492	.12869	.2173	.08504	.95278	.13226
.2124	.08364	.95488	.12876	.2174	.08507	.95274	.13233
.2125	.08367	.95484	.12883	.2175	.08510	.95270	.13240
.2126	.08370	.95480	.12890	.2176	.08513	.95266	.13247
.2127	.08373	.95476	.12897	.2177	.08515	.95260	.13255
.2128	.08376	.95472	.12904	.2178	.08518	.95256	.13262
.2129	.08379	.95468	.12911	.2179	.08521	.95252	.13269
.2130	.08382	.95464	.12918	.2180	.08524	.95248	.13276
.2131	.08384	.95458	.12926	.2181	.08527	.95244	.13283
.2132	.08387	.95454	.12933	.2182	.08529	.95238	.13291
.2133	.08390	.95450	.12940	.2183	.08532	.95234	.13298
.2134	.08393	.95446	.12947	.2184	.08535	.95230	.13305
.2135	.08396	.95442	.12954	.2185	.08538	.95226	.13312
.2136	.08399	.95438	.12961	.2186	.08541	.95222	.13319
.2137	.08402	.95434	.12968	.2187	.08544	.95218	.13326
.2138	.08404	.95428	.12976	.2188	.08546	.95212	.13334
.2139	.08407	.95424	.12983	.2189	.08549	.95208	.13341
.2140	.08410	.95420	.12990	.2190	.08552	.95204	.13348
.2141	.08413	.95416	.12997	.2191	.08555	.95200	.13355
.2142	.08416	.95412	.13004	.2192	.08558	.95196	.13362
.2143	.08419	.95408	.13011	.2193	.08560	.95190	.13370
.2144	.08422	.95404	.13018	.2194	.08563	.95186	.13377
.2145	.08424	.95398	.13026	.2195	.08566	.95182	.13384
.2146	.08427	.95394	.13033	.2196	.08569	.95178	.13391
.2147	.08430	.95390	.13040	.2197	.08572	.95174	.13398
.2148	.08433	.95386	.13047	.2198	.08574	.95168	.13406
.2149	.08436	.95382	.13054	.2199	.08577	.95164	.13413
.2150	.08439	.95378	.13061	.2200	.08580	.95160	.13420

TABLE II. THREE-POINT INTERPOLATION COEFFICIENTS

p	c_{-1} $-$	c_0 $+$	c_1 $+$	p	c_{-1} $-$	c_0 $+$	c_1 $+$
.2200	.08580	.95160	.13420	.2250	.08719	.94938	.13781
.2201	.08583	.95156	.13427	.2251	.08721	.94932	.13789
.2202	.08586	.95152	.13434	.2252	.08724	.94928	.13796
.2203	.08588	.95146	.13442	.2253	.08727	.94924	.13803
.2204	.08591	.95142	.13449	.2254	.08730	.94920	.13810
.2205	.08594	.95138	.13456	.2255	.08732	.94914	.13818
.2206	.08597	.95134	.13463	.2256	.08735	.94910	.13825
.2207	.08600	.95130	.13470	.2257	.08738	.94906	.13832
.2208	.08602	.95124	.13478	.2258	.08741	.94902	.13839
.2209	.08605	.95120	.13485	.2259	.08743	.94896	.13847
.2210	.08608	.95116	.13492	.2260	.08746	.94892	.13854
.2211	.08611	.95112	.13499	.2261	.08749	.94888	.13861
.2212	.08614	.95108	.13506	.2262	.08752	.94884	.13868
.2213	.08616	.95102	.13514	.2263	.08754	.94878	.13876
.2214	.08619	.95098	.13521	.2264	.08757	.94874	.13883
.2215	.08622	.95094	.13528	.2265	.08760	.94870	.13890
.2216	.08625	.95090	.13535	.2266	.08763	.94866	.13897
.2217	.08627	.95084	.13543	.2267	.08765	.94860	.13905
.2218	.08630	.95080	.13550	.2268	.08768	.94856	.13912
.2219	.08633	.95076	.13557	.2269	.08771	.94852	.13919
.2220	.08636	.95072	.13564	.2270	.08774	.94848	.13926
.2221	.08639	.95068	.13571	.2271	.08776	.94842	.13934
.2222	.08641	.95062	.13579	.2272	.08779	.94838	.13941
.2223	.08644	.95058	.13586	.2273	.08782	.94834	.13948
.2224	.08647	.95054	.13593	.2274	.08784	.94828	.13956
.2225	.08650	.95050	.13600	.2275	.08787	.94824	.13963
.2226	.08652	.95044	.13608	.2276	.08790	.94820	.13970
.2227	.08655	.95040	.13615	.2277	.08793	.94816	.13977
.2228	.08658	.95036	.13622	.2278	.08795	.94810	.13985
.2229	.08661	.95032	.13629	.2279	.08798	.94806	.13992
.2230	.08664	.95028	.13636	.2280	.08801	.94802	.13999
.2231	.08666	.95022	.13644	.2281	.08804	.94798	.14006
.2232	.08669	.95018	.13651	.2282	.08806	.94792	.14014
.2233	.08672	.95014	.13658	.2283	.08809	.94788	.14021
.2234	.08675	.95010	.13665	.2284	.08812	.94784	.14028
.2235	.08677	.95004	.13673	.2285	.08814	.94778	.14036
.2236	.08680	.95000	.13680	.2286	.08817	.94774	.14043
.2237	.08683	.94996	.13687	.2287	.08820	.94770	.14050
.2238	.08686	.94992	.13694	.2288	.08823	.94766	.14057
.2239	.08688	.94986	.13702	.2289	.08825	.94760	.14065
.2240	.08691	.94982	.13709	.2290	.08828	.94756	.14072
.2241	.08694	.94978	.13716	.2291	.08831	.94752	.14079
.2242	.08697	.94974	.13723	.2292	.08833	.94746	.14087
.2243	.08699	.94968	.13731	.2293	.08836	.94742	.14094
.2244	.08702	.94964	.13738	.2294	.08839	.94738	.14101
.2245	.08705	.94960	.13745	.2295	.08841	.94732	.14109
.2246	.08708	.94956	.13752	.2296	.08844	.94728	.14116
.2247	.08710	.94950	.13760	.2297	.08847	.94724	.14123
.2248	.08713	.94946	.13767	.2298	.08850	.94720	.14130
.2249	.08716	.94942	.13774	.2299	.08852	.94714	.14138
.2250	.08719	.94938	.13781	.2300	.08855	.94710	.14145

TABLE II. THREE-POINT INTERPOLATION COEFFICIENTS

p	c_{-1} −	c_0 +	c_1 +	p	c_{-1} −	c_0 +	c_1 +
.2300	.08855	.94710	.14145	.2350	.08989	.94478	.14511
.2301	.08858	.94706	.14152	.2351	.08991	.94472	.14519
.2302	.08860	.94700	.14160	.2352	.08994	.94468	.14526
.2303	.08863	.94696	.14167	.2353	.08997	.94464	.14533
.2304	.08866	.94692	.14174	.2354	.08999	.94458	.14541
.2305	.08868	.94686	.14182	.2355	.09002	.94454	.14548
.2306	.08871	.94682	.14189	.2356	.09005	.94450	.14555
.2307	.08874	.94678	.14196	.2357	.09007	.94444	.14563
.2308	.08877	.94674	.14203	.2358	.09010	.94440	.14570
.2309	.08879	.94668	.14211	.2359	.09013	.94436	.14577
.2310	.08882	.94664	.14218	.2360	.09015	.94430	.14585
.2311	.08885	.94660	.14225	.2361	.09018	.94426	.14592
.2312	.08887	.94654	.14233	.2362	.09020	.94420	.14600
.2313	.08890	.94650	.14240	.2363	.09023	.94416	.14607
.2314	.08893	.94646	.14247	.2364	.09026	.94412	.14614
.2315	.08895	.94640	.14255	.2365	.09028	.94406	.14622
.2316	.08898	.94636	.14262	.2366	.09031	.94402	.14629
.2317	.08901	.94632	.14269	.2367	.09034	.94398	.14636
.2318	.08903	.94626	.14277	.2368	.09036	.94392	.14644
.2319	.08906	.94622	.14284	.2369	.09039	.94388	.14651
.2320	.08909	.94618	.14291	.2370	.09042	.94384	.14658
.2321	.08911	.94612	.14299	.2371	.09044	.94378	.14666
.2322	.08914	.94608	.14306	.2372	.09047	.94374	.14673
.2323	.08917	.94604	.14313	.2373	.09049	.94368	.14681
.2324	.08920	.94600	.14320	.2374	.09052	.94364	.14688
.2325	.08922	.94594	.14328	.2375	.09055	.94360	.14695
.2326	.08925	.94590	.14335	.2376	.09057	.94354	.14703
.2327	.08928	.94586	.14342	.2377	.09060	.94350	.14710
.2328	.08930	.94580	.14350	.2378	.09063	.94346	.14717
.2329	.08933	.94576	.14357	.2379	.09065	.94340	.14725
.2330	.08936	.94572	.14364	.2380	.09068	.94336	.14732
.2331	.08938	.94566	.14372	.2381	.09070	.94330	.14740
.2332	.08941	.94562	.14379	.2382	.09073	.94326	.14747
.2333	.08944	.94558	.14386	.2383	.09076	.94322	.14754
.2334	.08946	.94552	.14394	.2384	.09078	.94316	.14762
.2335	.08949	.94548	.14401	.2385	.09081	.94312	.14769
.2336	.08952	.94544	.14408	.2386	.09084	.94308	.14776
.2337	.08954	.94538	.14416	.2387	.09086	.94302	.14784
.2338	.08957	.94534	.14423	.2388	.09089	.94298	.14791
.2339	.08960	.94530	.14430	.2389	.09091	.94292	.14799
.2340	.08962	.94524	.14438	.2390	.09094	.94288	.14806
.2341	.08965	.94520	.14445	.2391	.09097	.94284	.14813
.2342	.08968	.94516	.14452	.2392	.09099	.94278	.14821
.2343	.08970	.94510	.14460	.2393	.09102	.94274	.14828
.2344	.08973	.94506	.14467	.2394	.09104	.94268	.14836
.2345	.08975	.94500	.14475	.2395	.09107	.94264	.14843
.2346	.08978	.94496	.14482	.2396	.09110	.94260	.14850
.2347	.08981	.94492	.14489	.2397	.09112	.94254	.14858
.2348	.08983	.94486	.14497	.2398	.09115	.94250	.14865
.2349	.08986	.94482	.14504	.2399	.09117	.94244	.14873
.2350	.08989	.94478	.14511	.2400	.09120	.94240	.14880

TABLE II. THREE-POINT INTERPOLATION COEFFICIENTS

p	c_{-1} $-$	c_0 $+$	c_1 $+$	p	c_{-1} $-$	c_0 $+$	c_1 $+$
.2400	.09120	.94240	.14880	.2450	.09249	.93998	.15251
.2401	.09123	.94236	.14887	.2451	.09251	.93992	.15259
.2402	.09125	.94230	.14895	.2452	.09254	.93988	.15266
.2403	.09128	.94226	.14902	.2453	.09256	.93982	.15274
.2404	.09130	.94220	.14910	.2454	.09259	.93978	.15281
.2405	.09133	.94216	.14917	.2455	.09261	.93972	.15289
.2406	.09136	.94212	.14924	.2456	.09264	.93968	.15296
.2407	.09138	.94206	.14932	.2457	.09267	.93964	.15303
.2408	.09141	.94202	.14939	.2458	.09269	.93958	.15311
.2409	.09143	.94196	.14947	.2459	.09272	.93954	.15318
.2410	.09146	.94192	.14954	.2460	.09274	.93948	.15326
.2411	.09149	.94188	.14961	.2461	.09277	.93944	.15333
.2412	.09151	.94182	.14969	.2462	.09279	.93938	.15341
.2413	.09154	.94178	.14976	.2463	.09282	.93934	.15348
.2414	.09156	.94172	.14984	.2464	.09284	.93928	.15356
.2415	.09159	.94168	.14991	.2465	.09287	.93924	.15363
.2416	.09161	.94162	.14999	.2466	.09289	.93918	.15371
.2417	.09164	.94158	.15006	.2467	.09292	.93914	.15378
.2418	.09167	.94154	.15013	.2468	.09294	.93908	.15386
.2419	.09169	.94148	.15021	.2469	.09297	.93904	.15393
.2420	.09172	.94144	.15028	.2470	.09300	.93900	.15400
.2421	.09174	.94138	.15036	.2471	.09302	.93894	.15408
.2422	.09177	.94134	.15043	.2472	.09305	.93890	.15415
.2423	.09180	.94130	.15050	.2473	.09307	.93884	.15423
.2424	.09182	.94124	.15058	.2474	.09310	.93880	.15430
.2425	.09185	.94120	.15065	.2475	.09312	.93874	.15438
.2426	.09187	.94114	.15073	.2476	.09315	.93870	.15445
.2427	.09190	.94110	.15080	.2477	.09317	.93864	.15453
.2428	.09192	.94104	.15088	.2478	.09320	.93860	.15460
.2429	.09195	.94100	.15095	.2479	.09322	.93854	.15468
.2430	.09198	.94096	.15102	.2480	.09325	.93850	.15475
.2431	.09200	.94090	.15110	.2481	.09327	.93844	.15483
.2432	.09203	.94086	.15117	.2482	.09330	.93840	.15490
.2433	.09205	.94080	.15125	.2483	.09332	.93834	.15498
.2434	.09208	.94076	.15132	.2484	.09335	.93830	.15505
.2435	.09210	.94070	.15140	.2485	.09337	.93824	.15513
.2436	.09213	.94066	.15147	.2486	.09340	.93820	.15520
.2437	.09216	.94062	.15154	.2487	.09342	.93814	.15528
.2438	.09218	.94056	.15162	.2488	.09345	.93810	.15535
.2439	.09221	.94052	.15169	.2489	.09347	.93804	.15543
.2440	.09223	.94046	.15177	.2490	.09350	.93800	.15550
.2441	.09226	.94042	.15184	.2491	.09352	.93794	.15558
.2442	.09228	.94036	.15192	.2492	.09355	.93790	.15565
.2443	.09231	.94032	.15199	.2493	.09357	.93784	.15573
.2444	.09233	.94026	.15207	.2494	.09360	.93780	.15580
.2445	.09236	.94022	.15214	.2495	.09362	.93774	.15588
.2446	.09239	.94018	.15221	.2496	.09365	.93770	.15595
.2447	.09241	.94012	.15229	.2497	.09367	.93764	.15603
.2448	.09244	.94008	.15236	.2498	.09370	.93760	.15610
.2449	.09246	.94002	.15244	.2499	.09372	.93754	.15618
.2450	.09249	.93998	.15251	.2500	.09375	.93750	.15625

TABLE II. THREE-POINT INTERPOLATION COEFFICIENTS

p	c_{-1} −	c_0 +	c_1 +	p	c_{-1} −	c_0 +	c_1 +
.2500	.09375	.93750	.15625	.2550	.09499	.93498	.16001
.2501	.09377	.93744	.15633	.2551	.09501	.93492	.16009
.2502	.09380	.93740	.15640	.2552	.09504	.93488	.16016
.2503	.09382	.93734	.15648	.2553	.09506	.93482	.16024
.2504	.09385	.93730	.15655	.2554	.09509	.93478	.16031
.2505	.09387	.93724	.15663	.2555	.09511	.93472	.16039
.2506	:09390	.93720	.15670	.2556	.09513	.93466	.16047
.2507	.09392	.93714	.15678	.2557	.09516	.93462	.16054
.2508	.09395	.93710	.15685	.2558	.09518	.93456	.16062
.2509	.09397	.93704	.15693	.2559	.09521	.93452	.16069
.2510	.09400	.93700	.15700	.2560	.09523	.93446	.16077
.2511	.09402	.93694	.15708	.2561	.09526	.93442	.16084
.2512	.09405	.93690	.15715	.2562	.09528	.93436	.16092
.2513	.09407	.93684	.15723	.2563	.09531	.93432	.16099
.2514	.09410	.93680	.15730	.2564	.09533	.93426	.16107
.2515	.09412	.93674	.15738	.2565	.09535	.93420	.16115
.2516	.09415	.93670	.15745	.2566	.09538	.93416	.16122
.2517	.09417	.93664	.15753	.2567	.09540	.93410	.16130
.2518	.09420	.93660	.15760	.2568	.09543	.93406	.16137
.2519	.09422	.93654	.15768	.2569	.09545	.93400	.16145
.2520	.09425	.93650	.15775	.2570	.09548	.93396	.16152
.2521	.09427	.93644	.15783	.2571	.09550	.93390	.16160
.2522	.09430	.93640	.15790	.2572	.09552	.93384	.16168
.2523	.09432	.93634	.15798	.2573	.09555	.93380	.16175
.2524	.09435	.93630	.15805	.2574	.09557	.93374	.16183
.2525	.09437	.93624	.15813	.2575	.09560	.93370	.16190
.2526	.09440	.93620	.15820	.2576	.09562	.93364	.16198
.2527	.09442	.93614	.15828	.2577	.09565	.93360	.16205
.2528	.09445	.93610	.15835	.2578	.09567	.93354	.16213
.2529	.09447	.93604	.15843	.2579	.09569	.93348	.16221
.2530	.09450	.93600	.15850	.2580	.09572	.93344	.16228
.2531	.09452	.93594	.15858	.2581	.09574	.93338	.16236
.2532	.09454	.93588	.15866	.2582	.09577	.93334	.16243
.2533	.09457	.93584	.15873	.2583	.09579	.93328	.16251
.2534	.09459	.93578	.15881	.2584	.09581	.93322	.16259
.2535	.09462	.93574	.15888	.2585	.09584	.93318	.16266
.2536	.09464	.93568	.15896	.2586	.09586	.93312	.16274
.2537	.09467	.93564	.15903	.2587	.09589	.93308	.16281
.2538	.09469	.93558	.15911	.2588	.09591	.93302	.16289
.2539	.09472	.93554	.15918	.2589	.09594	.93298	.16296
.2540	.09474	.93548	.15926	.2590	.09596	.93292	.16304
.2541	.09477	.93544	.15933	.2591	.09598	.93286	.16312
.2542	.09479	.93538	.15941	.2592	.09601	.93282	.16319
.2543	.09482	.93534	.15948	.2593	.09603	.93276	.16327
.2544	.09484	.93528	.15956	.2594	.09606	.93272	.16334
.2545	.09486	.93522	.15964	.2595	.09608	.93266	.16342
.2546	.09489	.93518	.15971	.2596	.09610	.93260	.16350
.2547	.09491	.93512	.15979	.2597	.09613	.93256	.16357
.2548	.09494	.93508	.15986	.2598	.09615	.93250	.16365
.2549	.09496	.93502	.15994	.2599	.09618	.93246	.16372
.2550	.09499	.93498	.16001	.2600	.09620	.93240	.16380

TABLE II. THREE-POINT INTERPOLATION COEFFICIENTS

p	c_{-1} $-$	c_0 $+$	c_1 $+$	p	c_{-1} $-$	c_0 $+$	c_1 $+$
.2600	.09620	.93240	.16380	.2650	.09739	.92978	.16761
.2601	.09622	.93234	.16388	.2651	.09741	.92972	.16769
.2602	.09625	.93230	.16395	.2652	.09743	.92966	.16777
.2603	.09627	.93224	.16403	.2653	.09746	.92962	.16784
.2604	.09630	.93220	.16410	.2654	.09748	.92956	.16792
.2605	.09632	.93214	.16418	.2655	.09750	.92950	.16800
.2606	.09634	.93208	.16426	.2656	.09753	.92946	.16807
.2607	.09637	.93204	.16433	.2657	.09755	.92940	.16815
.2608	.09639	.93198	.16441	.2658	.09758	.92936	.16822
.2609	.09642	.93194	.16448	.2659	.09760	.92930	.16830
.2610	.09644	.93188	.16456	.2660	.09762	.92924	.16838
.2611	.09646	.93182	.16464	.2661	.09765	.92920	.16845
.2612	.09649	.93178	.16471	.2662	.09767	.92914	.16853
.2613	.09651	.93172	.16479	.2663	.09769	.92908	.16861
.2614	.09654	.93168	.16486	.2664	.09772	.92904	.16868
.2615	.09656	.93162	.16494	.2665	.09774	.92898	.16876
.2616	.09658	.93156	.16502	.2666	.09776	.92892	.16884
.2617	.09661	.93152	.16509	.2667	.09779	.92888	.16891
.2618	.09663	.93146	.16517	.2668	.09781	.92882	.16899
.2619	.09665	.93140	.16525	.2669	.09783	.92876	.16907
.2620	.09668	.93136	.16532	.2670	.09786	.92872	.16914
.2621	.09670	.93130	.16540	.2671	.09788	.92866	.16922
.2622	.09673	.93126	.16547	.2672	.09790	.92860	.16930
.2623	.09675	.93120	.16555	.2673	.09793	.92856	.16937
.2624	.09677	.93114	.16563	.2674	.09795	.92850	.16945
.2625	.09680	.93110	.16570	.2675	.09797	.92844	.16953
.2626	.09682	.93104	.16578	.2676	.09800	.92840	.16960
.2627	.09684	.93098	.16586	.2677	.09802	.92834	.16968
.2628	.09687	.93094	.16593	.2678	.09804	.92828	.16976
.2629	.09689	.93088	.16601	.2679	.09806	.92822	.16984
.2630	.09692	.93084	.16608	.2680	.09809	.92818	.16991
.2631	.09694	.93078	.16616	.2681	.09811	.92812	.16999
.2632	.09696	.93072	.16624	.2682	.09813	.92806	.17007
.2633	.09699	.93068	.06631	.2683	.09816	.92802	.17014
.2634	.09701	.93062	.16639	.2684	.09818	.92796	.17022
.2635	.09703	.93056	.16647	.2685	.09820	.92790	.17030
.2636	.09706	.93052	.16654	.2686	.09823	.92786	.17037
.2637	.09708	.93046	.16662	.2687	.09825	.92780	.17045
.2638	.09710	.93040	.16670	.2688	.09827	.92774	.17053
.2639	.09713	.93036	.16677	.2689	.09830	.92770	.17060
.2640	.09715	.93030	.16685	.2690	.09832	.92764	.17068
.2641	.09718	.93026	.16692	.2691	.09834	.92758	.17076
.2642	.09720	.93020	.16700	.2692	.09837	.92754	.17083
.2643	.09722	.93014	.16708	.2693	.09839	.92748	.17091
.2644	.09725	.93010	.16715	.2694	.09841	.92742	.17099
.2645	.09727	.93004	.16723	.2695	.09843	.92736	.17107
.2646	.09729	.92998	.16731	.2696	.09846	.92732	.17114
.2647	.09732	.92994	.16738	.2697	.09848	.92726	.17122
.2648	.09734	.92988	.16746	.2698	.09850	.92720	.17130
.2649	.09736	.92982	.16754	.2699	.09853	.92716	.17137
.2650	.09739	.92978	.16761	.2700	.09855	.92710	.17145

TABLE II. THREE-POINT INTERPOLATION COEFFICIENTS

p	c_{-1} $-$	c_0 $+$	c_1 $+$	p	c_{-1} $-$	c_0 $+$	c_1 $+$
.2700	.09855	.92710	.17145	.2750	.09969	.92438	.17531
.2701	.09857	.92704	.17153	.2751	.09971	.92432	.17539
.2702	.09860	.92700	.17160	.2752	.09973	.92426	.17547
.2703	.09862	.92694	.17168	.2753	.09975	.92420	.17555
.2704	.09864	.92688	.17176	.2754	.09978	.92416	.17562
.2705	.09866	.92682	.17184	.2755	.09980	.92410	.17570
.2706	.09869	.92678	.17191	.2756	.09982	.92404	.17578
.2707	.09871	.92672	.17199	.2757	.09984	.92398	.17586
.2708	.09873	.92666	.17207	.2758	.09987	.92394	.17593
.2709	.09876	.92662	.17214	.2759	.09989	.92388	.17601
.2710	.09878	.92656	.17222	.2760	.09991	.92382	.17609
.2711	.09880	.92650	.17230	.2761	.09993	.92376	.17617
.2712	.09883	.92646	.17237	.2762	.09996	.92372	.17624
.2713	.09885	.92640	.17245	.2763	.09998	.92366	.17632
.2714	.09887	.92634	.17253	.2764	.10000	.92360	.17640
.2715	.09889	.92628	.17261	.2765	.10002	.92354	.17648
.2716	.09892	.92624	.17268	.2766	.10005	.92350	.17655
.2717	.09894	.92618	.17276	.2767	.10007	.92344	.17663
.2718	.09896	.92612	.17284	.2768	.10009	.92338	.17671
.2719	.09899	.92608	.17291	.2769	.10011	.92332	.17679
.2720	.09901	.92602	.17299	.2770	.10014	.92328	.17686
.2721	.09903	.92596	.17307	.2771	.10016	.92322	.17694
.2722	.09905	.92590	.17315	.2772	.10018	.92316	.17702
.2723	.09908	.92586	.17322	.2773	.10020	.92310	.17710
.2724	.09910	.92580	.17330	.2774	.10022	.92304	.17718
.2725	.09912	.92574	.17338	.2775	.10025	.92300	.17725
.2726	.09914	.92568	.17346	.2776	.10027	.92294	.17733
.2727	.09917	.92564	.17353	.2777	.10029	.92288	.17741
.2728	.09919	.92558	.17361	.2778	.10031	.92282	.17749
.2729	.09921	.92552	.17369	.2779	.10034	.92278	.17756
.2730	.09924	.92548	.17376	.2780	.10036	.92272	.17764
.2731	.09926	.92542	.17384	.2781	.10038	.92266	.17772
.2732	.09928	.92536	.17392	.2782	.10040	.92260	.17780
.2733	.09930	.92530	.17400	.2783	.10042	.92254	.17788
.2734	.09933	.92526	.17407	.2784	.10045	.92250	.17795
.2735	.09935	.92520	.17415	.2785	.10047	.92244	.17803
.2736	.09937	.92514	.17423	.2786	.10049	.92238	.17811
.2737	.09939	.92508	.17431	.2787	.10051	.92232	.17819
.2738	.09942	.92504	.17438	.2788	.10054	.92228	.17826
.2739	.09944	.92498	.17446	.2789	.10056	.92222	.17834
.2740	.09946	.92492	.17454	.2790	.10058	.92216	.17842
.2741	.09948	.92486	.17462	.2791	.10060	.92210	.17850
.2742	.09951	.92482	.17469	.2792	.10062	.92204	.17858
.2743	.09953	.92476	.17477	.2793	.10065	.92200	.17865
.2744	.09955	.92470	.17485	.2794	.10067	.92194	.17873
.2745	.09957	.92464	.17493	.2795	.10069	.92188	.17881
.2746	.09960	.92460	.17500	.2796	.10071	.92182	.17889
.2747	.09962	.92454	.17508	.2797	.10073	.92176	.17897
.2748	.09964	.92448	.17516	.2798	.10076	.92172	.17904
.2749	.09966	.92442	.17524	.2799	.10078	.92166	.17912
.2750	.09969	.92438	.17531	.2800	.10080	.92160	.17920

TABLE II. THREE-POINT INTERPOLATION COEFFICIENTS

p	c_{-1} $-$	c_0 $+$	c_1 $+$	p	c_{-1} $-$	c_0 $+$	c_1 $+$
.2800	.10080	.92160	.17920	.2850	.10189	.91878	.18311
.2801	.10082	.92154	.17928	.2851	.10191	.91872	.18319
.2802	.10084	.92148	.17936	.2852	.10193	.91866	.18327
.2803	.10087	.92144	.17943	.2853	.10195	.91860	.18335
.2804	.10089	.92138	.17951	.2854	.10197	.91854	.18343
.2805	.10091	.92132	.17959	.2855	.10199	.91848	.18351
.2806	.10093	.92126	.17967	.2856	.10202	.91844	.18358
.2807	.10095	.92120	.17975	.2857	.10204	.91838	.18366
.2808	.10098	.92116	.17982	.2858	.10206	.91832	.18374
.2809	.10100	.92110	.17990	.2859	.10208	.91826	.18382
.2810	.10102	.92104	.17998	.2860	.10210	.91820	.18390
.2811	.10104	.92098	.18006	.2861	.10212	.91814	.18398
.2812	.10106	.92092	.18014	.2862	.10214	.91808	.18406
.2813	.10109	.92088	.18021	.2863	.10217	.91804	.18413
.2814	.10111	.92082	.18029	.2864	.10219	.91798	.18421
.2815	.10113	.92076	.18037	.2865	.10221	.91792	.18429
.2816	.10115	.92070	.18045	.2866	.10223	.91786	.18437
.2817	.10117	.92064	.18053	.2867	.10225	.91780	.18445
.2818	.10119	.92058	.18061	.2868	.10227	.91774	.18453
.2819	.10122	.92054	.18068	.2869	.10229	.91768	.18461
.2820	.10124	.92048	.18076	.2870	.10232	.91764	.18468
.2821	.10126	.92042	.18084	.2871	.10234	.91758	.18476
.2822	.10128	.92036	.18092	.2872	.10236	.91752	.18484
.2823	.10130	.92030	.18100	.2873	.10238	.91746	.18492
.2824	.10133	.92026	.18107	.2874	.10240	.91740	.18500
.2825	.10135	.92020	.18115	.2875	.10242	.91734	.18508
.2826	.10137	.92014	.18123	.2876	.10244	.91728	.18516
.2827	.10139	.92008	.18131	.2877	.10246	.91722	.18524
.2828	.10141	.92002	.18139	.2878	.10249	.91718	.18531
.2829	.10143	.91996	.18147	.2879	.10251	.91712	.18539
.2830	.10146	.91992	.18154	.2880	.10253	.91706	.18547
.2831	.10148	.91986	.18162	.2881	.10255	.91700	.18555
.2832	.10150	.91980	.18170	.2882	.10257	.91694	.18563
.2833	.10152	.91974	.18178	.2883	.10259	.91688	.18571
.2834	.10154	.91968	.18186	.2884	.10261	.91682	.18579
.2835	.10156	.91962	.18194	.2885	.10263	.91676	.18587
.2836	.10159	.91958	.18201	.2886	.10266	.91672	.18594
.2837	.10161	.91952	.18209	.2887	.10268	.91666	.18602
.2838	.10163	.91946	.18217	.2888	.10270	.91660	.18610
.2839	.10165	.91940	.18225	.2889	.10272	.91654	.18618
.2840	.10167	.91934	.18233	.2890	.10274	.91648	.18626
.2841	.10169	.91928	.18241	.2891	.10276	.91642	.18634
.2842	.10172	.91924	.18248	.2892	.10278	.91636	.18642
.2843	.10174	.91918	.18256	.2893	.10280	.91630	.18650
.2844	.10176	.91912	.18264	.2894	.10282	.91624	.18658
.2845	.10178	.91906	.18272	.2895	.10284	.91618	.18666
.2846	.10180	.91900	.18280	.2896	.10287	.91614	.18673
.2847	.10182	.91894	.18288	.2897	.10289	.91608	.18681
.2848	.10184	.91888	.18296	.2898	.10291	.91602	.18689
.2849	.10187	.91884	.18303	.2899	.10293	.91596	.18697
.2850	.10189	.91878	.18311	.2900	.10295	.91590	.18705

TABLE II. THREE-POINT INTERPOLATION COEFFICIENTS

p	c_{-1} −	c_0 +	c_1 +	p	c_{-1} −	c_0 +	c_1 +
.2900	.10295	.91590	.18705	.2950	.10399	.91298	.19101
.2901	.10297	.91584	.18713	.2951	.10401	.91292	.19109
.2902	.10299	.91578	.18721	.2952	.10403	.91286	.19117
.2903	.10301	.91572	.18729	.2953	.10405	.91280	.19125
.2904	.10303	.91566	.18737	.2954	.10407	.91274	.19133
.2905	.10305	.91560	.18745	.2955	.10409	.91268	.19141
.2906	.10308	.91556	.18752	.2956	.10411	.91262	.19149
.2907	.10310	.91550	.18760	.2957	.10413	.91256	.19157
.2908	.10312	.91544	.18768	.2958	.10415	.91250	.19165
.2909	.10314	.91538	.18776	.2959	.10417	.91244	.19173
.2910	.10316	.91532	.18784	.2860	.10419	.91238	.19181
.2911	.10318	.91526	.18792	.2961	.10421	.91232	.19189
.2912	.10320	.91520	.18800	.2962	.10423	.91226	.19197
.2913	.10322	.91514	.18808	.2963	.10425	.91220	.19205
.2914	.10324	.91508	.18816	.2964	.10427	.91214	.19213
.2915	.10326	.91502	.18824	.2965	.10429	.91208	.19221
.2916	.10328	.91496	.18832	.2966	.10431	.91202	.19229
.2917	.10331	.91492	.18839	.2967	.10433	.91196	.19237
.2918	.10333	.91486	.18847	.2968	.10435	.91190	.19245
.2919	.10335	.91480	.18855	.2969	.10438	.91186	.19252
.2920	.10337	.91474	.18863	.2970	.10440	.91180	.19260
.2921	.10339	.91468	.18871	.2971	.10442	.91174	.19268
.2922	.10341	.91462	.18879	.2972	.10444	.91168	.19276
.2923	.10343	.91456	.18887	.2973	.10446	.91162	.19284
.2924	.10345	.91450	.18895	.2974	.10448	.91156	.19292
.2925	.10347	.91444	.18903	.2975	.10450	.91150	.19300
.2926	.10349	.91438	.18911	.2976	.10452	.91144	.19308
.2927	.10351	.91432	.18919	.2977	.10454	.91138	.19316
.2928	.10353	.91426	.18927	.2978	.10456	.91132	.19324
.2929	.10355	.91420	.18935	.2979	.10458	.91126	.19332
.2930	.10358	.91416	.18942	.2980	.10460	.91120	.19340
.2931	.10360	.91410	.18950	.2981	.10462	.91114	.19348
.2932	.10362	.91404	.18958	.2982	.10464	.91108	.19356
.2933	.10364	.91398	.18966	.2983	.10466	.91102	.19364
.2934	.10366	.91392	.18974	.2984	.10468	.91096	.19372
.2935	.10368	.91386	.18982	.2985	.10470	.91090	.19380
.2936	.10370	.91380	.18990	.2986	.10472	.91084	.19388
.2937	.10372	.91374	.18998	.2987	.10474	.91078	.19396
.2738	.10374	.91368	.19006	.2988	.10476	.91072	.19404
.2939	.10376	.91362	.19014	.2989	.10478	.91066	.19412
.2940	.10378	.91356	.19022	.2990	.10480	.91060	.19420
.2941	.10380	.91350	.19030	.2991	.10482	.91054	.19428
.2942	.10382	.91344	.19038	.2992	.10484	.91048	.19436
.2943	.10384	.91338	.19046	.2993	.10486	.91042	.19444
.2944	.10386	.91332	.19054	.2994	.10488	.91036	.19452
.2945	.10388	.91326	.19062	.2995	.10490	.91030	.19460
.2946	.10391	.91322	.19069	.2996	.10492	.91024	.19468
.2947	.10393	.91316	.19077	.2997	.10494	.91018	.19476
.2948	.10395	.91310	.19085	.2998	.10496	.91012	.19484
.2949	.10397	.91304	.19093	.2999	.10498	.91006	.19492
.2950	.10399	.91298	.19101	.3000	.10500	.91000	.19500

TABLE II. THREE-POINT INTERPOLATION COEFFICIENTS

p	c_{-1} $-$	c_0 $+$	c_1 $+$	p	c_{-1} $-$	c_0 $+$	c_1 $+$
.3000	.10500	.91000	.19500	.3050	.10599	.90698	.19901
.3001	.10502	.90994	.19508	.3051	.10601	.90692	.19909
.3002	.10504	.90988	.19516	.3052	.10603	.90686	.19917
.3003	.10506	.90982	.19524	.3053	.10605	.90680	.19925
.3004	.10508	.90976	.19532	.3054	.10607	.90674	.19933
.3005	.10510	.90970	.19540	.3055	.10608	.90666	.19942
.3006	.10512	.90964	.19548	.3056	.10610	.90660	.19950
.3007	.10514	.90958	.19556	.3057	.10612	.90654	.19958
.3008	.10516	.90952	.19564	.3058	.10614	.90648	.19966
.3009	.10518	.90946	.19572	.3059	.10616	.90642	.19974
.3010	.10520	.90940	.19580	.3060	.10618	.90636	.19982
.3011	.10522	.90934	.19588	.3061	.10620	.90630	.19990
.3012	.10524	.90928	.19596	.3062	.10622	.90624	.19998
.3013	.10526	.90922	.19604	.3063	.10624	.90618	.20006
.3014	.10528	.90916	.19612	.3064	.10626	.90612	.20014
.3015	.10530	.90910	.19620	.3065	.10628	.90606	.20022
.3016	.10532	.90904	.19628	.3066	.10630	.90600	.20030
.3017	.10534	.90898	.19636	.3067	.10632	.90594	.20038
.3018	.10536	.90892	.19644	.3068	.10634	.90588	.20046
.3019	.10538	.90886	.19652	.3069	.10636	.90582	.20054
.3020	.10540	.90880	.19660	.3070	.10638	.90576	.20062
.3021	.10542	.90874	.19668	.3071	.10639	.90568	.20071
.3022	.10544	.90868	.19676	.3072	.10641	.90562	.20079
.3023	.10546	.90862	.19684	.3073	.10643	.90556	.20087
.3024	.10548	.90856	.19692	.3074	.10645	.90550	.20095
.3025	.10550	.90850	.19700	.3075	.10647	.90544	.20103
.3026	.10552	.90844	.19708	.3076	.10649	.90538	.20111
.3027	.10554	.90838	.19716	.3077	.10651	.90532	.20119
.3028	.10556	.90832	.19724	.3078	.10653	.90526	.20127
.3029	.10558	.90826	.19732	.3079	.10655	.90520	.20135
.3030	.10560	.90820	.19740	.3080	.10657	.90514	.20143
.3031	.10562	.90814	.19748	.3081	.10659	.90508	.20151
.3032	.10563	.90806	.19757	.3082	.10661	.90502	.20159
.3033	.10565	.90800	.19765	.3083	.10663	.90496	.20167
.3034	.10567	.90794	.19773	.3084	.10664	.90488	.20176
.3035	.10569	.90788	.19781	.3085	.10666	.90482	.20184
.3036	.10571	.90782	.19789	.3086	.10668	.90476	.20192
.3037	.10573	.90776	.19797	.3087	.10670	.90470	.20200
.3038	.10575	.90770	.19805	.3088	.10672	.90464	.20208
.3039	.10577	.90764	.19813	.3089	.10674	.90458	.20216
.3040	.10579	.90758	.19821	.3090	.10676	.90452	.20224
.3041	.10581	.90752	.19829	.3091	.10678	.90446	.20232
.3042	.10583	.90746	.19837	.3092	.10680	.90440	.20240
.3043	.10585	.90740	.19845	.3093	.10682	.90434	.20248
.3044	.10587	.90734	.19853	.3094	.10684	.90428	.20256
.3045	.10589	.90728	.19861	.3095	.10685	.90420	.20265
.3046	.10591	.90722	.19869	.3096	.10687	.90414	.20273
.3047	.10593	.90716	.19877	.3097	.10689	.90408	.20281
.3048	.10595	.90710	.19885	.3098	.10691	.90402	.20289
.3049	.10597	.90704	.19893	.3099	.10693	.90396	.20297
.3050	.10599	.90698	.19901	.3100	.10695	.90390	.20305

TABLE II. THREE-POINT INTERPOLATION COEFFICIENTS

p	c_{-1} −	c_0 +	c_1 +	p	c_{-1} −	c_0 +	c_1 +
.3100	.10695	.90390	.20305	.3150	.10789	.90078	.20711
.3101	.10697	.90384	.20313	.3151	.10791	.90072	.20719
.3102	.10699	.90378	.20321	.3152	.10792	.90064	.20728
.3103	.10701	.90372	.20329	.3153	.10794	.90058	.20736
.3104	.10703	.90366	.20337	.3154	.10796	.90052	.20744
.3105	.10704	.90358	.20346	.3155	.10798	.90046	.20752
.3106	.10706	.90352	.20354	.3156	.10800	.90040	.20760
.3107	.10708	.90346	.20362	.3157	.10802	.90034	.20768
.3108	.10710	.90340	.20370	.3158	.10804	.90028	.20776
.3109	.10712	.90334	.20378	.3159	.10805	.90020	.20785
.3110	.10714	.90328	.20386	.3160	.10807	.90014	.20793
.3111	.10716	.90322	.20394	.3161	.10809	.90008	.20801
.3112	.10718	.90316	.20402	.3162	.10811	.90002	.20809
.3113	.10720	.90310	.20410	.3163	.10813	.89996	.20817
.3114	.10722	.90304	.20418	.3164	.10815	.89990	.20825
.3115	.10723	.90296	.20427	.3165	.10816	.89982	.20834
.3116	.10725	.90290	.20435	.3166	.10818	.89976	.20842
.3117	.10727	.90284	.20443	.3167	.10820	.89970	.20850
.3118	.10729	.90278	.20451	.3168	.10822	.89964	.20858
.3119	.10731	.90272	.20459	.3169	.10824	.89958	.20866
.3120	.10733	.90266	.20467	.3170	.10826	.89952	.20874
.3121	.10735	.90260	.20475	.3171	.10827	.89944	.20883
.3122	.10737	.90254	.20483	.3172	.10829	.89938	.20891
.3123	.10738	.90246	.20492	.3173	.10831	.89932	.20899
.3124	.10740	.90240	.20500	.3174	.10833	.89926	.20907
.3125	.10742	.90234	.20508	.3175	.10835	.89920	.20915
.3126	.10744	.90228	.20516	.3176	.10837	.89914	.20923
.3127	.10746	.90222	.20524	.3177	.10838	.89906	.20932
.3128	.10748	.90216	.20532	.3178	.10840	.89900	.20940
.3129	.10750	.90210	.20540	.3179	.10842	.89894	.20948
.3130	.10752	.90204	.20548	.3180	.10844	.89888	.20956
.3131	.10753	.90196	.20557	.3181	.10846	.89882	.20964
.3132	.10755	.90190	.20565	.3182	.10847	.89874	.20973
.3133	.10757	.90184	.20573	.3183	.10849	.89868	.20981
.3134	.10759	.90178	.20581	.3184	.10851	.89862	.20989
.3135	.10761	.90172	.20589	.3185	.10853	.89856	.20997
.3136	.10763	.90166	.20597	.3186	.10855	.89850	.21005
.3137	.10765	.90160	.20605	.3187	.10857	.89844	.21013
.3138	.10766	.90152	.20614	.3188	.10858	.89836	.21022
.3139	.10768	.90146	.20622	.3189	.10860	.89830	.21030
.3140	.10770	.90140	.20630	.3190	.10862	.89824	.21038
.3141	.10772	.90134	.20638	.3191	.10864	.89818	.21046
.3142	.10774	.90128	.20646	.3192	.10866	.89812	.21054
.3143	.10776	.90122	.20654	.3193	.10867	.89804	.21063
.3144	.10778	.90116	.20662	.3194	.10869	.89798	.21071
.3145	.10779	.90108	.20671	.3195	.10871	.89792	.21079
.3146	.10781	.90102	.20679	.3196	.10873	.89786	.21087
.3147	.10783	.90096	.20687	.3197	.10875	.89780	.21095
.3148	.10785	.90090	.20695	.3198	.10876	.89772	.21104
.3149	.10787	.90084	.20703	.3199	.10878	.89766	.21112
.3150	.10789	.90078	.20711	.3200	.10880	.89760	.21120

TABLE II. THREE-POINT INTERPOLATION COEFFICIENTS

p	c_{-1} −	c_0 +	c_1 +		p	c_{-1} −	c_0 +	c_1 +
.3200	.10880	.89760	.21120		.3250	.10969	.89438	.21531
.3201	.10882	.89754	.21128		.3251	.10970	.89430	.21540
.3202	.10884	.89748	.21136		.3252	.10972	.89424	.21548
.3203	.10885	.89740	.21145		.3253	.10974	.89418	.21556
.3204	.10887	.89734	.21153		.3254	.10976	.89412	.21564
.3205	.10889	.89728	.21161		.3255	.10977	.89404	.21573
.3206	.10891	.89722	.21169		.3256	.10979	.89398	.21581
.3207	.10893	.89716	.21177		.3257	.10981	.89392	.21589
.3208	.10894	.89708	.21186		.3258	.10983	.89386	.21597
.3209	.10896	.89702	.21194		.3259	.10984	.89378	.21606
.3210	.10898	.89696	.21202		.3260	.10986	.89372	.21614
.3211	.10900	.89690	.21210		.3261	.10988	.89366	.21622
.3212	.10902	.89684	.21218		.3262	.10990	.89360	.21630
.3213	.10903	.89676	.21227		.3263	.10991	.89352	.21639
.3214	.10905	.89670	.21235		.3264	.10993	.89346	.21647
.3215	.10907	.89664	.21243		.3265	.10995	.89340	.21655
.3216	.10909	.89658	.21251		.3266	.10997	.89334	.21663
.3217	.10910	.89650	.21260		.3267	.10998	.89326	.21672
.3218	.10912	.89644	.21268		.3268	.11000	.89320	.21680
.3219	.10914	.89638	.21276		.3269	.11002	.89314	.21688
.3220	.10916	.89632	.21284		.3270	.11004	.89308	.21696
.3221	.10918	.89626	.21292		.3271	.11005	.89300	.21705
.3222	.10919	.89618	.21301		.3272	.11007	.89294	.21713
.3223	.10921	.89612	.21309		.3273	.11009	.89288	.21721
.3224	.10923	.89606	.21317		.3274	.11010	.89280	.21730
.3225	.10925	.89600	.21325		.3275	.11012	.89274	.21738
.3226	.10926	.89592	.21334		.3276	.11014	.89268	.21746
.3227	.10928	.89586	.21342		.3277	.11016	.89262	.21754
.3228	.10930	.89580	.21350		.3278	.11017	.89254	.21763
.3229	.10932	.89574	.21358		.3279	.11019	.89248	.21771
.3230	.10934	.89568	.21366		.3280	.11021	.89242	.21779
.3231	.10935	.89560	.21375		.3281	.11023	.89236	.21787
.3232	.10937	.89554	.21383		.3282	.11024	.89228	.21796
.3233	.10939	.89548	.21391		.3283	.11026	.89222	.21804
.3234	.10941	.89542	.21399		.3284	.11028	.89216	.21812
.3235	.10942	.89534	.21408		.3285	.11029	.89208	.21821
.3236	.10944	.89528	.21416		.3286	.11031	.89202	.21829
.3237	.10946	.89522	.21424		.3287	.11033	.89196	.21837
.3238	.10948	.89516	.21432		.3288	.11035	.89190	.21845
.3239	.10949	.89508	.21441		.3289	.11036	.89182	.21854
.3240	.10951	.89502	.21449		.3290	.11038	.89176	.21862
.3241	.10953	.89496	.21457		.3291	.11040	.89170	.21870
.3242	.10955	.89490	.21465		.3292	.11041	.89162	.21879
.3243	.10956	.89482	.21474		.3293	.11043	.89156	.21887
.3244	.10958	.89476	.21482		.3294	.11045	.89150	.21895
.3245	.10960	.89470	.21490		.3295	.11046	.89142	.21904
.3246	.10962	.89464	.21498		.3296	.11048	.89136	.21912
.3247	.10963	.89456	.21507		.3297	.11050	.89130	.21920
.3248	.10965	.89450	.21515		.3298	.11052	.89124	.21928
.3249	.10967	.89444	.21523		.3299	.11053	.89116	.21937
.3250	.10969	.89438	.21531		.3300	.11055	.89110	.21945

TABLE II. THREE-POINT INTERPOLATION COEFFICIENTS

p	c_{-1} −	c_0 +	c_1 +	p	c_{-1} −	c_0 +	c_1 +
.3300	.11055	.89110	.21945	.3350	.11139	.88778	.22361
.3301	.11057	.89104	.21953	.3351	.11140	.88770	.22370
.3302	.11058	.89096	.21962	.3352	.11142	.88764	.22378
.3303	.11060	.89090	.21970	.3353	.11144	.88758	.22386
.3304	.11062	.89084	.21978	.3354	.11145	.88750	.22395
.3305	.11063	.89076	.21987	.3355	.11147	.88744	.22403
.3306	.11065	.89070	.21995	.3356	.11149	.88738	.22411
.3307	.11067	.89064	.22003	.3357	.11150	.88730	.22420
.3308	.11069	.89058	.22011	.3358	.11152	.88724	.22428
.3309	.11070	.89050	.22020	.3359	.11154	.88718	.22436
.3310	.11072	.89044	.22028	.3360	.11155	.88710	.22445
.3311	.11074	.89038	.22036	.3361	.11157	.88704	.22453
.3312	.11075	.89030	.22045	.3362	.11158	.88696	.22462
.3313	.11077	.89024	.22053	.3363	.11160	.88690	.22470
.3314	.11079	.89018	.22061	.3364	.11162	.88684	.22478
.3315	.11080	.89010	.22070	.3365	.11163	.88676	.22487
.3316	.11082	.89004	.22078	.3366	.11165	.88670	.22495
.3317	.11084	.88998	.22086	.3367	.11167	.88664	.22503
.3318	.11085	.88990	.22095	.3368	.11168	.88656	.22512
.3319	.11087	.88984	.22103	.3369	.11170	.88650	.22520
.3320	.11089	.88978	.22111	.3370	.11172	.88644	.22528
.3321	.11090	.88970	.22120	.3371	.11173	.88636	.22537
.3322	.11092	.88964	.22128	.3372	.11175	.88630	.22545
.3323	.11094	.88958	.22136	.3373	.11176	.88622	.22554
.3324	.11096	.88952	.22144	.3374	.11178	.88616	.22562
.3325	.11097	.88944	.22153	.3375	.11180	.88610	.22570
.3326	.11099	.88938	.22161	.3376	.11181	.88602	.22579
.3327	.11101	.88932	.22169	.3377	.11183	.88596	.22587
.3328	.11102	.88924	.22178	.3378	.11185	.88590	.22595
.3329	.11104	.88918	.22186	.3379	.11186	.88582	.22604
.3330	.11106	.88912	.22194	.3380	.11188	.88576	.22612
.3331	.11107	.88904	.22203	.3381	.11189	.88568	.22621
.3332	.11109	.88898	.22211	.3382	.11191	.88562	.22629
.3333	.11111	.88892	.22219	.3383	.11193	.88556	.22637
.3334	.11112	.88884	.22228	.3384	.11194	.88548	.22646
.3335	.11114	.88878	.22236	.3385	.11196	.88542	.22654
.3336	.11116	.88872	.22244	.3386	.11198	.88536	.22662
.3337	.11117	.88864	.22253	.3387	.11199	.88528	.22671
.3338	.11119	.88858	.22261	.3388	.11201	.88522	.22679
.3339	.11121	.88852	.22269	.3389	.11202	.88514	.22688
.3340	.11122	.88844	.22278	.3390	.11204	.88508	.22696
.3341	.11124	.88838	.22286	.3391	.11206	.88502	.22704
.3342	.11126	.88832	.22294	.3392	.11207	.88494	.22713
.3343	.11127	.88824	.22303	.3393	.11209	.88488	.22721
.3344	.11129	.88818	.22311	.3394	.11210	.88480	.22730
.3345	.11130	.88810	.22320	.3395	.11212	.88474	.22738
.3346	.11132	.88804	.22328	.3396	.11214	.88468	.22746
.3347	.11134	.88798	.22336	.3397	.11215	.88460	.22755
.3348	.11135	.88790	.22345	.3398	.11217	.88454	.22763
.3349	.11137	.88784	.22353	.3399	.11218	.88446	.22772
.3350	.11139	.88778	.22361	.3400	.11220	.88440	.22780

TABLE II. THREE-POINT INTERPOLATION COEFFICIENTS

p	c_{-1} $-$	c_0 $+$	c_1 $+$	p	c_{-1} $-$	c_0 $+$	c_1 $+$
.3400	.11220	.88440	.22780	.3450	.11299	.88098	.23201
.3401	.11222	.88434	.22788	.3451	.11300	.88090	.23210
.3402	.11223	.88426	.22797	.3452	.11302	.88084	.23218
.3403	.11225	.88420	.22805	.3453	.11303	.88076	.23227
.3404	.11226	.88412	.22814	.3454	.11305	.88070	.23235
.3405	.11228	.88406	.22822	.3455	.11306	.88062	.23244
.3406	.11230	.88400	.22830	.3456	.11308	.88056	.23252
.3407	.11231	.88392	.22839	.3457	.11310	.88050	.23260
.3408	.11233	.88386	.22847	.3458	.11311	.88042	.23269
.3409	.11234	.88378	.22856	.3459	.11313	.88036	.23277
.3410	.11236	.88372	.22864	.3460	.11314	.88028	.23286
.3411	.11238	.88366	.22872	.3461	.11316	.88022	.23294
.3412	.11239	.88358	.22881	.3462	.11317	.88014	.23303
.3413	.11241	.88352	.22889	.3463	.11319	.88008	.23311
.3414	.11242	.88344	.22898	.3464	.11320	.88000	.23320
.3415	.11244	.88338	.22906	.3465	.11322	.87994	.23328
.3416	.11245	.88330	.22915	.3466	.11323	.87986	.23337
.3417	.11247	.88324	.22923	.3467	.11325	.87980	.23345
.3418	.11249	.88318	.22931	.3468	.11326	.87972	.23354
.3419	.11250	.88310	.22940	.3469	.11328	.87966	.23362
.3420	.11252	.88304	.22948	.3470	.11330	.87960	.23370
.3421	.11253	.88296	.22957	.3471	.11331	.87952	.23379
.3422	.11255	.88290	.22965	.3472	.11333	.87946	.23387
.3423	.11257	.88284	.22973	.3473	.11334	.87938	.23396
.3424	.11258	.88276	.22982	.3474	.11336	.87932	.23404
.3425	.11260	.88270	.22990	.3475	.11337	.87924	.23413
.3426	.11261	.88262	.22999	.3476	.11339	.87918	.23421
.3427	.11263	.88256	.23007	.3477	.11340	.87910	.23430
.3428	.11264	.88248	.23016	.3478	.11342	.87904	.23438
.3429	.11266	.88242	.23024	.3479	.11343	.87896	.23447
.3430	.11268	.88236	.23032	.3480	.11345	.87890	.23455
.3431	.11269	.88228	.23041	.3481	.11346	.87882	.23464
.3432	.11271	.88222	.23049	.3482	.11348	.87876	.23472
.3433	.11272	.88214	.23058	.3483	.11349	.87868	.23481
.3434	.11274	.88208	.23066	.3484	.11351	.87862	.23489
.3435	.11275	.88200	.23075	.3485	.11352	.87854	.23498
.3436	.11277	.88194	.23083	.3486	.11354	.87848	.23506
.3437	.11279	.88188	.23091	.3487	.11355	.87840	.23515
.3438	.11280	.88180	.23100	.3488	.11357	.87834	.23523
.3439	.11282	.88174	.23108	.3489	.11358	.87826	.23532
.3440	.11283	.88166	.23117	.3490	.11360	.87820	.23540
.3441	.11285	.88160	.23125	.3491	.11361	.87812	.23549
.3442	.11286	.88152	.23134	.3492	.11363	.87806	.23557
.3443	.11288	.88146	.23142	.3493	.11364	.87798	.23566
.3444	.11289	.88138	.23151	.3494	.11366	.87792	.23574
.3445	.11291	.88132	.23159	.3495	.11367	.87784	.23583
.3446	.11293	.88126	.23167	.3496	.11369	.87778	.23591
.3447	.11294	.88118	.23176	.3497	.11370	.87770	.23600
.3448	.11296	.88112	.23184	.3498	.11372	.87764	.23608
.3449	.11297	.88104	.23193	.3499	.11373	.87756	.23617
.3450	.11299	.88098	.23201	.3500	.11375	.87750	.23625

TABLE II. THREE-POINT INTERPOLATION COEFFICIENTS

p	c_{-1} $-$	c_0 $+$	c_1 $+$	p	c_{-1} $-$	c_0 $+$	c_1 $+$
.3500	.11375	.87750	.23625	.3550	.11449	.87398	.24051
.3501	.11376	.87742	.23634	.3551	.11450	.87390	.24060
.3502	.11378	.87736	.23642	.3552	.11452	.87384	.24068
.3503	.11379	.87728	.23651	.3553	.11453	.87376	.24077
.3504	.11381	.87722	.23659	.3554	.11455	.87370	.24085
.3505	.11382	.87714	.23668	.3555	.11456	.87362	.24094
.3506	.11384	.87708	.23676	.3556	.11457	.87354	.24103
.3507	.11385	.87700	.23685	.3557	.11459	.87348	.24111
.3508	.11387	.87694	.23693	.3558	.11460	.87340	.24120
.3509	.11388	.87686	.23702	.3559	.11462	.87334	.24128
.3510	.11390	.87680	.23710	.3560	.11463	.87326	.24137
.3511	.11391	.87672	.23719	.3561	.11465	.87320	.24145
.3512	.11393	.87666	.23727	.3562	.11466	.87312	.24154
.3513	.11394	.87658	.23736	.3563	.11468	.87306	.24162
.3514	.11396	.87652	.23744	.3564	.11469	.87298	.24171
.3515	.11397	.87644	.23753	.3565	.11470	.87290	.24180
.3516	.11399	.87638	.23761	.3566	.11472	.87284	.24188
.3517	.11400	.87630	.23770	.3567	.11473	.87276	.24197
.3518	.11402	.87624	.23778	.3568	.11475	.87270	.24205
.3519	.11403	.87616	.23787	.3569	.11476	.87262	.24214
.3520	.11405	.87610	.23795	.3570	.11478	.87256	.24222
.3521	.11406	.87602	.23804	.3571	.11479	.87248	.24231
.3522	.11408	.87596	.23812	.3572	.11480	.87240	.24240
.3523	.11409	.87588	.23821	.3573	.11482	.87234	.24248
.3524	.11411	.87582	.23829	.3574	.11483	.87226	.24257
.3525	.11412	.87574	.23838	.3575	.11485	.87220	.24265
.3526	.11414	.87568	.23846	.3576	.11486	.87212	.24274
.3527	.11415	.87560	.23855	.3577	.11488	.87206	.24282
.3528	.11417	.87554	.23863	.3578	.11489	.87198	.24291
.3529	.11418	.87546	.23872	.3579	.11490	.87190	.24300
.3530	.11420	.87540	.23880	.3580	.11492	.87184	.24308
.3531	.11421	.87532	.23889	.3581	.11493	.87176	.24317
.3532	.11422	.87524	.23898	.3582	.11495	.87170	.24325
.3533	.11424	.87518	.23906	.3583	.11496	.87162	.24334
.3534	.11425	.87510	.23915	.3584	.11497	.87154	.24343
.3535	.11427	.87504	.23923	.3585	.11499	.87148	.24351
.3536	.11428	.87496	.23932	.3586	.11500	.87140	.24360
.3537	.11430	.87490	.23940	.3587	.11502	.87134	.24368
.3538	.11431	.87482	.23949	.3588	.11503	.87126	.24377
.3539	.11433	.87476	.23957	.3589	.11505	.87120	.24385
.3540	.11434	.87468	.23966	.3590	.11506	.87112	.24394
.3541	.11436	.87462	.23974	.3591	.11507	.87104	.24403
.3542	.11437	.87454	.23983	.3592	.11509	.87098	.24411
.3543	.11439	.87448	.23991	.3593	.11510	.87090	.24420
.3544	.11440	.87440	.24000	.3594	.11512	.87084	.24428
.3545	.11441	.87432	.24009	.3595	.11513	.87076	.24437
.3546	.11443	.87426	.24017	.3596	.11514	.87068	.24446
.3547	.11444	.87418	.24026	.3597	.11516	.87062	.24454
.3548	.11446	.87412	.24034	.3598	.11517	.87054	.24463
.3549	.11447	.87404	.24043	.3599	.11519	.87048	.24471
.3550	.11449	.87398	.24051	.3600	.11520	.87040	.24480

TABLE II. THREE-POINT INTERPOLATION COEFFICIENTS

p	c_{-1} $-$	c_0 $+$	c_1 $+$	p	c_{-1} $-$	c_0 $+$	c_1 $+$
.3600	.11520	.87040	.24480	.3650	.11589	.86678	.24911
.3601	.11521	.87032	.24489	.3651	.11590	.86670	.24920
.3602	.11523	.87026	.24497	.3652	.11591	.86662	.24929
.3603	.11524	.87018	.24506	.3653	.11593	.86656	.24937
.3604	.11526	.87012	.24514	.3654	.11594	.86648	.24946
.3605	.11527	.87004	.24523	.3655	.11595	.86640	.24955
.3606	.11528	.86996	.24532	.3656	.11597	.86634	.24963
.3607	.11530	.86990	.24540	.3657	.11598	.86626	.24972
.3608	.11531	.86982	.24549	.3658	.11600	.86620	.24980
.3609	.11533	.86976	.24557	.3659	.11601	.86612	.24989
.3610	.11534	.86968	.24566	.3660	.11602	.86604	.24998
.3611	.11535	.86960	.24575	.3661	.11604	.86598	.25006
.3612	.11537	.86954	.24583	.3662	.11605	.86590	.25015
.3613	.11538	.86946	.24592	.3663	.11606	.86582	.25024
.3614	.11540	.86940	.24600	.3664	.11608	.86576	.25032
.3615	.11541	.86932	.24609	.3665	.11609	.86568	.25041
.3616	.11542	.86924	.24618	.3666	.11610	.86560	.25050
.3617	.11544	.86918	.24626	.3667	.11612	.86554	.25058
.3618	.11545	.86910	.24635	.3668	.11613	.86546	.25067
.3619	.11546	.86902	.24644	.3669	.11614	.86538	.25076
.3620	.11548	.86896	.24652	.3670	.11616	.86532	.25084
.3621	.11549	.86888	.24661	.3671	.11617	.86524	.25093
.3622	.11551	.86882	.24669	.3672	.11618	.86516	.25102
.3623	.11552	.86874	.24678	.3673	.11620	.86510	.25110
.3624	.11553	.86866	.24687	.3674	.11621	.86502	.25119
.3625	.11555	.86860	.24695	.3675	.11622	.86494	.25128
.3626	.11556	.86852	.24704	.3676	.11624	.86488	.25136
.3627	.11557	.86844	.24713	.3677	.11625	.86480	.25145
.3628	.11559	.86838	.24721	.3678	.11626	.86472	.25154
.3629	.11560	.86830	.24730	.3679	.11627	.86464	.25163
.3630	.11562	.86824	.24738	.3680	.11629	.86458	.25171
.3631	.11563	.86816	.24747	.3681	.11630	.86450	.25180
.3632	.11564	.86808	.24756	.3682	.11631	.86442	.25189
.3633	.11566	.86802	.24764	.3683	.11633	.86436	.25197
.3634	.11567	.86794	.24773	.3684	.11634	.86428	.25206
.3635	.11568	.86786	.24782	.3685	.11635	.86420	.25215
.3636	.11570	.86780	.24790	.3686	.11637	.86414	.25223
.3637	.11571	.86772	.24799	.3687	.11638	.86406	.25232
.3638	.11572	.86764	.24808	.3688	.11639	.86398	.25241
.3639	.11574	.86758	.24816	.3689	.11641	.86392	.25249
.3640	.11575	.86750	.24825	.3690	.11642	.86384	.25258
.3641	.11577	.86744	.24833	.3691	.11643	.86376	.25267
.3642	.11578	.86736	.24842	.3692	.11645	.86370	.25275
.3643	.11579	.86728	.24851	.3693	.11646	.86362	.25284
.3644	.11581	.86722	.24859	.3694	.11647	.86354	.25293
.3645	.11582	.86714	.24868	.3695	.11648	.86346	.25302
.3646	.11583	.86706	.24877	.3696	.11650	.86340	.25310
.3647	.11585	.86700	.24885	.3697	.11651	.86332	.25319
.3648	.11586	.86692	.24894	.3698	.11652	.86324	.25328
.3649	.11587	.86684	.24903	.3699	.11654	.86318	.25336
.3650	.11589	.86678	.24911	.3700	.11655	.86310	.25345

TABLE II. THREE-POINT INTERPOLATION COEFFICIENTS

p	c_{-1} $-$	c_0 $+$	c_1 $+$	p	c_{-1} $-$	c_0 $+$	c_1 $+$
.3700	.11655	.86310	.25345	.3750	.11719	.85938	.25781
.3701	.11656	.86302	.25354	.3751	.11720	.85930	.25790
.3702	.11658	.86296	.25362	.3752	.11721	.85922	.25799
.3703	.11659	.86288	.25371	.3753	.11722	.85914	.25808
.3704	.11660	.86280	.25380	.3754	.11724	.85908	.25816
.3705	.11661	.86272	.25389	.3755	.11725	.85900	.25825
.3706	.11663	.86266	.25397	.3756	.11726	.85892	.25834
.3707	.11664	.86258	.25406	.3757	.11727	.85884	.25843
.3708	.11665	.86250	.25415	.3758	.11729	.85878	.25851
.3709	.11667	.86244	.25423	.3759	.11730	.85870	.25860
.3710	.11668	.86236	.25432	.3760	.11731	.85862	.25869
.3711	.11669	.86228	.25441	.3761	.11732	.85854	.25878
.3712	.11671	.86222	.25449	.3762	.11734	.85848	.25886
.3713	.11672	.86214	.25458	.3763	.11735	.85840	.25895
.3714	.11673	.86206	.25467	.3764	.11736	.85832	.25904
.3715	.11674	.86198	.25476	.3765	.11737	.85824	.25913
.3716	.11676	.86192	.25484	.3766	.11739	.85818	.25921
.3717	.11677	.86184	.25493	.3767	.11740	.85810	.25930
.3718	.11678	.86176	.25502	.3768	.11741	.85802	.25939
.3719	.11680	.86170	.25510	.3769	.11742	.85794	.25948
.3720	.11681	.86162	.25519	.3770	.11744	.85788	.25956
.3721	.11682	.86154	.25528	.3771	.11745	.85780	.25965
.3722	.11683	.86146	.25537	.3772	.11746	.85772	.25974
.3723	.11685	.86140	.25545	.3773	.11747	.85764	.25983
.3724	.11686	.86132	.25554	.3774	.11748	.85756	.25992
.3725	.11687	.86124	.25563	.3775	.11750	.85750	.26000
.3726	.11688	.86116	.25572	.3776	.11751	.85742	.26009
.3727	.11690	.86110	.25580	.3777	.11752	.85734	.26018
.3728	.11691	.86102	.25589	.3778	.11753	.85726	.26027
.3729	.11692	.86094	.25598	.3779	.11755	.85720	.26035
.3730	.11694	.86088	.25606	.3780	.11756	.85712	.26044
.3731	.11695	.86080	.25615	.3781	.11757	.85704	.26053
.3732	.11696	.86072	.25624	.3782	.11758	.85696	.26062
.3733	.11697	.86064	.25633	.3783	.11759	.85688	.26071
.3734	.11699	.86058	.25641	.3784	.11761	.85682	.26079
.3735	.11700	.86050	.25650	.3785	.11762	.85674	.26088
.3736	.11701	.86042	.25659	.3786	.11763	.85666	.26097
.3737	.11702	.86034	.25668	.3787	.11764	.85658	.26106
.3738	.11704	.86028	.25676	.3788	.11766	.85652	.26114
.3739	.11705	.86020	.25685	.3789	.11767	.85644	.26123
.3740	.11706	.86012	.25694	.3790	.11768	.85636	.26132
.3741	.11707	.86004	.25703	.3791	.11769	.85628	.26141
.3742	.11709	.85998	.25711	.3792	.11770	.85620	.26150
.3743	.11710	.85990	.25720	.3793	.11772	.85614	.26158
.3744	.11711	.85982	.25729	.3794	.11773	.85606	.26167
.3745	.11712	.85974	.25738	.3795	.11774	.85598	.26176
.3746	.11714	.85968	.25746	.3796	.11775	.85590	.26185
.3747	.11715	.85960	.25755	.3797	.11776	.85582	.26194
.3748	.11716	.85952	.25764	.3798	.11778	.85576	.26202
.3749	.11717	.85944	.25773	.3799	.11779	.85568	.26211
.3750	.11719	.85938	.25781	.3800	.11780	.85560	.26220

TABLE II. THREE-POINT INTERPOLATION COEFFICIENTS

p	c_{-1} −	c_0 +	c_1 +	p	c_{-1} −	c_0 +	c_1 +
.3800	.11780	.85560	.26220	.3850	.11839	.85178	.26661
.3801	.11781	.85552	.26229	.3851	.11840	.85170	.26670
.3802	.11782	.85544	.26238	.3852	.11841	.85162	.26679
.3803	.11784	.85538	.26246	.3853	.11842	.85154	.26688
.3804	.11785	.85530	.26255	.3854	.11843	.85146	.26697
.3805	.11786	.85522	.26264	.3855	.11844	.85138	.26706
.3806	.11787	.85514	.26273	.3856	.11846	.85132	.26714
.3807	.11788	.85506	.26282	.3857	.11847	.85124	.26723
.3808	.11790	.85500	.26290	.3858	.11848	.85116	.26732
.3809	.11791	.85492	.26299	.3859	.11849	.85108	.26741
.3810	.11792	.85484	.26308	.3860	.11850	.85100	.26750
.3811	.11793	.85476	.26317	.3861	.11851	.85092	.26759
.3812	.11794	.85468	.26326	.3862	.11852	.85084	.26768
.3813	.11796	.85462	.26334	.3863	.11854	.85078	.26776
.3814	.11797	.85454	.26343	.3864	.11855	.85070	.26785
.3815	.11798	.85446	.26352	.3865	.11856	.85062	.26794
.3816	.11799	.85438	.26361	.3866	.11857	.85054	.26803
.3817	.11800	.85430	.26370	.3867	.11858	.85046	.26812
.3818	.11801	.85422	.26379	.3868	.11859	.85038	.26821
.3819	.11803	.85416	.26387	.3869	.11860	.85030	.26830
.3820	.11804	.85408	.26396	.3870	.11862	.85024	.26838
.3821	.11805	.85400	.26405	.3871	.11863	.85016	.26847
.3822	.11806	.85392	.26414	.3872	.11864	.85008	.26856
.3823	.11807	.85384	.26423	.3873	.11865	.85000	.26865
.3824	.11809	.85378	.26431	.3874	.11866	.84992	.26874
.3825	.11810	.85370	.26440	.3875	.11867	.84984	.26883
.3826	.11811	.85362	.26449	.3876	.11868	.84976	.26892
.3827	.11812	.85354	.26458	.3877	.11869	.84968	.26901
.3828	.11813	.85346	.26467	.3878	.11871	.84962	.26909
.3829	.11814	.85338	.26476	.3879	.11872	.84954	.26918
.3830	.11816	.85332	.26484	.3880	.11873	.84946	.26927
.3831	.11817	.85324	.26493	.3881	.11874	.84938	.26936
.3832	.11818	.85316	.26502	.3882	.11875	.84930	.26945
.3833	.11819	.85308	.26511	.3883	.11876	.84922	.26954
.3834	.11820	.85300	.26520	.3884	.11877	.84914	.26963
.3835	.11821	.85292	.26529	.3885	.11878	.84906	.26972
.3836	.11823	.85286	.26537	.3886	.11880	.84900	.26980
.3837	.11824	.85278	.26546	.3887	.11881	.84892	.26989
.3838	.11825	.85270	.26555	.3888	.11882	.84884	.26998
.3839	.11826	.85262	.26564	.3889	.11883	.84876	.27007
.3840	.11827	.85254	.26573	.3890	.11884	.84868	.27016
.3841	.11828	.85246	.26582	.3891	.11885	.84860	.27025
.3842	.11830	.85240	.26590	.3892	.11886	.84852	.27034
.3843	.11831	.85232	.26599	.3893	.11887	.84844	.27043
.3844	.11832	.85224	.26608	.3894	.11888	.84836	.27052
.3845	.11833	.85216	.26617	.3895	.11889	.84828	.27061
.3846	.11834	.85208	.26626	.3896	.11891	.84822	.27069
.3847	.11835	.85200	.26635	.3897	.11892	.84814	.27078
.3848	.11836	.85192	.26644	.3898	.11893	.84806	.27087
.3849	.11838	.85186	.26652	.3899	.11894	.84798	.27096
.3850	.11839	.85178	.26661	.3900	.11895	.84790	.27105

TABLE II. THREE-POINT INTERPOLATION COEFFICIENTS

p	c_{-1} $-$	c_0 $+$	c_1 $+$	p	c_{-1} $-$	c_0 $+$	c_1 $+$
.3900	.11895	.84790	.27105	.3950	.11949	.84398	.27551
.3901	.11896	.84782	.27114	.3951	.11950	.84390	.27560
.3902	.11897	.84774	.27123	.3952	.11951	..84382	.27569
.3903	.11898	.84766	.27132	.3953	.11952	.84374	.27578
.3904	.11899	.84758	.27141	.3954	.11953	.84366	.27587
.3905	.11900	.84750	.27150	.3955	.11954	.84358	.27596
.3906	.11902	.84744	.27158	.3956	.11955	.84350	.27605
.3907	.11903	.84736	.27167	.3957	.11956	.84342	.27614
.3908	.11904	.84728	.27176	.3958	.11957	.84334	.27623
.3909	.11905	.84720	.27185	.3959	.11958	.84326	.27632
.3910	.11906	.84712	.27194	.3960	.11959	.84318	.27641
.3911	.11907	.84704	.27203	.3961	.11960	.84310	.27650
.3912	.11908	.84696	.27212	.3962	.11961	.84302	.27659
.3913	.11909	.84688	.27221	.3963	.11962	.84294	.27668
.3914	.11910	.84680	.27230	.3964	.11963	.84286	.27677
.3915	.11911	.84672	.27239	.3965	.11964	.84278	.27686
.3916	.11912	.84664	.27248	.3966	.11965	.84270	.27695
.3917	.11914	.84658	.27256	.3967	.11966	.84262	.27704
.3918	.11915	.84650	.27265	.3968	.11967	.84254	.27713
.3919	.11916	.84642	.27274	.3969	.11969	.84248	.27721
.3920	.11917	.84634	.27283	.3970	.11970	.84240	.27730
.3921	.11918	.84626	.27292	.3971	.11971	.84232	.27739
.3922	.11919	.84618	.27301	.3972	.11972	.84224	.27748
.3923	.11920	.84610	.27310	.3973	.11973	.84216	.27757
.3924	.11921	.84602	.27319	.3974	.11974	.84208	.27766
.3925	.11922	.84594	.27328	.3975	.11975	.84200	.27775
.3926	.11923	.84586	.27337	.3976	.11976	.84192	.27784
.3927	.11924	.84578	.27346	.3977	.11977	.84184	.27793
.3928	.11925	.84570	.27355	.3978	.11978	.84176	.27802
.3929	.11926	.84562	.27364	.3979	.11979	.84168	.27811
.3930	.11928	.84556	.27372	.3980	.11980	.84160	.27820
.3931	.11929	.84548	.27381	.3981	.11981	.84152	.27829
.3932	.11930	.84540	.27390	.3982	.11982	.84144	.27838
.3933	.11931	.84532	.27399	.3983	.11983	.84136	.27847
.3934	.11932	.84524	.27408	.3984	.11984	.84128	.27856
.3935	.11933	.84516	.27417	.3985	.11985	.84120	.27865
.3936	.11934	.84508	.27426	.3986	.11986	.84112	.27874
.3937	.11935	.84500	.27435	.3987	.11987	.84104	.27883
.3938	.11936	.84492	.27444	.3988	.11988	.84096	.27892
.3939	.11937	.84484	.27453	.3989	.11989	.84088	.27901
.3940	.11938	.84476	.27462	.3990	.11990	.84080	.27910
.3941	.11939	.84468	.27471	.3991	.11991	.84072	.27919
.3942	.11940	.84460	.27480	.3992	.11992	.84064	.27928
.3943	.11941	.84452	.27489	.3993	.11993	.84056	.27937
.3944	.11942	.84444	.27498	.3994	.11994	.84048	.27946
.3945	.11943	.84436	.27507	.3995	.11995	.84040	.27955
.3946	.11945	.84430	.27515	.3996	.11996	.84032	.27964
.3947	.11946	.84422	.27524	.3997	.11997	.84024	.27973
.3948	.11947	.84414	.27533	.3998	.11998	.84016	.27982
.3949	.11948	.84406	.27542	.3999	.11999	.84008	.27991
.3950	.11949	.84398	.27551	.4000	.12000	.84000	.28000

TABLE II. THREE-POINT INTERPOLATION COEFFICIENTS

p	c_{-1} −	c_0 +	c_1 +	p	c_{-1} −	c_0 +	c_1 +
.4000	.12000	.84000	.28000	.4050	.12049	.83598	.28451
.4001	.12001	.83992	.28009	.4051	.12050	.83590	.28460
.4002	.12002	.83984	.28018	.4052	.12051	.83582	.28469
.4003	.12003	.83976	.28027	.4053	.12052	.83574	.28478
.4004	.12004	.83968	.28036	.4054	.12053	.83566	.28487
.4005	.12005	.83960	.28045	.4055	.12053	.83556	.28497
.4006	.12006	.83952	.28054	.4056	.12054	.83548	.28506
.4007	.12007	.83944	.28063	.4057	.12055	.83540	.28515
.4008	.12008	.83936	.28072	.4058	.12056	.83532	.28524
.4009	.12009	.83928	.28081	.4059	.12057	.83524	.28533
.4010	.12010	.83920	.28090	.4060	.12058	.83516	.28542
.4011	.12011	.83912	.28099	.4061	.12059	.83508	.28551
.4012	.12012	.83904	.28108	.4062	.12060	.83500	.28560
.4013	.12013	.83896	.28117	.4063	.12061	.83492	.28569
.4014	.12014	.83888	.28126	.4064	.12062	.83484	.28578
.4015	.12015	.83880	.28135	.4065	.12063	.83476	.28587
.4016	.12016	.83872	.28144	.4066	.12064	.83468	.28596
.4017	.12017	.83864	.28153	.4067	.12065	.83460	.28605
.4018	.12018	.83856	.28162	.4068	.12066	.83452	.28614
.4019	.12019	.83848	.28171	.4069	.12067	.83444	.28623
.4020	.12020	.83840	.28180	.4070	.12068	.83436	.28632
.4021	.12021	.83832	.28189	.4071	.12068	.83426	.28642
.4022	.12022	.83824	.28198	.4072	.12069	.83418	.28651
.4023	.12023	.83816	.28207	.4073	.12070	.83410	.28660
.4024	.12024	.83808	.28216	.4074	.12071	.83402	.28669
.4025	.12025	.83800	.28225	.4075	.12072	.83394	.28678
.4026	.12026	.83792	.28234	.4076	.12073	.83386	.28687
.4027	.12027	.83784	.28243	.4077	.12074	.83378	.28696
.4028	.12028	.83776	.28252	.4078	.12075	.83370	.28705
.4029	.12029	.83768	.28261	.4079	.12076	.83362	.28714
.4030	.12030	.83760	.28270	.4080	.12077	.83354	.28723
.4031	.12031	.83752	.28279	.4081	.12078	.83346	.28732
.4032	.12031	.83742	.28289	.4082	.12079	.83338	.28741
.4033	.12032	.83734	.28298	.4083	.12080	.83330	.28750
.4034	.12033	.83726	.28307	.4084	.12080	.83320	.28760
.4035	.12034	.83718	.28316	.4085	.12081	.83312	.28769
.4036	.12035	.83710	.28325	.4086	.12082	.83304	.28778
.4037	.12036	.83702	.28334	.4087	.12083	.83296	.28787
.4038	.12037	.83694	.28343	.4088	.12084	.83288	.28796
.4039	.12038	.83686	.28352	.4089	.12085	.83280	.28805
.4040	.12039	.83678	.28361	.4090	.12086	.83272	.28814
.4041	.12040	.83670	.28370	.4091	.12087	.83264	.28823
.4042	.12041	.83662	.28379	.4092	.12088	.83256	.28832
.4043	.12042	.83654	.28388	.4093	.12089	.83248	.28841
.4044	.12043	.83646	.28397	.4094	.12090	.83240	.28850
.4045	.12044	.83638	.28406	.4095	.12090	.83230	.28860
.4046	.12045	.83630	.28415	.4096	.12091	.83222	.28869
.4047	.12046	.83622	.28424	.4097	.12092	.83214	.28878
.4048	.12047	.83614	.28433	.4098	.12093	.83206	.28887
.4049	.12048	.83606	.28442	.4099	.12094	.83198	.28896
.4050	.12049	.83598	.28451	.4100	.12095	.83190	.28905

TABLE II. THREE-POINT INTERPOLATION COEFFICIENTS

p	c_{-1} −	c_0 +	c_1 +	p	c_{-1} −	c_0 +	c_1 +
.4100	.12095	.83190	.28905	.4150	.12139	.82778	.29361
.4101	.12096	.83182	.28914	.4151	.12140	.82770	.29370
.4102	.12097	.83174	.28923	.4152	.12140	.82760	.29380
.4103	.12098	.83166	.28932	.4153	.12141	.82752	.29389
.4104	.12099	.83158	.28941	.4154	.12142	.82744	.29398
.4105	.12099	.83148	.28951	.4155	.12143	.82736	.29407
.4106	.12100	.83140	.28960	.4156	.12144	.82728	.29416
.4107	.12101	.83132	.28969	.4157	.12145	.82720	.29425
.4108	.12102	.83124	.28978	.4158	.12146	.82712	.29434
.4109	.12103	.83116	.28987	.4159	.12146	.82702	.29444
.4110	.12104	.83108	.28996	.4160	.12147	.82694	.29453
.4111	.12105	.83100	.29005	.4161	.12148	.82686	.29462
.4112	.12106	.83092	.29014	.4162	.12149	.82678	.29471
.4113	.12107	.83084	.29023	.4163	.12150	.82670	.29480
.4114	.12108	.83076	.29032	.4164	.12151	.82662	.29489
.4115	.12108	.83066	.29042	.4165	.12151	.82652	.29499
.4116	.12109	.83058	.29051	.4166	.12152	.82644	.29508
.4117	.12110	.83050	.29060	.4167	.12153	.82636	.29517
.4118	.12111	.83042	.29069	.4168	.12154	.82628	.29526
.4119	.12112	.83034	.29078	.4169	.12155	.82620	.29535
.4120	.12113	.83026	.29087	.4170	.12156	.82612	.29544
.4121	.12114	.83018	.29096	.4171	.12156	.82602	.29554
.4122	.12115	.83010	.29105	.4172	.12157	.82594	.29563
.4123	.12115	.83000	.29115	.4173	.12158	.82586	.29572
.4124	.12116	.82992	.29124	.4174	.12159	.82578	.29581
.4125	.12117	.82984	.29133	.4175	.12160	.82570	.29590
.4126	.12118	.82976	.29142	.4176	.12161	.82562	.29599
.4127	.12119	.82968	.29151	.4177	.12161	.82552	.29609
.4128	.12120	.82960	.29160	.4178	.12162	.82544	.29618
.4129	.12121	.82952	.29169	.4179	.12163	.82536	.29627
.4130	.12122	.82944	.29178	.4180	.12164	.82528	.29636
.4131	.12122	.82934	.29188	.4181	.12165	.82520	.29645
.4132	.12123	.82926	.29197	.4182	.12165	.82510	.29655
.4133	.12124	.82918	.29206	.4183	.12166	.82502	.29664
.4134	.12125	.82910	.29215	.4184	.12167	.82494	.29673
.4135	.12126	.82902	.29224	.4185	.12168	.82486	.29682
.4136	.12127	.82894	.29233	.4186	.12169	.82478	.29691
.4137	.12128	.82886	.29242	.4187	.12170	.82470	.29700
.4138	.12128	.82876	.29252	.4188	.12170	.82460	.29710
.4139	.12129	.82868	.29261	.4189	.12171	.82452	.29719
.4140	.12130	.82860	.29270	.4190	.12172	.82444	.29728
.4141	.12131	.82852	.29279	.4191	.12173	.82436	.29737
.4142	.12132	.82844	.29288	.4192	.12174	.82428	.29746
.4143	.12133	.82836	.29297	.4193	.12174	.82418	.29756
.4144	.12134	.82828	.29306	.4194	.12175	.82410	.29765
.4145	.12134	.82818	.29316	.4195	.12176	.82402	.29774
.4146	.12135	.82810	.29325	.4196	.12177	.82394	.29783
.4147	.12136	.82802	.29334	.4197	.12178	.82386	.29792
.4148	.12137	.82794	.29343	.4198	.12178	.82376	.29802
.4149	.12138	.82786	.29352	.4199	.12179	.82368	.29811
.4150	.12139	.82778	.29361	.4200	.12180	.82360	.29820

TABLE II. THREE-POINT INTERPOLATION COEFFICIENTS

p	c₋₁ −	c₀ +	c₁ +	p	c₋₁ −	c₀ +	c₁ +
.4200	.12180	.82360	.29820	.4250	.12219	.81938	.30281
.4201	.12181	.82352	.29829	.4251	.12219	.81928	.30291
.4202	.12182	.82344	.29838	.4252	.12220	.81920	.30300
.4203	.12182	.82334	.29848	.4253	.12221	.81912	.30309
.4204	.12183	.82326	.29857	.4254	.12222	.81904	.30318
.4205	.12184	.82318	.29866	.4255	.12222	.81894	.30328
.4206	.12185	.82310	.29875	.4256	.12223	.81886	.30337
.4207	.12186	.82302	.29884	.4257	.12224	.81878	.30346
.4208	.12186	.82292	.29894	.4258	.12225	.81870	.30355
.4209	.12187	.82284	.29903	.4259	.12225	.81860	.30365
.4210	.12188	.82276	.29912	.4260	.12226	.81852	.30374
.4211	.12189	.82268	.29921	.4261	.12227	.81844	.30383
.4212	.12190	.82260	.29930	.4262	.12228	.81836	.30392
.4213	.12190	.82250	.29940	.4263	.12228	.81826	.30402
.4214	.12191	.82242	.29949	.4264	.12229	.81818	.30411
.4215	.12192	.82234	.29958	.4265	.12230	.81810	.30420
.4216	.12193	.82226	.29967	.4266	.12231	.81802	.30429
.4217	.12193	.82216	.29977	.4267	.12231	.81792	.30439
.4218	.12194	.82208	.29986	.4268	.12232	.81784	.30448
.4219	.12195	.82200	.29995	.4269	.12233	.81776	.30457
.4220	.12196	.82192	.30004	.4270	.12234	.81768	.30466
.4221	.12197	.82184	.30013	.4271	.12234	.81758	.30476
.4222	.12197	.82174	.30023	.4272	.12235	.81750	.30485
.4223	.12198	.82166	.30032	.4273	.12236	.81742	.30494
.4224	.12199	.82158	.30041	.4274	.12236	.81732	.30504
.4225	.12200	.82150	.30050	.4275	.12237	.81724	.30513
.4226	.12200	.82140	.30060	.4276	.12238	.81716	.30522
.4227	.12201	.82132	.30069	.4277	.12239	.81708	.30531
.4228	.12202	.82124	.30078	.4278	.12239	.81698	.30541
.4229	.12203	.82116	.30087	.4279	.12240	.81690	.30550
.4230	.12204	.82108	.30096	.4280	.12241	.81682	.30559
.4231	.12204	.82098	.30106	.4281	.12242	.81674	.30568
.4232	.12205	.82090	.30115	.4282	.12242	.81664	.30578
.4233	.12206	.82082	.30124	.4283	.12243	.81656	.30587
.4234	.12207	.82074	.30133	.4284	.12244	.81648	.30596
.4235	.12207	.82064	.30143	.4285	.12244	.81638	.30606
.4236	.12208	.82056	.30152	.4286	.12245	.81630	.30615
.4237	.12209	.82048	.30161	.4287	.12246	.81622	.30624
.4238	.12210	.82040	.30170	.4288	.12247	.81614	.30633
.4239	.12210	.82030	.30180	.4289	.12247	.81604	.30643
.4240	.12211	.82022	.30189	.4290	.12248	.81596	.30652
.4241	.12212	.82014	.30198	.4291	.12249	.81588	.30661
.4242	.12213	.82006	.30207	.4292	.12249	.81578	.30671
.4243	.12213	.81996	.30217	.4293	.12250	.81570	.30680
.4244	.12214	.81988	.30226	.4294	.12251	.81562	.30689
.4245	.12215	.81980	.30235	.4295	.12251	.81552	.30699
.4246	.12216	.81972	.30244	.4296	.12252	.81544	.30708
.4247	.12216	.81962	.30254	.4297	.12253	.81536	.30717
.4248	.12217	.81954	.30263	.4298	.12254	.81528	.30726
.4249	.12218	.81946	.30272	.4299	.12254	.81518	.30736
.4250	.12219	.81938	.30281	.4300	.12255	.81510	.30745

TABLE II. THREE-POINT INTERPOLATION COEFFICIENTS

p	c_{-1} $-$	c_0 $+$	c_1 $+$	p	c_{-1} $-$	c_0 $+$	c_1 $+$
.4300	.12255	.81510	.30745	.4350	.12289	.81078	.31211
.4301	.12256	.81502	.30754	.4351	.12289	.81068	.31221
.4302	.12256	.81492	.30764	.4352	.12290	.81060	.31230
.4303	.12257	.81484	.30773	.4353	.12291	.81052	.31239
.4304	.12258	.81476	.30782	.4354	.12291	.81042	.31249
.4305	.12258	.81466	.30792	.4355	.12292	.81034	.31258
.4306	.12259	.81458	.30801	.4356	.12293	.81026	.31267
.4307	.12260	.81450	.30810	.4357	.12293	.81016	.31277
.4308	.12261	.81442	.30819	.4358	.12294	.81008	.31286
.4309	.12261	.81432	.30829	.4359	.12295	.81000	.31295
.4310	.12262	.81424	.30838	.4360	.12295	.80990	.31305
.4311	.12263	.81416	.30847	.4361	.12296	.80982	.31314
.4312	.12263	.81406	.30857	.4362	.12296	.80972	.31324
.4313	.12264	.81398	.30866	.4363	.12297	.80964	.31333
.4314	.12265	.81390	.30875	.4364	.12298	.80956	.31342
.4315	.12265	.81380	.30885	.4365	.12298	.80946	.31352
.4316	.12266	.81372	.30894	.4366	.12299	.80938	.31361
.4317	.12267	.81364	.30903	.4367	.12300	.80930	.31370
.4318	.12267	.81354	.30913	.4368	.12300	.80920	.31380
.4319	.12268	.81346	.30922	.4369	.12301	.80912	.31389
.4320	.12269	.81338	.30931	.4370	.12302	.80904	.31398
.4321	.12269	.81328	.30941	.4371	.12302	.80894	.31408
.4322	.12270	.81320	.30950	.4372	.12303	.80886	.31417
.4323	.12271	.81312	.30959	.4373	.12303	.80876	.31427
.4324	.12272	.81304	.30968	.4374	.12304	.80868	.31436
.4325	.12272	.81294	.30978	.4375	.12305	.80860	.31445
.4326	.12273	.81286	.30987	.4376	.12305	.80850	.31455
.4327	.12274	.81278	.30996	.4377	.12306	.80842	.31464
.4328	.12274	.81268	.31006	.4378	.12307	.80834	.31473
.4329	.12275	.81260	.31015	.4379	.12307	.80824	.31483
.4330	.12276	.81252	.31024	.4380	.12308	.80816	.31492
.4331	.12276	.81242	.31034	.4381	.12308	.80806	.31502
.4332	.12277	.81234	.31043	.4382	.12309	.80798	.31511
.4333	.12278	.81226	.31052	.4383	.12310	.80790	.31520
.4334	.12278	.81216	.31062	.4384	.12310	.80780	.31530
.4335	.12279	.81208	.31071	.4385	.12311	.80772	.31539
.4336	.12280	.81200	.31080	.4386	.12312	.80764	.31548
.4337	.12280	.81190	.31090	.4387	.12312	.80754	.31558
.4338	.12281	.81182	.31099	.4388	.12313	.80746	.31567
.4339	.12282	.81174	.31108	.4389	.12313	.80736	.31577
.4340	.12282	.81164	.31118	.4390	.12314	.80728	.31586
.4341	.12283	.81156	.31127	.4391	.12315	.80720	.31595
.4342	.12284	.81148	.31136	.4392	.12315	.80710	.31605
.4343	.12284	.81138	.31146	.4393	.12316	.80702	.31614
.4344	.12285	.81130	.31155	.4394	.12316	.80692	.31624
.4345	.12285	.81120	.31165	.4395	.12317	.80684	.31633
.4346	.12286	.81112	.31174	.4396	.12318	.80676	.31642
.4347	.12287	.81104	.31183	.4397	.12318	.80666	.31652
.4348	.12287	.81094	.31193	.4398	.12319	.80658	.31661
.4349	.12288	.81086	.31202	.4399	.12319	.80648	.31671
.4350	.12289	.81078	.31211	.4400	.12320	.80640	.31680

TABLE II. THREE-POINT INTERPOLATION COEFFICIENTS

p	c₋₁ −	c₀ +	c₁ +	p	c₋₁ −	c₀ +	c₁ +
.4400	.12320	.80640	.31680	.4450	.12349	.80198	.32151
.4401	.12321	.80632	.31689	.4451	.12349	.80188	.32161
.4402	.12321	.80622	.31699	.4452	.12350	.80180	.32170
.4403	.12322	.80614	.31708	.4453	.12350	.80170	.32180
.4404	.12322	.80604	.31718	.4454	.12351	.80162	.32189
.4405	.12323	.80596	.31727	.4455	.12351	.80152	.32199
.4406	.12324	.80588	.31736	.4456	.12352	.80144	.32208
.4407	.12324	.80578	.31746	.4457	.12353	.80136	.32217
.4408	.12325	.80570	.31755	.4458	.12353	.80126	.32227
.4409	.12325	.80560	.31765	.4459	.12354	.80118	.32236
.4410	.12326	.80552	.31774	.4460	.12354	.80108	.32246
.4411	.12327	.80544	.31783	.4461	.12355	.80100	.32255
.4412	.12327	.80534	.31793	.4462	.12355	.80090	.32265
.4413	.12328	.80526	.31802	.4463	.12356	.80082	.32274
.4414	.12328	.80516	.31812	.4464	.12356	.80072	.32284
.4415	.12329	.80508	.31821	.4465	.12357	.80064	.32293
.4416	.12329	.80498	.31831	.4466	.12357	.80054	.32303
.4417	.12330	.80490	.31840	.4467	.12358	.80046	.32312
.4418	.12331	.80482	.31849	.4468	.12358	.80036	.32322
.4419	.12331	.80472	.31859	.4469	.12359	.80028	.32331
.4420	.12332	.80464	.31868	.4470	.12360	.80020	.32340
.4421	.12332	.80454	.31878	.4471	.12360	.80010	.32350
.4422	.12333	.80446	.31887	.4472	.12361	.80002	.32359
.4423	.12334	.80438	.31896	.4473	.12361	.79992	.32369
.4424	.12334	.80428	.31906	.4474	.12362	.79984	.32378
.4425	.12335	.80420	.31915	.4475	.12362	.79974	.32388
.4426	.12335	.80410	.31925	.4476	.12363	.79966	.32397
.4427	.12336	.80402	.31934	.4477	.12363	.79956	.32407
.4428	.12336	.80392	.31944	.4478	.12364	.79948	.32416
.4429	.12337	.80384	.31953	.4479	.12364	.79938	.32426
.4430	.12338	.80376	.31962	.4480	.12365	.79930	.32435
.4431	.12338	.80366	.31972	.4481	.12365	.79920	.32445
.4432	.12339	.80358	.31981	.4482	.12366	.79912	.32454
.4433	.12339	.80348	.31991	.4483	.12366	.79902	.32464
.4434	.12340	.80340	.32000	.4484	.12367	.79894	.32473
.4435	.12340	.80330	.32010	.4485	.12367	.79884	.32483
.4436	.12341	.80322	.32019	.4486	.12368	.79876	.32492
.4437	.12342	.80314	.32028	.4487	.12368	.79866	.32502
.4438	.12342	.80304	.32038	.4488	.12369	.79858	.32511
.4439	.12343	.80296	.32047	.4489	.12369	.79848	.32521
.4440	.12343	.80286	.32057	.4490	.12370	.79840	.32530
.4441	.12344	.80278	.32066	.4491	.12370	.79830	.32540
.4442	.12344	.80268	.32076	.4492	.12371	.79822	.32549
.4443	.12345	.80260	.32085	.4493	.12371	.79812	.32559
.4444	.12345	.80250	.32095	.4494	.12372	.79804	.32568
.4445	.12346	.80242	.32104	.4495	.12372	.79794	.32578
.4446	.12347	.80234	.32113	.4496	.12373	.79786	.32587
.4447	.12347	.80224	.32123	.4497	.12373	.79776	.32597
.4448	.12348	.80216	.32132	.4498	.12374	.79768	.32606
.4449	.12348	.80206	.32142	.4499	.12374	.79758	.32616
.4450	.12349	.80198	.32151	.4500	.12375	.79750	.32625

TABLE II. THREE-POINT INTERPOLATION COEFFICIENTS

p	c_{-1} −	c_0 +	c_1 +	p	c_{-1} −	c_0 +	c_1 +
.4500	.12375	.79750	.32625	.4550	.12399	.79298	.33101
.4501	.12375	.79740	.32635	.4551	.12399	.79288	.33111
.4502	.12376	.79732	.32644	.4552	.12400	.79280	.33120
.4503	.12376	.79722	.32654	.4553	.12400	.79270	.33130
.4504	.12377	.79714	.32663	.4554	.12401	.79262	.33139
.4505	.12377	.79704	.32673	.4555	.12401	.79252	.33149
.4506	.12378	.79696	.32682	.4556	.12401	.79242	.33159
.4507	.12378	.79686	.32692	.4557	.12402	.79234	.33168
.4508	.12379	.79678	.32701	.4558	.12402	.79224	.33178
.4509	.12379	.79668	.32711	.4559	.12403	.79216	.33187
.4510	.12380	.79660	.32720	.4560	.12403	.79206	.33197
.4511	.12380	.79650	.32730	.4561	.12404	.79198	.33206
.4512	.12381	.79642	.32739	.4562	.12404	.79188	.33216
.4513	.12381	.79632	.32749	.4563	.12405	.79180	.33225
.4514	.12382	.79624	.32758	.4564	.12405	.79170	.33235
.4515	.12382	.79614	.32768	.4565	.12405	.79160	.33245
.4516	.12383	.79606	.32777	.4566	.12406	.79152	.33254
.4517	.12383	.79596	.32787	.4567	.12406	.79142	.33264
.4518	.12384	.79588	.32796	.4568	.12407	.79134	.33273
.4519	.12384	.79578	.32806	.4569	.12407	.79124	.33283
.4520	.12385	.79570	.32815	.4570	.12408	.79116	.33292
.4521	.12385	.79560	.32825	.4571	.12408	.79106	.33302
.4522	.12386	.79552	.32834	.4572	.12408	.79096	.33312
.4523	.12386	.79542	.32844	.4573	.12409	.79088	.33321
.4524	.12387	.79534	.32853	.4574	.12409	.79078	.33331
.4525	.12387	.79524	.32863	.4575	.12410	.79070	.33340
.4526	.12388	.79516	.32872	.4576	.12410	.79060	.33350
.4527	.12388	.79506	.32882	.4577	.12411	.79052	.33359
.4528	.12389	.79498	.32891	.4578	.12411	.79042	.33369
.4529	.12389	.79488	.32901	.4579	.12411	.79032	.33379
.4530	.12390	.79480	.32910	.4580	.12412	.79024	.33388
.4531	.12390	.79470	.32920	.4581	.12412	.79014	.33398
.4532	.12390	.79460	.32930	.4582	.12413	.79006	.33407
.4533	.12391	.79452	.32939	.4583	.12413	.78996	.33417
.4534	.12391	.79442	.32949	.4584	.12413	.78986	.33427
.4535	.12392	.79434	.32958	.4585	.12414	.78978	.33436
.4536	.12392	.79424	.32968	.4586	.12414	.78968	.33446
.4537	.12393	.79416	.32977	.4587	.12415	.78960	.33455
.4538	.12393	.79406	.32987	.4588	.12415	.78950	.33465
.4539	.12394	.79398	.32996	.4589	.12416	.78942	.33474
.4540	.12394	.79388	.33006	.4590	.12416	.78932	.33484
.4541	.12395	.79380	.33015	.4591	.12416	.78922	.33494
.4542	.12395	.79370	.33025	.4592	.12417	.78914	.33503
.4543	.12396	.79362	.33034	.4593	.12417	.78904	.33513
.4544	.12396	.79352	.33044	.4594	.12418	.78896	.33522
.4545	.12396	.79342	.33054	.4595	.12418	.78886	.33532
.4546	.12397	.79334	.33063	.4596	.12418	.78876	.33542
.4547	.12397	.79324	.33073	.4597	.12419	.78868	.33551
.4548	.12398	.79316	.33082	.4598	.12419	.78858	.33561
.4549	.12398	.79306	.33092	.4599	.12420	.78850	.33570
.4550	.12399	.79298	.33101	.4600	.12420	.78840	.33580

TABLE II. THREE-POINT INTERPOLATION COEFFICIENTS

p	c_{-1} −	c_0 +	c_1 +	p	c_{-1} −	c_0 +	c_1 +
.4600	.12420	.78840	.33580	.4650	.12439	.78378	.34061
.4601	.12420	.78830	.33590	.4651	.12439	.78368	.34071
.4602	.12421	.78822	.33599	.4652	.12439	.78358	.34081
.4603	.12421	.78812	.33609	.4653	.12440	.78350	.34090
.4604	.12422	.78804	.33618	.4654	.12440	.78340	.34100
.4605	.12422	.78794	.33628	.4655	.12440	.78330	.34110
.4606	.12422	.78784	.33638	.4656	.12441	.78322	.34119
.4607	.12423	.78776	.33647	.4657	.12441	.78312	.34129
.4608	.12423	.78766	.33657	.4658	.12442	.78304	.34138
.4609	.12424	.78758	.33666	.4659	.12442	.78294	.34148
.4610	.12424	.78748	.33676	.4660	.12442	.78284	.34158
.4611	.12424	.78738	.33686	.4661	.12443	.78276	.34167
.4612	.12425	.78730	.33695	.4662	.12443	.78266	.34177
.4613	.12425	.78720	.33705	.4663	.12443	.78256	.34187
.4614	.12426	.78712	.33714	.4664	.12444	.78248	.34196
.4615	.12426	.78702	.33724	.4665	.12444	.78238	.34206
.4616	.12426	.78692	.33734	.4666	.12444	.78228	.34216
.4617	.12427	.78684	.33743	.4667	.12445	.78220	.34225
.4618	.12427	.78674	.33753	.4668	.12445	.78210	.34235
.4619	.12427	.78664	.33763	.4669	.12445	.78200	.34245
.4620	.12428	.78656	.33772	.4670	.12446	.78192	.34254
.4621	.12428	.78646	.33782	.4671	.12446	.78182	.34264
.4622	.12429	.78638	.33791	.4672	.12446	.78172	.34274
.4623	.12429	.78628	.33801	.4673	.12447	.78164	.34283
.4624	.12429	.78618	.33811	.4674	.12447	.78154	.34293
.4625	.12430	.78610	.33820	.4675	.12447	.78144	.34303
.4626	.12430	.78600	.33830	.4676	.12448	.78136	.34312
.4627	.12430	.78590	.33840	.4677	.12448	.78126	.34322
.4628	.12431	.78582	.33849	.4678	.12448	.78116	.34332
.4629	.12431	.78572	.33859	.4679	.12448	.78106	.34342
.4630	.12432	.78564	.33868	.4680	.12449	.78098	.34351
.4631	.12432	.78554	.33878	.4681	.12449	.78088	.34361
.4632	.12432	.78544	.33888	.4682	.12449	.78078	.34371
.4633	.12433	.78536	.33897	.4683	.12450	.78070	.34380
.4634	.12433	.78526	.33907	.4684	.12450	.78060	.34390
.4635	.12433	.78516	.33917	.4685	.12450	.78050	.34400
.4636	.12434	.78508	.33926	.4686	.12451	.78042	.34409
.4637	.12434	.78498	.33936	.4687	.12451	.78032	.34419
.4638	.12434	.78488	.33946	.4688	.12451	.78022	.34429
.4639	.12435	.78480	.33955	.4689	.12452	.78014	.34438
.4640	.12435	.78470	.33965	.4690	.12452	.78004	.34448
.4641	.12436	.78462	.33974	.4691	.12452	.77994	.34458
.4642	.12436	.78452	.33984	.4692	.12453	.77986	.34467
.4643	.12436	.78442	.33994	.4693	.12453	.77976	.34477
.4644	.12437	.78434	.34003	.4694	.12453	.77966	.34487
.4645	.12437	.78424	.34013	.4695	.12453	.77956	.34497
.4646	.12437	.78414	.34023	.4696	.12454	.77948	.34506
.4647	.12438	.78406	.34032	.4697	.12454	.77938	.34516
.4648	.12438	.78396	.34042	.4698	.12454	.77928	.34526
.4649	.12438	.78386	.34052	.4699	.12455	.77920	.34535
.4650	.12439	.78378	.34061	.4700	.12455	.77910	.34545

TABLE II. THREE-POINT INTERPOLATION COEFFICIENTS

p	c_{-1} −	c_0 +	c_1 +	p	c_{-1} −	c_0 +	c_1 +
.4700	.12455	.77910	.34545	.4750	.12469	.77438	.35031
.4701	.12455	.77900	.34555	.4751	.12469	.77428	.35041
.4702	.12456	.77892	.34564	.4752	.12469	.77418	.35051
.4703	.12456	.77882	.34574	.4753	.12469	.77408	.35061
.4704	.12456	.77872	.34584	.4754	.12470	.77400	.35070
.4705	.12456	.77862	.34594	.4755	.12470	.77390	.35080
.4706	.12457	.77854	.34603	.4756	.12470	.77380	.35090
.4707	.12457	.77844	.34613	.4757	.12470	.77370	.35100
.4708	.12457	.77834	.34623	.4758	.12471	.77362	.35109
.4709	.12458	.77826	.34632	.4759	.12471	.77352	.35119
.4710	.12458	.77816	.34642	.4760	.12471	.77342	.35129
.4711	.12458	.77806	.34652	.4761	.12471	.77332	.35139
.4712	.12459	.77798	.34661	.4762	.12472	.77324	.35148
.4713	.12459	.77788	.34671	.4763	.12472	.77314	.35158
.4714	.12459	.77778	.34681	.4764	.12472	.77304	.35168
.4715	.12459	.77768	.34691	.4765	.12472	.77294	.35178
.4716	.12460	.77760	.34700	.4766	.12473	.77286	.35187
.4717	.12460	.77750	.34710	.4767	.12473	.77276	.35197
.4718	.12460	.77740	.34720	.4768	.12473	.77266	.35207
.4719	.12461	.77732	.34729	.4769	.12473	.77256	.35217
.4720	.12461	.77722	.34739	.4770	.12474	.77248	.35226
.4721	.12461	.77712	.34749	.4771	.12474	.77238	.35236
.4722	.12461	.77702	.34759	.4772	.12474	.77228	.35246
.4723	.12462	.77694	.34768	.4773	.12474	.77218	.35256
.4724	.12462	.77684	.34778	.4774	.12474	.77208	.35266
.4725	.12462	.77674	.34788	.4775	.12475	.77200	.35275
.4726	.12462	.77664	.34798	.4776	.12475	.77190	.35285
.4727	.12463	.77656	.34807	.4777	.12475	.77180	.35295
.4728	.12463	.77646	.34817	.4778	.12475	.77170	.35305
.4729	.12463	.77636	.34827	.4779	.12476	.77162	.35314
.4730	.12464	.77628	.34836	.4780	.12476	.77152	.35324
.4731	.12464	.77618	.34846	.4781	.12476	.77142	.35334
.4732	.12464	.77608	.34856	.4782	.12476	.77132	.35344
.4733	.12464	.77598	.34866	.4783	.12476	.77122	.35354
.4734	.12465	.77590	.34875	.4784	.12477	.77114	.35363
.4735	.12465	.77580	.34885	.4785	.12477	.77104	.35373
.4736	.12465	.77570	.34895	.4786	.12477	.77094	.35383
.4737	.12465	.77560	.34905	.4787	.12477	.77084	.35393
.4738	.12466	.77552	.34914	.4788	.12478	.77076	.35402
.4739	.12466	.77542	.34924	.4789	.12478	.77066	.35412
.4740	.12466	.77532	.34934	.4790	.12478	.77056	.35422
.4741	.12466	.77522	.34944	.4791	.12478	.77046	.35432
.4742	.12467	.77514	.34953	.4792	.12478	.77036	.35442
.4743	.12467	.77504	.34963	.4793	.12479	.77028	.35451
.4744	.12467	.77494	.34973	.4794	.12479	.77018	.35461
.4745	.12467	.77484	.34983	.4795	.12479	.77008	.35471
.4746	.12468	.77476	.34992	.4796	.12479	.76998	.35481
.4747	.12468	.77466	.35002	.4797	.12479	.76988	.35491
.4748	.12468	.77456	.35012	.4798	.12480	.76980	.35500
.4749	.12468	.77446	.35022	.4799	.12480	.76970	.35510
.4750	.12469	.77438	.35031	.4800	.12480	.76960	.35520

TABLE II. THREE-POINT INTERPOLATION COEFFICIENTS

p	c_{-1} $-$	c_0 $+$	c_1 $+$	p	c_{-1} $-$	c_0 $+$	c_1 $+$
.4800	.12480	.76960	.35520	.4850	.12489	.76478	.36011
.4801	.12480	.76950	.35530	.4851	.12489	.76468	.36021
.4802	.12480	.76940	.35540	.4852	.12489	.76458	.36031
.4803	.12481	.76932	.35549	.4853	.12489	.76448	.36041
.4804	.12481	.76922	.35559	.4854	.12489	.76438	.36051
.4805	.12481	.76912	.35569	.4855	.12489	.76428	.36061
.4806	.12481	.76902	.35579	.4856	.12490	.76420	.36070
.4807	.12481	.76892	.35589	.4857	.12490	.76410	.36080
.4808	.12482	.76884	.35598	.4858	.12490	.76400	.36090
.4809	.12482	.76874	.35608	.4859	.12490	.76390	.36100
.4810	.12482	.76864	.35618	.4860	.12490	.76380	.36110
.4811	.12482	.76854	.35628	.4861	.12490	.76370	.36120
.4812	.12482	.76844	.35638	.4862	.12490	.76360	.36130
.4813	.12483	.76836	.35647	.4863	.12491	.76352	.36139
.4814	.12483	.76826	.35657	.4864	.12491	.76342	.36149
.4815	.12483	.76816	.35667	.4865	.12491	.76332	.36159
.4816	.12483	.76806	.35677	.4866	.12491	.76322	.36169
.4817	.12483	.76796	.35687	.4867	.12491	.76312	.36179
.4818	.12483	.76786	.35697	.4868	.12491	.76302	.36189
.4819	.12484	.76778	.35706	.4869	.12491	.76292	.36199
.4820	.12484	.76768	.35716	.4870	.12492	.76284	.36208
.4821	.12484	.76758	.35726	.4871	.12492	.76274	.36218
.4822	.12484	.76748	.35736	.4872	.12492	.76264	.36228
.4823	.12484	.76738	.35746	.4873	.12492	.76254	.36238
.4824	.12485	.76730	.35755	.4874	.12492	.76244	.36248
.4825	.12485	.76720	.35765	.4875	.12492	.76234	.36258
.4826	.12485	.76710	.35775	.4876	.12492	.76224	.36268
.4827	.12485	.76700	.35785	.4877	.12492	.76214	.36278
.4828	.12485	.76690	.35795	.4878	.12493	.76206	.36287
.4829	.12485	.76680	.35805	.4879	.12493	.76196	.36297
.4830	.12486	.76672	.35814	.4880	.12493	.76186	.36307
.4831	.12486	.76662	.35824	.4881	.12493	.76176	.36317
.4832	.12486	.76652	.35834	.4882	.12493	.76166	.36327
.4833	.12486	.76642	.35844	.4883	.12493	.76156	.36337
.4834	.12486	.76632	.35854	.4884	.12493	.76146	.36347
.4835	.12486	.76622	.35864	.4885	.12493	.76136	.36357
.4836	.12487	.76614	.35873	.4886	.12494	.76128	.36366
.4837	.12487	.76604	.35883	.4887	.12494	.76118	.36376
.4838	.12487	.76594	.35893	.4888	.12494	.76108	.36386
.4839	.12487	.76584	.35903	.4889	.12494	.76098	.36396
.4840	.12487	.76574	.35913	.4890	.12494	.76088	.36406
.4841	.12487	.76564	.35923	.4891	.12494	.76078	.36416
.4842	.12488	.76556	.35932	.4892	.12494	.76068	.36426
.4843	.12488	.76546	.35942	.4893	.12494	.76058	.36436
.4844	.12488	.76536	.35952	.4894	.12494	.76048	.36446
.4845	.12488	.76526	.35962	.4895	.12494	.76038	.36456
.4846	.12488	.76516	.35972	.4896	.12495	.76030	.36465
.4847	.12488	.76506	.35982	.4897	.12495	.76020	.36475
.4848	.12488	.76496	.35992	.4898	.12495	.76010	.36485
.4849	.12489	.76488	.36001	.4899	.12495	.76000	.36495
.4850	.12489	.76478	.36011	.4900	.12495	.75990	.36505

TABLE II. THREE-POINT INTERPOLATION COEFFICIENTS

p	c_{-1} −	c_0 +	c_1 +	p	c_{-1} −	c_0 +	c_1 +
.4900	.12495	.75990	.36505	.4950	.12499	.75498	.37001
.4901	.12495	.75980	.36515	.4951	.12499	.75488	.37011
.4902	.12495	.75970	.36525	.4952	.12499	.75478	.37021
.4903	.12495	.75960	.36535	.4953	.12499	.75468	.37031
.4904	.12495	.75950	.36545	.4954	.12499	.75458	.37041
.4905	.12495	.75940	.36555	.4955	.12499	.75448	.37051
.4906	.12496	.75932	.36564	.4956	.12499	.75438	.37061
.4907	.12496	.75922	.36574	.4957	.12499	.75428	.37071
.4908	.12496	.75912	.36584	.4958	.12499	.75418	.37081
.4909	.12496	.75902	.36594	.4959	.12499	.75408	.37091
.4910	.12496	.75892	.36604	.4960	.12499	.75398	.37101
.4911	.12496	.75882	.36614	.4961	.12499	.75388	.37111
.4912	.12496	.75872	.36624	.4962	.12499	.75378	.37121
.4913	.12496	.75862	.36634	.4963	.12499	.75368	.37131
.4914	.12496	.75852	.36644	.4964	.12499	.75358	.37141
.4915	.12496	.75842	.36654	.4965	.12499	.75348	.37151
.4916	.12496	.75832	.36664	.4966	.12499	.75338	.37161
.4917	.12497	.75824	.36673	.4967	.12499	.75328	.37171
.4918	.12497	.75814	.36683	.4968	.12499	.75318	.37181
.4919	.12497	.75804	.36693	.4969	.12500	.75310	.37190
.4920	.12497	.75794	.36703	.4970	.12500	.75300	.37200
.4921	.12497	.75784	.36713	.4971	.12500	.75290	.37210
.4922	.12497	.75774	.36723	.4972	.12500	.75280	.37220
.4923	.12497	.75764	.36733	.4973	.12500	.75270	.37230
.4924	.12497	.75754	.36743	.4974	.12500	.75260	.37240
.4925	.12497	.75744	.36753	.4975	.12500	.75250	.37250
.4926	.12497	.75734	.36763	.4976	.12500	.75240	.37260
.4927	.12497	.75724	.36773	.4977	.12500	.75230	.37270
.4928	.12497	.75714	.36783	.4978	.12500	.75220	.37280
.4929	.12497	.75704	.36793	.4979	.12500	.75210	.37290
.4930	.12498	.75696	.36802	.4980	.12500	.75200	.37300
.4931	.12498	.75686	.36812	.4981	.12500	.75190	.37310
.4932	.12498	.75676	.36822	.4982	.12500	.75180	.37320
.4933	.12498	.75666	.36832	.4983	.12500	.75170	.37330
.4934	.12498	.75656	.36842	.4984	.12500	.75160	.37340
.4935	.12498	.75646	.36852	.4985	.12500	.75150	.37350
.4936	.12498	.75636	.36862	.4986	.12500	.75140	.37360
.4937	.12498	.75626	.36872	.4987	.12500	.75130	.37370
.4938	.12498	.75616	.36882	.4988	.12500	.75120	.37380
.4939	.12498	.75606	.36892	.4989	.12500	.75110	.37390
.4940	.12498	.75596	.36902	.4990	.12500	.75100	.37400
.4941	.12498	.75586	.36912	.4991	.12500	.75090	.37410
.4942	.12498	.75576	.36922	.4992	.12500	.75080	.37420
.4943	.12498	.75566	.36932	.4993	.12500	.75070	.37430
.4944	.12498	.75556	.36942	.4994	.12500	.75060	.37440
.4945	.12498	.75546	.36952	.4995	.12500	.75050	.37450
.4946	.12499	.75538	.36961	.4996	.12500	.75040	.37460
.4947	.12499	.75528	.36971	.4997	.12500	.75030	.37470
.4948	.12499	.75518	.36981	.4998	.12500	.75020	.37480
.4949	.12499	.75508	.36991	.4999	.12500	.75010	.37490
.4950	.12499	.75498	.37001	.5000	.12500	.75000	.37500

TABLE III. FOUR-POINT INTERPOLATION COEFFICIENTS

p ($p<.5$)	c_{-1} −	c_0 +	c_1 +	c_2 −	p ($p>.5$)
.000	.00000 00	1.00000 00	.00000 00	.00000 00	1.000
.001	.00033 28	.99949 89	.00100 06	.00016 67	.999
.002	.00066 47	.99899 61	.00200 19	.00033 33	.998
.003	.00099 55	.99849 10	.00300 45	.00050 00	.997
.004	.00132 53	.99798 39	.00400 81	.00066 67	.996
.005	.00165 42	.99747 51	.00501 24	.00083 33	.995
.006	.00198 20	.99696 40	.00601 80	.00100 00	.994
.007	.00230 89	.99645 12	.00702 43	.00116 66	.993
.008	.00263 48	.99593 64	.00803 16	.00133 32	.992
.009	.00295 96	.99541 93	.00904 02	.00149 99	.991
.010	.00328 35	.99490 05	.01004 95	.00166 65	.990
.011	.00360 64	.99437 97	.01105 98	.00183 31	.989
.012	.00392 83	.99385 69	.01207 11	.00199 97	.988
.013	.00424 92	.99333 21	.01308 34	.00216 63	.987
.014	.00456 91	.99280 53	.01409 67	.00233 29	.986
.015	.00488 81	.99227 68	.01511 07	.00249 94	.985
.016	.00520 60	.99174 60	.01612 60	.00266 60	.984
.017	.00552 30	.99121 35	.01714 20	.00283 25	.983
.018	.00583 90	.99067 90	.01815 90	.00299 90	.982
.019	.00615 40	.99014 25	.01917 70	.00316 55	.981
.020	.00646 80	.98960 40	.02019 60	.00333 20	.980
.021	.00678 10	.98906 35	.02121 60	.00349 85	.979
.022	.00709 31	.98852 13	.02223 67	.00366 49	.978
.023	.00740 42	.98797 71	.02325 84	.00383 13	.977
.024	.00771 43	.98743 09	.02428 11	.00399 77	.976
.025	.00802 34	.98688 27	.02530 48	.00416 41	.975
.026	.00833 16	.98633 28	.02632 92	.00433 04	.974
.027	.00863 88	.98578 09	.02735 46	.00449 67	.973
.028	.00894 50	.98522 70	.02838 10	.00466 30	.972
.029	.00925 02	.98467 11	.02940 84	.00482 93	.971
.030	.00955 45	.98411 35	.03043 65	.00499 55	.970
.031	.00985 78	.98355 39	.03146 56	.00516 17	.969
.032	.01016 01	.98299 23	.03249 57	.00532 79	.968
.033	.01046 15	.98242 90	.03352 65	.00549 40	.967
.034	.01076 19	.98186 37	.03455 83	.00566 01	.966
.035	.01106 13	.98129 64	.03559 11	.00582 62	.965
.036	.01135 98	.98072 74	.03662 46	.00599 22	.964
.037	.01165 73	.98015 64	.03765 91	.00615 82	.963
.038	.01195 38	.97958 34	.03869 46	.00632 42	.962
.039	.01224 94	.97900 87	.03973 08	.00649 01	.961
.040	.01254 40	.97843 20	.04076 80	.00665 60	.960
.041	.01283 77	.97785 36	.04180 59	.00682 18	.959
.042	.01313 03	.97727 29	.04284 51	.00698 77	.958
.043	.01342 21	.97669 08	.04388 47	.00715 34	.957
.044	.01371 29	.97610 67	.04492 53	.00731 91	.956
.045	.01400 27	.97552 06	.04596 69	.00748 48	.955
.046	.01429 16	.97493 28	.04700 92	.00765 04	.954
.047	.01457 95	.97434 30	.04805 25	.00781 60	.953
.048	.01486 64	.97375 12	.04909 68	.00798 16	.952
.049	.01515 24	.97315 77	.05014 18	.00814 71	.951
.050	.01543 75	.97256 25	.05118 75	.00831 25	.950
	− c_2	+ c_1	+ c_0	− c_{-1}	

TABLE III. FOUR-POINT INTERPOLATION COEFFICIENTS

p ($p<.5$)	c_{-1} $-$	c_0 $+$	c_1 $+$	c_2 $-$	p ($p>.5$)
.050	.01543 75	.97256 25	.05118 75	.00831 25	.950
.051	.01572 16	.97196 53	.05223 42	.00847 79	.949
.052	.01600 48	.97136 64	.05328 16	.00864 32	.948
.053	.01628 70	.97076 55	.05433 00	.00880 85	.947
,054	.01656 82	.97016 26	.05537 94	.00897 38	.946
.055	.01684 86	.96955 83	.05642 92	.00913 89	.945
.056	.01712 79	.96895 17	.05738 03	.00930 41	.944
.057	.01740 64	.96834 37	.05853 18	.00946 91	.943
.058	.01768 39	.96773 37	.05958 43	.00963 41	.942
.059	.01796 04	.96712 17	.06063 78	.00979 91	.941
.060	.01823 60	.96650 80	.06169 20	.00996 40	.940
.061	.01851 07	.96589 26	.06274 69	.01012 88	.939
.062	.01878 44	.96527 52	.06380 28	.01029 36	.938
.063	.01905 72	.96465 61	.06485 94	.01045 83	.937
.064	.01932 90	.96403 50	.06591 70	.01062 30	.936
.065	.01959 99	.96341 22	.06697 53	.01078 76	.935
.066	.01986 99	.96278 77	.06803 43	.01095 21	.934
.067	.02013 90	.96216 15	.06909 40	.01111 65	.933
.068	.02040 71	.96153 33	.07015 47	.01128 09	.932
.069	.02067 43	.96090 34	.07121 61	.01144 52	.931
.070	.02094 05	.96027 15	.07227 85	.01160 95	.930
.071	.02120 58	.95963 79	.07334 16	.01177 37	.929
.072	.02147 02	.95900 26	.07440 54	.01193 78	.928
.073	.02173 37	.95836 56	.07546 99	.01210 18	.927
.074	.02199 62	.95772 66	.07653 54	.01226 58	.926
.075	.02225 78	.95708 59	.07760 16	.01242 97	.925
.076	.02251 85	.95644 35	.07866 85	.01259 35	.924
.077	.02277 83	.95579 94	.07973 61	.01275 72	.923
.078	.02303 71	.95515 33	.08080 47	.01292 09	.922
.079	.02329 50	.95450 55	.08187 40	.01308 45	.921
.080	.02355 20	.95385 60	.08294 40	.01324 80	.920
.081	.02380 81	.95320 48	.08401 47	.01341 14	.919
.082	.02406 32	.95255 16	.08508 64	.01357 48	.918
.083	.02431 75	.95189 70	.08615 85	.01373 80	.917
.084	.02457 08	.95124 04	.08723 16	.01390 12	.916
.085	.02482 32	.95058 21	.08830 54	.01406 43	.915
.086	.02507 47	.94992 21	.08937 99	.01422 73	.914
.087	.02532 53	.94926 04	.09045 51	.01439 02	.913
.088	.02557 49	.94859 67	.09153 13	.01455 31	.912
.089	.02582 37	.94793 16	.09260 79	.01471 58	.911
.090	.02607 15	.94726 45	.09368 55	.01487 85	.910
.091	.02631 84	.94659 57	.09476 38	.01504 11	.909
.092	.02656 44	.94592 52	.09584 28	.01520 36	.908
.093	.02680 96	.94525 33	.09692 22	.01536 59	.907
.094	.02705 38	.94457 94	.09800 26	.01552 82	.906
.095	.02729 71	.94390 38	.09908 37	.01569 04	.905
.096	.02753 95	.94322 65	.10016 55	.01585 25	.904
.097	.02778 09	.94254 72	.10124 83	.01601 46	.903
.098	.02802 15	.94186 65	.10233 15	.01617 65	.902
.099	.02826 12	.94118 41	.10341 54	.01633 83	.901
.100	.02850 00	.94050 00	.10450 00	.01650 00	.900
	$-$ c_2	$+$ c_1	$+$ c_0	$-$ c_{-1}	

TABLE III. FOUR-POINT INTERPOLATION COEFFICIENTS

p ($p<.5$)	c_{-1} −	c_0 +	c_1 +	c_2 −	p ($p>.5$)
.100	.02850 00	.94050 00	.10450 00	.01650 00	.900
.101	.02873 79	.93981 42	.10558 53	.01666 16	.899
.102	.02897 49	.93912 67	.10667 13	.01682 31	.898
.103	.02921 10	.93843 75	.10775 80	.01698 45	.897
.104	.02944 61	.93774 63	.10884 57	.01714 59	.896
.105	.02968 04	.93705 37	.10993 38	.01730 71	.895
.106	.02991 38	.93635 94	.11102 26	.01746 82	.894
.107	.03014 63	.93566 34	.11211 21	.01762 92	.893
.108	.03037 80	.93496 60	.11320 20	.01779 00	.892
.109	.03060 87	.93426 66	.11429 29	.01795 08	.891
.110	.03083 85	.93356 55	.11538 45	.01811 15	.890
.111	.03106 74	.93286 27	.11647 68	.01827 21	.889
.112	.03129 55	.93215 85	.11756 95	.01843 25	.888
.113	.03152 26	.93145 23	.11866 32	.01859 29	.887
.114	.03174 89	.93074 47	.11975 73	.01875 31	.886
.115	.03197 43	.93003 54	.12085 21	.01891 32	.885
.116	.03219 88	.92932 44	.12194 76	.01907 32	.884
.117	.03242 24	.92861 17	.12304 38	.01923 31	.883
.118	.03264 52	.92789 76	.12414 04	.01939 28	.882
.119	.03286 70	.92718 15	.12523 80	.01955 25	.881
.120	.03308 80	.92646 40	.12633 60	.01971 20	.880
.021	.03330 81	.92574 48	.12743 47	.01987 14	.879
.121	.03352 73	.92502 39	.12853 41	.02003 07	.878
.123	.03374 56	.92430 13	.12963 42	.02018 99	.877
.124	.03396 31	.92357 73	.13073 47	.02034 89	.876
.125	.03417 97	.92285 16	.13183 59	.02050 78	.875
.126	.03439 54	.92212 42	.13293 78	.02066 66	.874
.127	.03461 02	.92139 51	.13404 04	.02082 53	.873
.128	.03482 42	.92066 46	.13514 34	.02098 38	.872
.129	.03503 73	.91993 24	.13624 71	.02114 22	.871
.130	.03524 95	.91919 85	.13735 15	.02130 05	.870
.131	.03546 08	.91846 29	.13845 66	.02145 87	.869
.132	.03567 13	.91772 59	.13956 21	.02161 67	.968
.133	.03588 09	.91698 72	.14066 83	.02177 46	.867
.134	.03608 97	.91624 71	.14177 49	.02193 23	.866
.135	.03629 76	.91550 53	.14288 22	.02208 99	.865
.136	.03650 46	.91476 18	.14399 02	.02224 74	.864
.137	.03671 07	.91401 66	.14509 89	.02240 48	.863
.138	.03691 60	.91327 00	.14620 80	.02256 20	.862
.139	.03712 04	.91252 17	.14731 78	.02271 91	.861
.140	.03732 40	.91177 20	.14842 80	.02287 60	.860
.141	.03752 67	.91102 06	.14953 89	.02303 28	.859
.142	.03772 85	.91026 75	.15065 05	.02318 95	.858
.143	.03792 95	.90951 30	.15176 25	.02334 60	.857
.144	.03812 97	.90875 71	.15287 49	.02350 23	.856
.145	.03832 89	.90799 92	.15398 83	.02365 86	.855
.146	.03852 74	.90724 02	.15510 18	.02381 46	.854
.147	.03872 49	.90647 92	.15621 63	.02397 06	.853
.148	.03892 16	.90571 68	.15733 12	.02412 64	.852
.149	.03911 75	.90495 30	.15844 65	.02428 20	.851
.150	.03931 25	.90418 75	.15956 25	.02443 75	.850
	− c_2	+ c_1	+ c_0	− c_{-1}	

TABLE III. FOUR-POINT INTERPOLATION COEFFICIENTS

p ($p<.5$)	c_{-1} $-$	c_0 $+$	c_1 $+$	c_2 $-$	p ($p>.5$)
.150	.03931 25	.90418 75	.15956 25	.02443 75	.850
.151	.03950 67	.90342 06	.16067 89	.02459 28	.849
.152	.03970 00	.90265 20	.16179 60	.02474 80	.848
.153	.03989 24	.90188 17	.16291 38	.02490 31	.847
.154	.04008 40	.90111 00	.16403 20	.02505 80	.846
.155	.04027 48	.90033 69	.16515 06	.02521 27	.845
.156	.04046 47	.89956 21	.16626 99	.02536 73	.844
.157	.04065 38	.89878 59	.16738 96	.02552 17	.843
.158	.04084 21	.89800 83	.16850 97	.02567 59	.842
.159	.04102 94	.89722 87	.16963 08	.02583 01	.841
.160	.04121 60	.89644 80	.17075 20	.02598 40	.840
.161	.04140 17	.89566 56	.17187 39	.02613 78	.839
.162	.04158 66	.89488 18	.17299 62	.02629 14	.838
.163	.04177 06	.89409 63	.17411 92	.02644 49	.837
.164	.04195 38	.89330 94	.17524 26	.02659 82	.836
.165	.04213 62	.89252 11	.17636 64	.02675 13	.835
.166	.04231 77	.89173 11	.17749 09	.02690 43	.834
.167	.04249 84	.89093 97	.17861 58	.02705 71	.833
.168	.04267 83	.89014 69	.17974 11	.02720 97	.832
.169	.04285 73	.88935 24	.18086 71	.02736 22	.831
.170	.04303 55	.88855 65	.18199 35	.02751 45	.830
.171	.04321 29	.88775 92	.18312 03	.02766 66	.829
.172	.04338 94	.88696 02	.18424 78	.02781 86	.828
.173	.04356 51	.88615 98	.18537 57	.02797 04	.827
.174	.04374 00	.88535 80	.18650 40	.02812 20	.826
.175	.04391 41	.88455 48	.18763 27	.02827 34	.825
.176	.04408 73	.88374 99	.18876 21	.02842 47	.824
.177	.04425 97	.88294 36	.18989 19	.02857 58	.823
.178	.04443 13	.88213 59	.19102 21	.02872 67	.822
.179	.04460 21	.88132 68	.19215 27	.02887 74	.821
.180	.04477 20	.88051 60	.19328 40	.02902 80	.820
.181	.04494 11	.87970 38	.19441 57	.02917 84	.819
.182	.04510 94	.87889 02	.19554 78	.02932 86	.818
.183	.04527 69	.87807 52	.19668 03	.02947 86	.817
.184	.04544 36	.87725 88	.19781 32	.02962 84	.816
.185	.04560 94	.87644 07	.19894 68	.02977 81	.815
.186	.04577 45	.87562 15	.20008 05	.02992 75	.814
.187	.04593 87	.87480 06	.20121 49	.03007 68	.813
.188	.04610 21	.87397 83	.20234 97	.03022 59	.812
.189	.04626 47	.87315 46	.20348 49	.03037 48	.811
.190	.04642 65	.87232 95	.20462 05	.03052 35	.810
.191	.04658 75	.87150 30	.20575 65	.03067 20	.809
.192	.04674 76	.87067 48	.20689 32	.03082 04	.808
.193	.04690 70	.86984 55	.20803 00	.03096 85	.807
.194	.04706 56	.86901 48	.20916 72	.03111 64	.806
.195	.04722 33	.86818 24	.21030 51	.03126 42	.805
.196	.04738 03	.86734 89	.21144 31	.03141 17	.804
.197	.04753 64	.86651 37	.21258 18	.03155 91	.803
.198	.04769 17	.86567 71	.21372 09	.03170 63	.802
.199	.04784 63	.86483 94	.21486 01	.03185 32	.801
.200	.04800 00	.86400 00	.21600 00	.33200 00	.800
	$-$ c_2	$+$ c_1	$+$ c_0	$-$ c_{-1}	

TABLE III. FOUR-POINT INTERPOLATION COEFFICIENTS

p ($p<.5$)	c_{-1} $-$	c_0 $+$	c_1 $+$	c_2 $-$	p ($p>.5$)
.200	.04800 00	.86400 00	.21600 00	.03200 00	.800
.201	.04815 29	.86315 92	.21714 03	.03214 66	.799
.202	.04830 51	.86231 73	.21828 07	.03229 29	.798
.203	.04845 64	.86147 37	.21942 18	.03243 91	.797
.204	.04860 69	.86062 87	.22056 33	.03258 51	.796
.205	.04875 67	.85978 26	.22170 49	.03273 08	.795
.206	.04890 56	.85893 48	.22284 72	.03287 64	.794
.207	.04905 38	.85808 59	.22398 96	.03302 17	.793
.208	.04920 12	.85723 56	.22513 24	.03316 68	.792
.209	.04934 77	.85638 36	.22627 59	.03331 18	.791
.210	.04949 35	.85553 05	.22741 95	.03345 65	.790
.211	.04963 85	.85467 60	.22856 35	.03360 10	.789
.212	.04978 27	.85382 01	.22970 79	.03374 53	.788
.213	.04992 61	.85296 28	.23085 27	.03388 94	.787
.214	.05006 87	.85210 41	.23199 79	.03403 33	.786
.215	.05021 06	.85124 43	.23314 32	.03417 69	.785
.216	.05035 16	.85038 28	.23428 92	.03432 04	.784
.217	.05049 19	.84952 02	.23543 53	.03446 36	.783
.218	.05063 14	.84865 62	.23658 18	.03460 66	.782
.219	.05077 01	.84779 08	.23772 87	.03474 94	.781
.220	.05090 80	.84692 40	.23887 60	.03489 20	.780
.221	.05104 51	.84605 58	.24002 37	.03503 44	.779
.222	.05118 15	.84518 65	.24117 15	.03517 65	.778
.223	.05131 71	.84431 58	.24231 97	.03531 84	.777
.224	.05145 19	.84344 37	.24346 83	.03546 01	.776
.225	.05158 59	.84257 02	.24461 73	.03560 16	.775
.226	.05171 92	.84169 56	.24576 64	.03574 28	.774
.227	.05185 17	.84081 96	.24691 59	.03588 38	.773
.228	.05198 34	.83994 22	.24806 58	.03602 46	.772
.229	.05211 43	.83906 34	.24921 61	.03616 52	.771
.230	.05224 45	.83818 35	.25036 65	.03630 55	.770
.231	.05237 39	.83730 22	.25151 73	.03644 56	.769
.232	.05250 25	.83641 95	.25266 85	.03658 55	.768
.233	.05263 04	.83553 57	.25381 98	.03672 51	.767
.234	.05275 75	.83465 05	.25497 15	.03686 45	.766
.235	.05288 38	.83376 39	.25612 36	.03700 37	.765
.236	.05300 94	.83287 62	.25727 58	.03714 26	.764
.237	.05313 42	.83198 71	.25842 84	.03728 13	.863
.238	.05325 82	.83109 66	.25958 14	.03741 98	.762
.239	.05338 15	.83020 50	.26073 45	.03755 80	.761
.240	.05350 40	.82931 20	.26188 80	.03769 60	.760
.241	.05362 58	.82841 79	.26304 16	.03783 37	.759
.242	.05374 67	.82752 21	.26419 59	.03797 13	.758
.243	.05386 70	.82662 55	.26535 00	.03810 85	.757
.244	.05398 65	.82572 75	.26650 45	.03824 55	.756
.245	.05410 52	.82482 81	.26765 94	.03838 23	.755
.246	.05422 32	.82392 76	.26881 44	.03851 88	.754
.247	.05434 04	.82302 57	.26996 98	.03865 51	.753
.248	.05445 68	.82212 24	.27112 56	.03879 12	.752
.249	.05457 25	.82121 80	.27228 15	.03892 70	.751
.250	.05468 75	.82031 25	.27343 75	.03906 25	.750
	$-$ c_2	$+$ c_1	$+$ c_0	$-$ c_{-1}	

TABLE III. FOUR-POINT INTERPOLATION COEFFICIENTS

p ($p<.5$)	c_{-1} −	c_0 +	c_1 +	c_2 −	p ($p>.5$)
.250	.05468 75	.82031 25	.27343 75	.03906 25	.750
.251	.05480 17	.81940 56	.27459 39	.03919 78	.749
.252	.05491 52	.81849 76	.27575 04	.03933 28	.748
.253	.05502 79	.81758 82	.27690 73	.03946 76	.747
.254	.05513 98	.81667 74	.27806 46	.03960 22	.746
.255	.05525 11	.81576 58	.27922 17	.03973 64	.745
.256	.05536 15	.81485 25	.28037 95	.03987 05	.744
.257	.05547 13	.81393 84	.28153 71	.04000 42	.743
.258	.05558 03	.81302 29	.28269 51	.04013 77	.742
.259	.05568 85	.81210 60	.28385 35	.04027 10	.741
.260	.05579 60	.81118 80	.28501 20	.04040 40	.740
.261	.05590 28	.81026 89	.28617 06	.04053 67	.739
.262	.05600 88	.80934 84	.28732 96	.04066 92	.738
.263	.05611 41	.80842 68	.28848 87	.04080 14	.737
.264	.05621 86	.80750 38	.28964 82	.04093 34	.736
.265	.05632 24	.80657 97	.29080 78	.04106 51	.735
.266	.05642 55	.80565 45	.29196 75	.04119 65	.734
.267	.05652 79	.80472 82	.29312 73	.04132 76	.733
.268	.05662 95	.80380 05	.29428 75	.04145 85	.732
.269	.05673 04	.80287 17	.29544 78	.04158 91	.731
.270	.05683 05	.80194 15	.29660 85	.04171 95	.730
.271	.05692 99	.80101 02	.29776 93	.04184 96	.729
.272	.05702 86	.80007 78	.29893 02	.04197 94	.728
.273	.05712 66	.79914 43	.30009 12	.04210 89	.727
.274	.05722 38	.79820 94	.30125 26	.04223 82	.726
.275	.05732 03	.79727 34	.30241 41	.04236 72	.725
.276	.05741 61	.79633 63	.30357 57	.04249 59	.724
.277	.05751 12	.79539 81	.30473 74	.04262 43	.723
.278	.05760 55	.79445 85	.30589 95	.04275 25	.722
.279	.05769 91	.79351 78	.30706 17	.04288 04	.721
.280	.05779 20	.79257 60	.30822 40	.04300 80	.720
.281	.05788 42	.79163 31	.30938 64	.04313 53	.719
.282	.05797 56	.79068 88	.31054 92	.04326 24	.718
.283	.05806 64	.78974 37	.31171 18	.04338 91	.717
.284	.05815 64	.78879 72	.31287 48	.04351 56	.716
.285	.05824 57	.78784 96	.31403 79	.04364 18	.715
.286	.05833 43	.78690 09	.31520 11	.04376 77	.714
.287	.05842 22	.78595 11	.31636 44	.04389 33	.713
.288	.05850 93	.78499 99	.31752 81	.04401 87	.712
.289	.05859 58	.78404 79	.31869 16	.04414 37	.711
.290	.05868 15	.78309 45	.31985 55	.04426 85	.710
.291	.05876 65	.78214 00	.32101 95	.04439 30	.709
.292	.05885 08	.78118 44	.32218 36	.04451 72	.708
.293	.05893 45	.78022 80	.32334 75	.04464 10	.707
.294	.05901 74	.77927 02	.32451 18	.04476 46	.706
.295	.05909 96	.77831 13	.32567 62	.04488 79	.705
.296	.05918 11	.77735 13	.32684 07	.04501 09	.704
.297	.05926 18	.77638 99	.32800 56	.04513 37	.703
.298	.05934 19	.77542 77	.32917 03	.04525 61	.702
.299	.05942 13	.77446 44	.33033 51	.04537 82	.701
.300	.05950 00	.77350 00	.33150 00	.04550 00	.700
	− c_2	+ c_1	+ c_0	− c_{-1}	

TABLE III. FOUR-POINT INTERPOLATION COEFFICIENTS

p ($p<.5$)	c_{-1} −	c_0 +	c_1 +	c_2 −	p ($p>.5$)
.300	.05950 00	.77350 00	.33150 00	.04550 00	**.700**
.301	.05957 80	.77253 45	.33266 50	.04562 15	.699
.302	.05965 53	.77156 79	.33383 01	.04574 27	.698
.303	.05973 19	.77060 02	.33499 53	.04586 36	.697
.304	.05980 77	.76963 11	.33616 09	.04598 43	.696
.305	.05988 29	.76866 12	.33732 63	.04610 46	.695
.306	.05995 74	.76769 02	.33849 18	.04622 46	.694
.307	.06003 12	.76671 81	.33965 74	.04634 43	.693
.308	.06010 44	.76574 52	.34082 28	.04646 36	.692
.309	.06017 68	.76477 09	.34198 86	.04658 27	.691
.310	.06024 85	.76379 55	.34315 45	.04670 15	**.690**
.311	.06031 95	.76281 90	.34432 05	.04682 00	.689
.312	.06038 99	.76184 17	.34548 63	.04693 81	.688
.313	.06045 95	.76086 30	.34665 25	.04705 60	.687
.314	.06052 85	.75988 35	.34781 85	.04717 35	.686
.315	.06059 68	.75890 29	.34898 46	.04729 07	.685
.316	.06066 44	.75792 12	.35015 08	.04740 76	.684
.317	.06073 13	.75693 84	.35131 71	.04752 42	.683
.318	.06079 76	.75595 48	.35248 32	.04764 04	.682
.319	.06086 31	.75496 98	.35364 97	.04775 64	.681
.320	.06092 80	.75398 40	.35481 60	.04787 20	**.680**
.321	.06099 22	.75299 71	.35598 24	.04798 73	.679
.322	.06105 57	.75200 91	.35714 89	.04810 23	.678
.323	.06111 85	.75102 00	.35831 55	.04821 70	.677
.324	.06118 07	.75003 01	.35948 19	.04833 13	.676
.325	.06124 22	.74903 91	.36064 84	.04844 53	.675
.326	.06130 30	.74804 70	.36181 50	.04855 90	.674
.327	.06136 31	.74705 38	.36298 17	.04867 24	.673
.328	.06142 26	.74605 98	.36414 82	.04878 54	.672
.329	.06148 14	.74506 47	.36531 48	.04889 81	.671
.330	.06153 95	.74406 85	.36648 15	.04901 05	**.670**
.331	.06159 69	.74307 12	.36764 83	.04912 26	.669
.332	.06165 37	.74207 31	.36881 49	.04923 43	.668
.333	.06170 98	.74107 39	.36998 16	.04934 57	.667
.334	.06176 53	.74007 39	.37114 81	.04945 67	.666
.335	.06182 01	.73907 28	.37231 47	.04956 74	.665
.336	.06187 42	.73807 06	.37348 14	.04967 78	.664
.337	.06192 76	.73706 73	.37464 82	.04978 79	.663
.338	.06198 04	.73606 32	.37581 48	.04989 76	.662
.339	.06203 25	.73505 80	.37698 15	.05000 70	.661
.340	.06208 40	.73405 20	.37814 80	.05011 60	**.660**
.341	.06213 48	.73304 49	.37931 46	.05022 47	.659
.342	.06218 49	.73203 67	.38048 13	.05033 31	.658
.343	.06223 44	.73102 77	.38164 78	.05044 11	.657
.344	.06228 33	.73001 79	.38281 41	.05054 87	.656
.345	.06233 14	.72900 67	.38398 08	.05065 61	.655
.346	.06237 90	.72799 50	.38514 70	.05076 30	.654
.347	.06242 58	.72698 19	.38631 36	.05086 97	.653
.348	.06247 20	.72596 80	.38748 00	.05097 60	.652
.349	.06251 76	.72495 33	.38864 62	.05108 19	.651
.350	.06256 25	.72393 75	.38981 25	.05118 75	**.650**
	− c_2	+ c_1	+ c_0	− c_{-1}	

TABLE III. FOUR-POINT INTERPOLATION COEFFICIENTS

p $(p<.5)$	c_{-1} $-$	c_0 $+$	c_1 $+$	c_2 $-$	p $(p>.5)$
.350	.06256 25	.72393 75	.38981 25	.05118 75	.650
.351	.06260 68	.72292 09	.39097 86	.05129 27	.649
.352	.06265 04	.72190 32	.39214 48	.05139 76	.648
.353	.06269 33	.72088 44	.39331 11	.05150 22	.647
.354	.06273 56	.71986 48	.39447 72	.05160 64	.646
.355	.06277 73	.71884 44	.39564 31	.05171 02	.645
.356	.06281 83	.71782 29	.39680 91	.05181 37	.644
.357	.06285 87	.71680 06	.39797 49	.05191 68	.643
.358	.06289 85	.71577 75	.39914 05	.05201 95	.642
.359	.06293 75	.71475 30	.40030 65	.05212 20	.641
.360	.06297 60	.71372 80	.40147 20	.05222 40	.640
.361	.06301 38	.71270 19	.40263 76	.05232 57	.639
.362	.06305 10	.71167 50	.40380 30	.05242 70	.638
.363	.06308 75	.71064 70	.40496 85	.05252 80	.637
.364	.06312 34	.70961 82	.40613 38	.05262 86	.636
.365	.06315 87	.70858 86	.40729 89	.05272 88	.635
.366	.06319 33	.70755 79	.40846 41	.05282 87	.634
.367	.06322 73	.70652 64	.40962 91	.05292 82	.633
.368	.06326 07	.70549 41	.41079 39	.05302 73	.632
.369	.06329 34	.70446 07	.41195 88	.05312 61	.631
.370	.06332 55	.70342 65	.41312 35	.05322 45	.630
.371	.06335 70	.70239 15	.41428 80	.05332 25	.629
.372	.06338 78	.70135 54	.41545 26	.05342 02	.628
.373	.06341 80	.70031 85	.41661 70	.05351 75	.627
.374	.06344 76	.69928 08	.41778 12	.05361 44	.626
.375	.06347 66	.69824 23	.41894 52	.05371 09	.625
.376	.06350 49	.69720 27	.42010 93	.05380 71	.624
.377	.06353 26	.69616 23	.42127 32	.05390 29	.623
.378	.06355 97	.69512 11	.42243 69	.05399 83	.622
.379	.06358 62	.69407 91	.42360 04	.05409 33	.621
.380	.06361 20	.69303 60	.42476 40	.05418 80	.620
.381	.06363 72	.69199 21	.42592 74	.05428 23	.619
.382	.06366 18	.69094 74	.42709 06	.05437 62	.618
.383	.06368 58	.68990 19	.42825 36	.05446 97	.617
.384	.06370 92	.68885 56	.42941 64	.05456 28	.616
.385	.06373 19	.68780 82	.43057 93	.05465 56	.615
.386	.06375 41	.68676 03	.43174 17	.05474 79	.614
.387	.06377 56	.68571 13	.43290 42	.05483 99	.613
.388	.06379 65	.68466 15	.43406 65	.05493 15	.612
.389	.06381 68	.68361 09	.43522 86	.05502 27	.611
.390	.06383 65	.68255 95	.43639 05	.05511 35	.610
.391	.06385 56	.68150 73	.43755 22	.05520 39	.609
.392	.06387 40	.68045 40	.43871 40	.05529 40	.608
.393	.06389 19	.67940 02	.43987 53	.05538 36	.607
.394	.06390 92	.67834 56	.44103 64	.05547 28	.606
.395	.06392 58	.67728 99	.44219 76	.05556 17	.605
.396	.06394 19	.67623 37	.44335 83	.05565 01	.604
.397	.06395 73	.67517 64	.44451 91	.05573 82	.603
.398	.06397 21	.67411 83	.44567 97	.05582 59	.602
.399	.06398 64	.67305 97	.44683 98	.05591 31	.601
.400	.06400 00	.67200 00	.44800 00	.05600 00	.600
	$-$ c_2	$+$ c_1	$+$ c_0	$-$ c_{-1}	

TABLE III. FOUR-POINT INTERPOLATION COEFFICIENTS

p ($p<.5$)	c_{-1} −	c_0 +	c_1 +	c_2 −	p ($p>.5$)
.400	.06400 00	.67200 00	.44800 00	.05600 00	.600
.401	.06401 30	.67093 95	.44916 00	.05608 65	.599
.402	.06402 55	.66987 85	.45031 95	.05617 25	.598
.403	.06403 73	.66881 64	.45147 91	.05625 82	.597
.404	.06404 85	.66775 35	.45263 85	.05634 35	.596
.405	.06405 92	.66669 01	.45379 74	.05642 83	.595
.406	.06406 92	.66562 56	.45495 64	.05651 28	.594
.407	.06407 87	.66456 06	.45611 49	.05659 68	.593
.408	.06408 76	.66349 48	.45727 32	.05668 04	.592
.409	.06409 58	.66242 79	.45843 16	.05676 37	.591
.410	.06410 35	.66136 05	.45958 95	.05684 65	.590
.411	.06411 06	.66029 23	.46074 72	.05692 89	.589
.412	.06411 71	.65922 33	.46190 47	.05701 09	.588
.413	.06412 30	.65815 35	.46306 20	.05709 25	.587
.414	.06412 83	.65708 29	.46421 91	.05717 37	.586
.415	.06413 31	.65601 18	.46537 57	.05725 44	.585
.416	.06413 72	.65493 96	.46653 24	.05733 48	.584
.417	.06414 08	.65386 69	.46768 86	.05741 47	.583
.418	.06414 38	.65279 34	.46884 46	.05749 42	.582
.419	.06414 62	.65171 91	.47000 04	.05757 33	.581
.420	.06414 80	.65064 40	.47115 60	.05765 20	.580
.421	.06414 92	.64956 81	.47231 14	.05773 03	.579
.422	.06414 99	.64849 17	.47346 63	.05780 81	.578
.423	.06415 00	.64741 45	.47462 10	.05788 55	.577
.424	.06414 95	.64633 65	.47577 55	.05796 25	.576
.425	.06414 84	.64525 77	.47692 98	.05803 91	.575
.426	.06414 68	.64417 84	.47808 36	.05811 52	.574
.427	.06414 46	.64309 83	.47923 72	.05819 09	.573
.428	.06414 18	.64201 74	.48039 06	.05826 62	.572
.429	.06413 84	.64093 57	.48154 38	.05834 11	.571
.430	.06413 45	.63985 35	.48269 65	.05841 55	.570
.431	.06413 00	.63877 05	.48384 90	.05848 95	.569
.432	.06412 49	.63768 67	.48500 13	.05856 31	.568
.433	.06411 93	.63660 24	.48615 31	.05863 62	.567
.434	.06411 31	.63551 73	.48730 47	.05870 89	.566
.435	.06410 63	.63443 14	.48845 61	.05878 12	.565
.436	.06409 90	.63334 50	.48960 70	.05885 30	.564
.437	.06409 11	.63225 78	.49075 77	.05892 44	.563
.438	.06408 26	.63116 98	.49190 82	.05899 54	.562
.439	.06407 36	.63008 13	.49305 82	.05906 59	.561
.440	.06406 40	.62899 20	.49420 80	.05913 60	.560
.441	.06405 39	.62790 22	.49535 73	.05920 56	.559
.442	.06404 31	.62681 13	.49650 67	.05927 49	.558
.443	.06403 19	.62572 02	.49765 53	.05934 36	.557
.444	.06402 01	.62462 83	.49880 37	.05941 19	.556
.445	.06400 77	.62353 56	.49995 19	.05947 98	.555
.446	.06399 48	.62244 24	.50109 96	.05954 72	.554
.447	.06398 13	.62134 84	.50224 71	.05961 42	.553
.448	.06396 72	.62025 36	.50339 44	.05968 08	.552
.449	.06395 26	.61915 83	.50454 12	.05974 69	.551
.450	.06393 75	.61806 25	.50568 75	.05981 25	.550
	− c_2	+ c_1	+ c_0	− c_{-1}	

TABLE III. FOUR-POINT INTERPOLATION COEFFICIENTS

p ($p<.5$)	c_{-1} $-$	c_0 $+$	c_1 $+$	c_2 $-$	p ($p>.5$)
.450	.06393 75	.61806 25	.50568 75	.05981 25	.550
.451	.06392 18	.61696 59	.50683 36	.05987 77	.549
.452	.06390 56	.61586 88	.50797 92	.05994 24	.548
.453	.06388 88	.61477 09	.50912 46	.06000 67	.547
.454	.06387 14	.61367 22	.51026 98	.06007 06	.546
.455	.06385 36	.61257 33	.51141 42	.06013 39	.545
.456	.06383 51	.61147 33	.51255 87	.06019 69	.544
.457	.06381 62	.61037 31	.51370 24	.06025 93	.543
.458	.06379 67	.60927 21	.51484 59	.06032 13	.542
.459	.06377 66	.60817 03	.51598 92	.06038 29	.541
.460	.06375 60	.60706 80	.51713 20	.06044 40	.540
.461	.06373 49	.60596 52	.51827 43	.06050 46	.539
.462	.06371 32	.60486 16	.51941 64	.06056 48	.538
.463	.06369 10	.60375 75	.52055 80	.06062 45	.537
.464	.06366 82	.60265 26	.52169 94	.06068 38	.536
.465	.06364 49	.60154 72	.52284 03	.06074 26	.535
.466	.06362 11	.60044 13	.52398 07	.06080 09	.534
.467	.06359 68	.59933 49	.52512 06	.06085 87	.533
.468	.06357 19	.59822 77	.52626 03	.06091 61	.532
.469	.06354 65	.59712 00	.52739 95	.06097 30	.531
.470	.06352 05	.59601 15	.52853 85	.06102 95	.530
.471	.06349 40	.59490 25	.52967 70	.06108 55	.529
.472	.06346 70	.59379 30	.53081 50	.06114 10	.528
.473	.06343 95	.59268 30	.53195 25	.06119 60	.527
.474	.06341 14	.59157 22	.53308 98	.06125 06	.526
.475	.06338 28	.59046 09	.53422 66	.06130 47	.525
.476	.06335 37	.58934 91	.53536 29	.06135 83	.524
.477	.06332 41	.58823 68	.53649 87	.06141 14	.523
.478	.06329 39	.58712 37	.53763 43	.06146 41	.522
.479	.06326 32	.58601 01	.53876 94	.06151 63	.521
.480	.06323 20	.58489 60	.53990 40	.06156 80	.520
.481	.06320 03	.58378 14	.54103 81	.06161 92	.519
.482	.06316 80	.58266 60	.54217 20	.06167 00	.518
.483	.06313 53	.58155 04	.54330 51	.06172 02	.517
.484	.06310 20	.58043 40	.54443 80	.06177 00	.516
.485	.06306 82	.57931 71	.54557 04	.06181 93	.515
.486	.06303 39	.57819 97	.54670 23	.06186 81	.514
.487	.06299 91	.57708 18	.54783 37	.06191 64	.513
.488	.06296 37	.57596 31	.54896 49	.06196 43	.512
.489	.06292 79	.57484 42	.55009 53	.06201 16	.511
.490	.06289 15	.57372 45	.55122 55	.06205 85	.510
.491	.06285 46	.57260 43	.55235 52	.06210 49	.509
.492	.06281 72	.57148 36	.55348 44	.06215 08	.508
.493	.06277 94	.57036 27	.55461 28	.06219 61	.507
.494	.06274 10	.56924 10	.55574 10	.06224 10	.506
.495	.06270 21	.56811 88	.55686 87	.06228 54	.505
.496	.06266 27	.56699 61	.55799 59	.06232 93	.504
.497	.06262 27	.56587 26	.55912 29	.06237 28	.503
.498	.06258 23	.56474 89	.56024 91	.06241 57	.502
.499	.06254 14	.56362 47	.56137 48	.06245 81	.501
.500	.06250 00	.56250 00	.56250 00	.06250 00	.500
	$-$ c_2	$+$ c_1	$+$ c_0	$-$ c_{-1}	

TABLE IV
SIX-POINT INTERPOLATION COEFFICIENTS

p (p<.5)	c_{-2} +	c_{-1} -	c_0 +	c_1 +	c_2 -	c_3 +	p (p>.5)
.00	.00000 00000	.00000 00000	1.00000 00000	.00000 00000	.00000 00000	.00000 00000	1.00
.01	.00049 57921	.00493 33767	.99654 20858	.01006 60817	.00250 38746	.00033 32917	.99
.02	.00098 30066	.00973 36932	.99283 67064	.02026 19736	.00501 43268	.00066 63334	.98
.03	.00146 14086	.01440 12590	.98888 64505	.03058 41170	.00752 95923	.00099 88752	.97
.04	.00193 07725	.01893 64224	.98469 39648	.04102 89152	.01004 78976	.00133 06675	.96
.05	.00239 08828	.02333 95703	.98026 19531	.05159 27344	.01256 74609	.00166 14609	.95
.06	.00284 15335	.02761 11276	.97559 31752	.06227 19048	.01508 64924	.00199 10065	.94
.07	.00328 25281	.03175 15567	.97069 04458	.07306 27217	.01760 31946	.00231 90557	.93
.08	.00371 36794	.03576 13568	.96555 66336	.08396 14464	.02011 57632	.00264 53606	.92
.09	.00413 48096	.03964 10640	.96019 46604	.09496 43071	.02262 23873	.00296 96742	.91
.10	.00454 57500	.04339 12500	.95460 75000	.10606 75000	.02512 12500	.00329 17500	.90
.11	.00494 63412	.04701* 25223	.94879 81771	.11726 71904	.02761 05290	.00361 13426	.89
.12	.00533 64326	.05050 55232	.94276 97664	.12855 95136	.03008 83968	.00392 82074	.88
.13	.00571 58827	.05387 09296	.93652 53917	.13994 05758	.03255 30217	.00424 21011	.87
.14	.00608 45585	.05710 94524	.93006 82248	.15140 64552	.03500 25676	.00455 27815	.86
.15	.00644 23359	.06022 18359	.92340 14844	.16295 32031	.03743 51953	.00486 00078	.85
.16	.00678 90995	.06320 88576	.91652 84352	.17457 68448	.03984 90624	.00516 35405	.84
.17	.00712 47422	.06607 13273	.90945 23870	.18627 33805	.04224 23240	.00546 31416	.83
.18	.00744 91654	.06881 00868	.90217 66936	.19803 87864	.04461 31332	.00575 85746	.82
.19	.00776 22787	.07142 60096	.89470 47517	.20986 90158	.04695 96417	.00604 96051	.81
.20	.00806 40000	.07392 00000	.88704 00000	.22176 00000	.04928 00000	.00633 60000	.80
.21	.00835 42553	.07629 29929	.87918 59183	.23370 76492	.05157 23583	.00661 75284	.79
.22	.00863 29786	.07854 59532	.87114 60264	.24570 78536	.05383 48668	.00689 39614	.78
.23	.00890 01118	.08067 98752	.86292 38830	.25775 64845	.05606 56760	.00716 50719	.77

	$+$ c_{-2}	$-$ c_{-1}	$+$ c_0	$+$ c_1	$-$ c_2	$+$ c_3	
.76	.00743 06355	.05826 29376	.26984 93952	.85452 30848	.08269 57824	.00915 56045	.24
.75	.00769 04297	.06042 48047	.28198 24219	.84594 72656	.08459 47266	.00939 94141	.25
.74	.00794 42345	.06254 94324	.29415 13848	.83720 00952	.08637 77876	.00963 15055	.26
.73	.00819 18324	.06463 49783	.30635 20892	.82828 52783	.08804 60729	.00985 18513	.27
.72	.00843 30086	.06667 96032	.31858 03264	.81920 65536	.08960 07168	.01006 04314	.28
.71	.00866 75509	.06868 14710	.33083 18746	.80996 76929	.09104 28802	.01025 72328	.29
.70	.00889 52500	.07063 87500	.34310 25000	.80057 25000	.09237 37500	.01044 22500	**.30**
.69	.00911 58993	.07254 96127	.35538 79579	.79102 48096	.09359 45385	.01061 54844	.31
.68	.00932 92954	.07441 22368	.36768 39936	.78132 84864	.09470 64832	.01077 69446	.32
.67	.00953 52378	.07622 48054	.37998 63433	.77148 74242	.09571 08458	.01092 66459	.33
.66	.00973 35295	.07798 55076	.39229 07352	.76150 55448	.09660 89124	.01106 46105	.34
.65	.00992 39766	.07969 25391	.40459 28906	.75138 67969	.09740 19922	.01119 08672	.35
.64	.01010 63885	.08134 41024	.41688 85248	.74113 51552	.09809 14176	.01130 54515	.36
.63	.01028 05783	.08293 84077	.42917 33480	.73075 46195	.09867 85435	.01140 84054	.37
.62	.01044 63626	.08447 36732	.44144 30664	.72024 92136	.09916 47468	.01149 97774	.38
.61	.01060 35618	.08594 81254	.45369 33833	.70962 29842	.09955 14258	.01157 96219	.39
.60	.01075 20000	.08736 00000	.46592 00000	.69888 00000	.09984 00000	.01164 80000	**.40**
.59	.01089 15052	.08870 75421	.47811 86167	.68802 43508	.10003 19092	.01170 49786	.41
.58	.01102 19094	.08998 90068	.49028 49336	.67706 01464	.10012 86132	.01175 06306	.42
.57	.01114 30487	.09120 26598	.50241 46520	.66599 15155	.10013 15915	.01178 50351	.43
.56	.01125 47635	.09234 67776	.51450 34752	.65482 26048	.10004 23424	.01180 82765	.44
.55	.01135 68984	.09341 96484	.52654 71094	.64355 75781	.09986 23828	.01182 04453	.45
.54	.01144 93025	.09441 95724	.53854 12648	.63220 06152	.09959 32476	.01182 16375	.46
.53	.01153 18292	.09534 48621	.55048 16567	.62075 59108	.09923 64892	.01181 19546	.47
.52	.01160 43366	.09619 38432	.56236 40064	.60922 76736	.09879 36768	.01179 15034	.48
.51	.01166 66877	.09696 48548	.57418 40421	.59762 01254	.09826 63965	.01176 03961	.49
.50	.01171 87500	.09765 62500	.58593 75000	.58593 75000	.09765 62500	.01171 87500	**.50**

TABLE V
EIGHT–POINT INTERPOLATION COEFFICIENTS

p	c_{-3} −	c_{-2} +	c_{-1} −	c_0 +	c_1 +	c_2 −	c_3 +	c_4 −
.1	.00088 64212 5	.00915 96862 5	.05246 00212 5	.96176 70562 5	.10686 30062 5	.03037 15912 5	.00663 28762 5	.00070 45912 5
.2	.00160 51200 0	.01634 30400 0	.08988 67200 0	.89886 72000 0	.22471 68000 0	.05992 44800 0	.01284 09600 0	.00135 16800 0
.3	.00211 57987 5	.02124 99787 5	.11278 83487 5	.81458 25187 5	.34910 67937 5	.08624 99137 5	.01810 18337 5	.00188 70637 5
.4	.00239 61600 0	.02376 19200 0	.12220 41600 0	.71285 76000 0	.47523 84000 0	.10692 86400 0	.02193 40800 0	.00226 30400 0
.5	.00244 14062 5	.02392 57812 5	.11962 89062 5	.59814 45312 5	.59814 45312 5	.11962 89062 5	.02392 57812 5	.00244 14062 5
.6	.00226 30400 0	.02193 40800 0	.10692 86400 0	.47523 84000 0	.71285 76000 0	.12220 41600 0	.02376 19200 0	.00239 61600 0
.7	.00188 70637 5	.01810 18337 5	.08624 99137 5	.34910 67937 5	.81458 25187 5	.11278 83487 5	.02124 99787 5	.00211 57987 5
.8	.00135 16800 0	.01284 09600 0	.05992 44800 0	.22471 68000 0	.89886 72000 0	.08988 67200 0	.01634 30400 0	.00160 51200 0
.9	.00070 45912 5	.00663 28762 5	.03037 15912 5	.10686 30062 5	.96176 70562 5	.05246 00212 5	.00915 96862 5	.00088 64212 5

TABLE VI SUPPLEMENT

Inverse interpolation maximal errors in $\chi/\sqrt{n}$ when P, approximately as recorded, is the argument for n's as shown

	1	**2**	**3**	**4**	**5**	**6**	**7**
	P=.58	P=.74	P=.82	P=.87	P=.91	P=.93	P=.95
$E^{-ii}=$	.0007	.0006	.0021	.0037	.0055	.0075	.0097
$E^{-iii}=$	.0001	.0003	.0005	.0018	.0035	.0058	large
	P=.30	P=.33	P=.35	P=.35	P=.36	P=.36	P=.36
$E^{-ii}=$	.0013	.0015	.0017	.0019	.0021	.0023	.0025
$E^{-iii}=$	.0002	.0004	.0005	.0007	.0009	.0012	.0013
	P=.12	P=.09	P=.07	P=.05	P=.03	P=.03	P=.02
$E^{-ii}=$	.0018	.0028	.0037	.0045	.0053	.0060	.0066
$E^{-iii}=$	.0003	.0007	.0012	.0016	.0021	.0026	.0028

	8	**9**	**10**	**12**	**15**	**19**	**24**	**30**
	P=.80	P=.82	P=.84	P=.86	P=.89	P=.92	P=.94	P=.96
$E^{-ii}=$	.0033	.0040	.0047	.0063	.0089	.013	.019	.027
$E^{-iii}=$	.0019	.0024	.0031	.0048	large	large	large	large
	P=.36	P=.36	P=.36	P=.36	P=.35	P=.35	P=.34	P=.33
$E^{-ii}=$	.0027	.0028	.0030	.0034	.0038	.0044	.0051	.0060
$E^{-iii}=$	.0016	.0018	.0020	.0024	.0031	.0040	large	large
	P=.07	P=.06	P=.05	P=.04	P=.03	P=.02	P=.01	P=.007
$E^{-ii}=$	.0057	.0062	.0066	.0073	.0083	.0092	.0099	.011
$E^{-iii}=$	.0028	.0032	.0035	.0038	.0039	.0032	.0004	.0016

Giving P, — the probability that, for a given n, a diver-

$\chi/\sqrt{n}$	χ^2/n	1	2	3	4	5	6	7
					n, — Number of de-			
.0	.00	1.0000	1.0000	1.0000	1.0000	1.0000	1.0000	1.0000
.1	.01	.9203	.9900	.9986	.9998	1.0000−	1.0000−	1.0000−
.2	.04	.8415	.9608	.9893	.9970	.9991	.9997	.9999
.3	.09	.7642	.9139	.9656	.9856	.9938	.9973	.9988
.4	.16	.6892	.8521	.9233	.9585	.9770	.9871	.9927
.5	.25	.6171	.7788	.8614	.9098	.9400	.9595	.9724
.6	.36	.5485	.6977	.7819	.8372	.8761	.9044	.9256
.7	.49	.4839	.6126	.6892	.7431	.7840	.8164	.8426
.8	.64	.4237	.5273	.5892	.6339	.6692	.6983	.7232
.9	.81	.3681	.4449	.4881	.5185	.5423	.5619	.5788
1.0	1.00	.3173	.3679	.3916	.4060	.4159	.4232	.4289
1.1	1.21	.2713	.2982	.3043	.3041	.3014	.2975	.2930
1.2	1.44	.2301	.2369	.2289	.2178	.2062	.1949	.1841
1.3	1.69	.1936	.1845	.1667	.1491	.1331	.1189	.1063
1.4	1.96	.1615	.1409	.1176	.0976	.0811	.0675	.0564
1.5	2.25	.1336	.1054	.0803	.0611	.0466	.0358	.0275
1.6	2.56	.1096	.0773	.0531	.0366	.0253	.0176	.0123
1.7	2.89	.0891	.0556	.0340	.0209	.0130	.0081	.0051
1.8	3.24	.0719	.0392	.0211	.0115	.0063	.0035	.0019
1.9	3.61	.0574	.0271	.0127	.0060	.0029	.0014	.0007
2.0	4.00	.0455	.0183	.0074	.0030	.0013	.0005	.0002
2.1	4.41	.0357	.0122	.0042	.0014	.0005	.0002	.0001
2.2	4.84	.0278	.0079	.0023	.0007	.0002	.0001	.0000+
2.3	5.29	.0214	.0050	.0012	.0003	.0001	.0000+	
2.4	5.76	.0164	.0032	.0006	.0001	.0000+		
2.5	6.25	.0124	.0019	.0003	.0001			
2.6	6.76	.0093	.0012	.0001	.0000+			
2.7	7.29	.0069	.0007	.0001				
2.8	7.84	.0051	.0004	.0000+				
2.9	8.41	.0037	.0002					
3.0	9.00	.0027	.0001					
3.1	9.61	.0019	.0001					
3.2	10.24	.0014	.0000+					
3.3	10.89	.0010						
3.4	11.56	.0007						
3.5	12.25	.0005						
3.6	12.96	.0003						
3.7	13.69	.0002						
3.8	14.44	.0001						
3.9	15.21	.0001						
4.0	16.00	.0001						
4.1	16.81	.0000+						

$\chi/\sqrt{n} = .5$	$E^{ii} =$	.0008	.0005	.0017	.0027	.0035	.0041	.0045
	$E^{iii} =$	.000+	0002	.0002	.0002	.0001	.0003	.0006
$\chi/\sqrt{n} = 1.0$	$E^{ii} =$	.0006	.0011	.0015	.0019	.0024	.0029	.0034
	$F^{iii} =$	.000+	.0001	.0002	.0003	.0005	.0006	.0008
$\chi/\sqrt{n} = 1.5$	$E^{ii} =$	.0004	.0008	.0010	.0011	.0011	.0011	.0010
	$E^{iii} =$	.000+	.0001	.0001	.0002	.0003	.0003	.0004

gence as great as χ^2 will arise as a matter of chance

grees of freedom

8	9	10	12	15	19	24	30
1.0000	1.0000	1.0000	1.0000	1.0000	1.0000	1.0000	1.0000
1.0000−	1.0000−	1.0000−	1.0000−	1.0000−	1.0000−	1.0000−	1.0000−
1.0000−	1.0000−	1.0000−	1.0000−	1.0000−	1.0000−	1.0000−	1.0000−
.9995	.9998	.9999	1.0000−	1.0000−	1.0000−	1.0000−	1.0000−
.9958	.9976	.9986	.9995	.9999	1.0000−	1.0000−	1.0000−
.9810	.9869	.9909	.9956	.9985	.9996	.9999	1.0000−
.9417	.9540	.9636	.9770	.9882	.9950	.9983	.9995
.8643	.8824	.8978	.9221	.9472	.9680	.9825	.9913
.7447	.7637	.7806	.8096	.8441	.8787	.9098	.9357
.5937	.6070	.6191	.6406	.6676	.6975	.7281	.7583
.4335	.4373	.4405	.4457	.4514	.4568	.4616	.4657
.2882	.2833	.2784	.2687	.2549	.2378	.2187	.1984
.1740	.1644	.1555	.1394	.1188	.0965	.0752	.0589
.0952	.0853	.0766	.0620	.0454	.0304	.0186	.0118
.0472	.0395	.0332	.0236	.0143	.0074	.0035	.0013
.0212	.0164	.0128	.0077	.0037	.0014	.0004	.0001
.0087	.0061	.0043	.0022	.0008	.0002	.0000+	.0000+
.0032	.0020	.0013	.0005	.0001	.0000+		
.0011	.0006	.0004	.0001	.0000+			
.0003	.0002	.0001	.0000+				
.0001	.0000+	.0000+					
.0000+							

		8	9	10	12	15	19	24	30
$\chi/\sqrt{n}=.7$	$E^{ii}=$	.0038	.0047	.0055	.0071	.0092	.011	.014	.015
	$E^{iii}=$	.0007	.0006	.0004	.0001	.0007	.0018	.0033	.0046
$\chi/\sqrt{n}=1.0$	$E^{ii}=$	.0039	.0044	.0049	.0060	.0075	.0097	.012	.016
	$E^{iii}=$	.0010	.0012	.0014	.0019	.0025	.0035	.0047	.0064
$\chi/\sqrt{n}=1.3$	$E^{ii}=$	.0027	.0028	.0029	.0028	.0028	.0021	.0015	.0011
	$E^{iii}=$	.0011	.0007	.0008	.0010	.0014	.0016	large	large

For inverse two- and three-point interpolation errors, see page 201

TABLE VII. SQUARE AND CUBE ROOTS AND NATURAL LOGARITHMS

N	$\sqrt{N}$	$\sqrt{10N}$	$\sqrt[3]{N}$	$\sqrt[3]{10N}$	$\sqrt[3]{100N}$	$\log_e N$
1.00	1.0000 000	3.1622 777	1.0000 000	2.1544 347	4.6415 888	.0000 0000
1.01	1.0049 876	3.1780 497	1.0033 223	2.1615 923	4.6570 095	.0099 5033
1.02	1.0099 505	3.1937 439	1.0066 227	2.1687 029	4.6723 287	.0198 0263
1.03	1.0148 892	3.2093 613	1.0099 016	2.1757 671	4.6875 481	.0295 5880
1.04	1.0198 039	3.2249 031	1.0131 594	2.1827 858	4.7026 694	.0392 2071
1.05	1.0246 951	3.2403 703	1.0163 964	2.1897 596	4.7176 940	.0487 9016
1.06	1.0295 630	3.2557 641	1.0196 128	2.1966 892	4.7326 235	.0582 6891
1.07	1.0344 080	3.2710 854	1.0228 091	2.2035 755	4.7474 594	.0676 5865
1.08	1.0392 305	3.2863 353	1.0259 856	2.2104 189	4.7622 032	.0769 6104
1.09	1.0440 307	3.3015 148	1.0291 425	2.2172 202	4.7768 562	.0861 7770
1.10	1.0488 088	3.3166 248	1.0322 801	2.2239 801	4.7914 199	.0953 1018
1.11	1.0535 654	3.3316 662	1.0353 988	2.2306 991	4.8058 955	.1043 6002
1.12	1.0583 005	3.3466 401	1.0384 988	2.2373 779	4.8202 845	.1133 2869
1.13	1.0630 146	3.3615 473	1.0415 804	2.2440 170	4.8345 881	.1222 1763
1.14	1.0677 078	3.3763 886	1.0446 439	2.2506 171	4.8488 076	.1310 2826
1.15	1.0723 805	3.3911 650	1.0476 896	2.2571 787	4.8629 441	.1397 6194
1.16	1.0770 330	3.4058 773	1.0507 176	2.2637 024	4.8769 990	.1484 2001
1.17	1.0816 654	3.4205 263	1.0537 282	2.2701 887	4.8909 732	.1570 0375
1.18	1.0862 780	3.4351 128	1.0567 218	2.2766 381	4.9048 681	.1655 1444
1 19	1.0908 712	3.4496 377	1.0596 985	2.2830 512	4.9186 847	.1739 5331
1.20	1.0954 451	3.4641 016	1.0626 586	2.2894 285	4.9324 241	.1823 2156
1.21	1.1000 000	3.4785 054	1.0656 022	2.2957 704	4.9460 874	.1906 2036
1.22	1.1045 361	3.4928 498	1.0685 297	2.3020 775	4.9596 757	.1988 5086
1.23	1.1090 537	3.5071 356	1.0714 413	2.3083 502	4.9731 898	.2070 1417
1.24	1.1135 529	3.5213 634	1.0743 371	2.3145 891	4.9866 310	.2151 1138
1.25	1.1180 340	3.5355 339	1.0772 173	2.3207 944	5.0000 000	.2231 4355
1.26	1.1224 972	3.5496 479	1.0800 823	2.3269 668	5.0132 979	.2311 1172
1.27	1.1269 428	3.5637 059	1.0829 321	2.3331 066	5.0265 257	.2390 1690
1.28	1.1313 708	3.5777 088	1.0857 670	2.3392 142	5.0396 842	.2468 6008
1.29	1.1357 817	3 5916 570	1.0885 872	2.3452 901	5.0527 743	.2546 4222
1.30	1.1401 754	3.6055 513	1.0913 929	2.3513 347	5.0657 970	.2623 6426
1.31	1.1445 523	3.6193 922	1.0941 842	2.3573 484	5.0787 531	.2700 2714
1.32	1.1489 125	3.6331 804	1.0969 613	2.3633 315	5.0916 434	.2776 3174
1.33	1.1532 563	3.6469 165	1.0997 244	2.3692 845	5.1044 687	.2851 7 94
1.34	1.1575 837	3.6606 010	1.1024 738	2.3752 077	5.1172 299	.2926 6961
1.35	1.1618 950	3.6742 346	1.1052 094	2.3811 016	5.1299 278	.3038 0145
1.36	1.1661 904	3.6878 178	1.1079 317	2.3869 664	5.1425 632	.3074 8470
1.37	1.1704 700	3.7013 511	1.1106 405	2.3928 025	5.1551 367	.3148 1074
1.38	1.1747 340	3.7148 351	1.1133 363	2.3986 103	5.1676 493	.3220 8350
1.39	1.1789 826	3.7282 704	1.1160 190	2.4043 901	5.1801 015	.3293 0375
1.40	1.1832 160	3.7416 574	1.1186 889	2.4101 423	5.1924 941	.3364 7224
1.41	1.1874 342	3.7549 967	1.1213 462	2.4158 671	5.2048 279	.3435 8970
1.42	1.1916 375	3.7682 887	1.1239 909	2.4215 649	5.2171 034	.3506 5687
1.43	1.1958 261	3.7815 341	1.1266 232	2.4272 360	5.2293 215	.3576 7444
1.44	1.2000 000	3.7947 332	1.1292 432	2.4328 806	5.2414 828	.3646 4311
1.45	1.2041 595	3.8078 866	1.1318 512	2.4384 995	5.2535 879	.3715 6356
1.46	1.2083 046	3.8209 946	1.1344 472	2.4440 924	5.2656 374	.3784 3644
1.47	1.2124 356	3.8340 579	1.1370 314	2.4496 598	5.2776 321	.3852 6240
1.48	1.2165 525	3.8470 768	1.1396 038	2.4552 021	5.2895 725	.3920 4209
1.49	1.2206 556	3.8600 518	1.1421 648	2.4607 194	5.3014 592	.3987 7612
1.50	1.2247 449	3.8729 833	1.1447 142	2.4662 121	5.3132 928	.4054 6511
$\frac{Eii}{Eiii}$	.0000 022	.0000 071 / 1	.0000 019	.0000 041	.0000 089 / 1	.0000 0800 / 6

TABLE VII. SQUARE AND CUBE ROOTS AND NATURAL LOGARITHMS

N	$\sqrt{N}$	$\sqrt{10N}$	$\sqrt[3]{N}$	$\sqrt[3]{10N}$	$\sqrt[3]{100N}$	$\log_e N$
1.50	1.2247 449	3.8729 833	1.1447 142	2.4662 121	5.3132 928	.4054 6511
1.51	1.2288 206	3.8858 718	1.1472 524	2.4716 804	5.3250 740	.4121 0965
1.52	1.2328 828	3.8987 177	1.1497 794	2.4771 247	5.3368 033	.4187 1033
1.53	1.2369 317	3.9115 214	1.1522 954	2.4825 451	5.3484 812	.4252 6774
1.54	1.2409 674	3.9242 834	1.1548 004	2.4879 419	5.3601 084	.4317 8242
1.55	1.2449 900	3.9370 039	1.1572 945	2.4933 155	5.3716 854	.4382 5493
1.56	1.2489 996	3.9496 835	1.1597 780	2.4986 660	5.3832 126	.4446 8582
1.57	1.2529 964	3.9623 226	1.1622 509	2.5039 936	5.3946 907	.4510 7562
1.58	1.2569 805	3.9749 214	1.1647 133	2.5092 987	5.4061 202	.4574 2485
1.59	1.2609 520	3.9874 804	1.1671 653	2.5145 815	5.4175 015	.4637 3402
1.60	1.2649 111	4.0000 000	1.1696 071	2.5198 421	5.4288 352	.4700 0363
1.61	1.2688 578	4.0124 805	1.1720 387	2.5250 809	5.4401 218	.4762 3418
1.62	1.2727 922	4.0249 224	1.1744 603	2.5302 980	5.4513 618	.4824 2615
1.63	1.2767 145	4.0373 258	1.1768 719	2.5354 937	5.4625 556	.4885 8001
1.64	1.2806 248	4.0496 913	1.1792 737	2.5406 682	5.4737 037	.4946 9624
1.65	1.2845 233	4.0620 192	1.1816 658	2.5458 217	5.4848 066	.5007 7529
1.66	1.2884 099	4.0743 098	1.1840 481	2.5509 544	5.4958 647	.5068 1760
1.67	1.2922 848	4.0865 633	1.1864 210	2.5560 666	5.5068 784	.5128 2363
1.68	1.2961 481	4.0987 803	1.1887 844	2.5611 583	5.5178 484	.5187 9379
1.69	1.3000 000	4.1109 610	1.1911 384	2.5662 299	5.5287 748	.5247 2853
1.70	1.3038 405	4.1231 056	1.1934 832	2.5712 816	5.5396 583	.5306 2825
1.71	1.3076 697	4.1352 146	1.1958 188	2.5763 135	5.5504 991	.5364 9337
1.72	1.3114 877	4.1472 883	1.1981 453	2.5813 258	5.5612 978	.5423 2429
1.73	1.3152 946	4.1593 269	1.2004 628	2.5863 187	5.5720 547	.5481 2141
1.74	1.3190 906	4.1713 307	1.2027 714	2.5912 924	5.5827 702	.5538 8511
1.75	1.3228 757	4.1833 001	1.2050 711	2.5962 471	5.5934 447	.5596 1579
1.76	1.3266 499	4.1952 354	1.2073 621	2.6011 829	5.6040 787	.5653 1381
1.77	1.3304 135	4.2071 368	1.2096 445	2.6061 001	5.6146 724	.5709 7955
1.78	1.3341 664	4.2190 046	1.2119 183	2.6109 988	5.6252 263	.5766 1336
1.79	1.3379 088	4.2308 392	1.2141 835	2.6158 792	5.6357 408	.5822 1562
1.80	1.3416 408	4.2426 407	1.2164 404	2.6207 414	5.6462 162	.5877 8666
1.81	1.3453 624	4.2544 095	1.2186 889	2.6255 857	5.6566 528	.5933 2685
1.82	1.3490 738	4.2661 458	1.2209 291	2.6304 121	5.6670 511	.5988 3650
1.83	1.3527 749	4.2778 499	1.2231 612	2.6352 209	5.6774 114	.6043 1597
1.84	1.3564 660	4.2895 221	1.2253 851	2.6400 122	5.6877 340	.6097 6557
1.85	1.3601 471	4.3011 626	1.2276 010	2.6447 862	5.6980 192	.6151 8564
1.86	1.3638 182	4.3127 717	1.2298 089	2.6495 431	5.7082 675	.6205 7649
1.87	1.3674 794	4.3243 497	1.2320 090	2.6542 829	5.7184 791	.6259 3843
1.88	1.3711 309	4.3358 967	1.2342 012	2.6590 058	5.7286 543	.6312 7178
1.89	1.3747 727	4.3474 130	1.2363 856	2.6637 120	5.7387 935	.6365 7683
1.90	1.3784 049	4.3588 989	1.2385 623	2.6684 016	5.7488 971	.6418 5389
1.91	1.3820 275	4.3703 547	1.2407 314	2.6730 749	5.7589 652	.6471 0324
1.92	1.3856 406	4.3817 805	1.2428 930	2.6777 318	5.7689 983	.6523 2519
1.93	1.3892 444	4.3931 765	1.2450 471	2.6823 726	5.7789 966	.6575 2000
1.94	1.3928 388	4.4045 431	1.2471 937	2.6869 974	5.7889 604	.6626 8797
1.95	1.3964 240	4.4158 804	1.2493 330	2.6916 063	5.7988 900	.6678 2937
1.96	1.4000 000	4.4271 887	1.2514 649	2.6961 995	5.8087 857	.6729 4447
1.97	1.4035 669	4.4384 682	1.2535 897	2.7007 771	5.8186 478	.6780 3354
1.98	1.4071 247	4.4497 191	1.2557 072	2.7053 392	5.8284 767	.6830 9684
1.99	1.4106 736	4.4609 416	1.2578 177	2.7098 860	5.8382 725	.6881 3464
2.00	1.4142 136	4.4720 360	1.2599 210	2.7144 176	5.8480 355	.6931 4718
$\begin{matrix} \text{Eii} \\ \text{Eiii} \end{matrix}=$	.0000 014	.0000 04	.0000 011	.0000 024	.0000 051	.0000 0408

2

TABLE VII. SQUARE AND CUBE ROOTS AND NATURAL LOGARITHMS

N	$\sqrt{N}$	$\sqrt{10N}$	$\sqrt[3]{N}$	$\sqrt[3]{10N}$	$\sqrt[3]{100N}$	$\log_e N$
2.00	1.4142 136	4.4721 360	1.2599 210	2.7144 176	5.8480 355	.6931 4718
2.01	1.4177 447	4.4833 024	1.2620 174	2.7189 341	5.8577 660	.6981 3472
2.02	1.4212 670	4.4944 410	1.2641 069	2.7234 357	5.8674 643	.7030 9751
2.03	1.4247 807	4.5055 521	1.2661 894	2.7279 224	5.8771 307	.7080 3579
2.04	1.4282 857	4.5166 359	1.2682 651	2.7323 944	5.8867 653	.7129 4981
2.05	1.4317 821	4.5276 926	1.2703 341	2.7368 518	5.8963 685	.7178 3979
2.06	1.4352 700	4.5387 223	1.2723 963	2.7412 948	5.9059 406	.7227 0598
2.07	1.4387 495	4.5497 253	1.2744 519	2.7457 234	5.9154 817	.7275 4861
2.08	1.4422 205	4.5607 017	1.2765 009	2.7501 377	5.9249 921	.7323 6789
2.09	1.4456 832	4.5716 518	1.2785 433	2.7545 380	5.9344 721	.7371 6407
2.10	1.4491 377	4.5825 757	1.2805 792	2.7589 242	5.9439 220	.7419 3734
2.11	1.4525 839	4.5934 736	1.2826 086	2.7632 965	5.9533 418	.7466 8795
2.12	1.4560 220	4.6043 458	1.2846 317	2.7676 550	5.9627 320	.7514 1609
2.13	1.4594 520	4.6151 923	1.2866 484	2.7719 998	5.9720 926	.7561 2198
2.14	1.4628 739	4.6260 134	1.2886 587	2.7763 311	5.9814 240	.7608 0583
2.15	1.4662 878	4.6368 092	1.2906 629	2.7806 489	5.9907 264	.7654 6784
2.16	1.4696 938	4.6475 800	1.2926 608	2.7849 533	6.0000 000	.7701 0822
2.17	1.4730 920	4.6583 259	1.2946 526	2.7892 445	6.0092 450	.7747 2717
2.18	1.4764 823	4.6690 470	1.2966 383	2.7935 224	6.0184 617	.7793 2488
2.19	1.4798 649	4.6797 436	1.2986 179	2.7977 874	6.0276 502	.7839 0154
2.20	1.4832 397	4.6904 158	1.3005 914	2.8020 393	6.0368 107	.7884 5736
2.21	1.4866 069	4.7010 637	1.3025 591	2.8062 784	6.0459 436	.7929 9252
2.22	1.4899 664	4.7116 876	1.3045 208	2.8105 048	6.0550 489	.7975 0720
2.23	1.4933 185	4.7222 876	1.3064 766	2.8147 184	6.0641 270	.8020 0159
2.24	1.4966 630	4.7328 638	1.3084 265	2.8189 195	6.0731 779	.8064 7587
2.25	1.5000 000	4.7434 165	1.3103 707	2.8231 081	6.0822 020	.8109 3022
2.26	1.5033 296	4.7539 457	1.3123 091	2.8272 843	6.0911 993	.8153 6481
2.27	1.5066 519	4.7644 517	1.3142 418	2.8314 482	6.1001 702	.8197 7983
2.28	1.5099 669	4.7749 346	1.3161 689	2.8355 999	6.1091 147	.8241 7544
2.29	1.5132 746	4.7853 944	1.3180 903	2.8397 394	6.1180 332	.8285 5182
2.30	1.5165 751	4.7958 315	1.3200 061	2.8438 670	6.1269 257	.8329 0912
2.31	1.5198 684	4.8062 459	1.3219 164	2.8479 826	6.1357 924	.8372 4752
2.32	1.5231 546	4.8166 378	1.3238 212	2.8520 863	6.1446 337	.8415 6719
2.33	1.5264 338	4.8270 074	1.3257 205	2.8561 782	6.1534 495	.8458 6827
2.34	1.5297 059	4.8373 546	1.3276 144	2.8602 585	6.1622 401	.8501 5093
2.35	1.5329 710	4.8476 799	1.3295 029	2.8643 272	6.1710 058	.8544 1533
2.36	1.5362 291	4.8579 831	1.3313 860	2.8683 843	6.1797 466	.8586 6152
2.37	1.5394 804	4.8682 646	1.3332 639	2.8724 300	6.1884 628	.8628 8996
2.38	1.5427 249	4.8785 244	1.3351 364	2.8764 643	6.1971 544	.8671 0049
2.39	1.5459 625	4.8887 626	1.3370 038	2.8804 873	6.2058 218	.8712 9337
2.40	1.5491 933	4.8989 795	1.3388 659	2.8844 991	6.2144 650	.8754 6874
2.41	1.5524 175	4.9091 751	1.3407 229	2.8884 998	6.2230 843	.8796 2675
2.42	1.5556 349	4.9193 496	1.3425 747	2.8924 895	6.2316 797	.8837 6754
2.43	1.5588 457	4.9295 030	1.3444 214	2.8964 682	6.2402 515	.8878 9126
2.44	1.5620 499	4.9396 356	1.3462 631	2.9004 359	6.2487 998	.8919 9804
2.45	1.5652 476	4.9497 475	1.3480 997	2.9043 928	6.2573 247	.8960 8802
2.46	1.5684 387	4.9598 387	1.3499 314	2.9083 391	6.2658 266	.9001 6135
2.47	1.5716 234	4.9699 095	1.3517 581	2.9122 746	6.2743 054	.9042 1815
2.48	1.5748 016	4.9799 598	1.3535 799	2.9161 995	6.2827 613	.9082 5856
2.49	1.5779 734	4.9899 900	1.3553 968	2.9201 138	6.2911 946	.9122 8271
2.50	1.5811 388	5.0000 000	1.3572 088	2.9240 177	6.2996 052	.9162 9073
$\mathrm{E^{ii}}$ $\mathrm{E^{iii}}$	.0000 009	.0000 030	.0000 007	.0000 016	.0000 034	.0000 0247

1

TABLE VII. SQUARE AND CUBE ROOTS AND NATURAL LOGARITHMS

N	$\sqrt{N}$	$\sqrt{10N}$	$\sqrt[3]{N}$	$\sqrt[3]{10N}$	$\sqrt[3]{100N}$	$\log_e N$
2.50	1.5811 388	5.0000 000	1.3572 088	2.9240 177	6.2996 052	.9162 9073
2.51	1.5842 980	5.0099 900	1.3590 160	2.9279 112	6.3079 935	.9202 8275
2.52	1.5874 508	5.0199 602	1.3608 184	2.9317 944	6.3163 596	.9242 5890
2.53	1.5905 974	5.0299 105	1.3626 161	2.9356 673	6.3247 035	.9282 1930
2.54	1.5937 377	5.0398 413	1.3644 090	2.9395 301	6.3330 255	.9321 6408
2.55	1.5968 719	5.0497 525	1.3661 972	2.9433 827	6.3413 257	.9360 9336
2.56	1.6000 000	5.0596 443	1.3679 808	2.9472 252	6.3496 042	.9400 0726
2.57	1.6031 220	5.0695 167	1.3697 597	2.9510 577	6.3578 612	.9439 0590
2.58	1.6062 378	5.0793 700	1.3715 340	2.9548 804	6.3660 968	.9477 8940
2.59	1.6093 477	5.0892 043	1.3733 037	2.9586 931	6.3743 111	.9516 5788
2.60	1.6124 515	5.0990 195	1.3750 689	2.9624 961	6.3825 043	.9555 1145
2.61	1.6155 494	5.1088 159	1.3768 295	2.9662 893	6.3906 765	.9593 5022
2.62	1.6186 414	5.1185 936	1.3785 857	2.9700 728	6.3988 279	.9631 7432
2.63	1.6217 275	5.1283 526	1.3803 374	2.9738 467	6.4069 586	.9669 8385
2.64	1.6248 077	5.1380 930	1.3820 846	2.9776 111	6.4150 687	.9707 7892
2.65	1.6278 821	5.1478 151	1.3838 275	2.9813 660	6.4231 583	.9745 5964
2.66	1.6309 506	5.1575 188	1.3855 660	2.9851 114	6.4312 276	.9783 2612
2.67	1.6340 135	5.1672 043	1.3873 001	2.9888 475	6.4392 767	.9820 7847
2.68	1.6370 706	5.1768 716	1.3890 299	2.9925 742	6.4473 057	.9858 1679
2.69	1.6401 219	5.1865 210	1.3907 554	2.9962 917	6.4553 148	.9895 4119
2.70	1.6431 677	5.1961 524	1.3924 767	3.0000 000	6.4633 041	.9932 5177
2.71	1.6462 078	5.2057 660	1.3941 936	3.0036 991	6.4712 736	.9969 4863
2.72	1.6492 423	5.2153 619	1.3959 064	3.0073 892	6.4792 236	1.0006 3188
2.73	1.6522 712	5.2249 402	1.3976 150	3.0110 702	6.4871 541	1.0043 0161
2.74	1.6552 945	5.2345 009	1.3993 194	3.0147 423	6.4950 653	1.0079 5792
2.75	1.6583 124	5.2440 442	1.4010 197	3.0184 054	6.5029 572	1.0116 0091
2.76	1.6613 248	5.2535 702	1.4027 158	3.0220 596	6.5108 301	1.0152 3068
2.77	1.6643 317	5.2630 789	1.4044 079	3.0257 050	6.5186 839	1.0188 4732
2.78	1.6673 332	5.2725 705	1.4060 959	3.0293 417	6.5265 189	1.0224 5093
2.79	1.6703 293	5.2820 451	1.4077 798	3.0329 697	6.5343 351	1.0260 4160
2.80	1.6733 201	5.2915 026	1.4094 597	3.0365 890	6.5421 326	1.0296 1942
2.81	1.6763 055	5.3009 433	1.4111 357	3.0401 997	6.5499 116	1.0331 8448
2.82	1.6792 856	5.3103 672	1.4128 076	3.0438 018	6.5576 722	1.0367 3688
2.83	1.6822 604	5.3197 744	1.4144 757	3.0473 954	6.5654 144	1.0402 7671
2.84	1.6852 300	5.3291 650	1.4161 398	3.0509 806	6.5731 385	1.0438 0405
2.85	1.6881 943	5.3385 391	1.4177 999	3.0545 574	6.5808 444	1.0473 1899
2.86	1.6911 535	5.3478 968	1.4194 562	3.0581 258	6.5885 323	1.0508 2162
2.87	1.6941 074	5.3572 381	1.4211 087	3.0616 859	6.5962 023	1.0543 1203
2.88	1.6970 563	5.3665 631	1.4227 573	3.0652 377	6.6038 545	1.0577 9029
2.89	1.7000 000	5.3758 720	1.4244 021	3.0687 814	6.6114 890	1.0612 5650
2.90	1.7029 386	5.3851 648	1.4260 431	3.0723 168	6.6191 059	1.0647 1074
2.91	1.7058 722	5.3944 416	1.4276 804	3.0758 442	6.6267 054	1.0681 5308
2.92	1.7088 007	5.4037 024	1.4293 139	3.0793 634	6.6342 874	1.0715 8362
2.93	1.7117 243	5.4129 474	1.4309 437	3.0828 747	6.6418 522	1.0750 0242
2.94	1.7146 428	5.4221 767	1.4325 698	3.0863 780	6.6493 998	1.0784 0958
2.95	1.7175 564	5.4313 902	1.4341 921	3.0898 733	6.6569 302	1.0818 0517
2.96	1.7204 651	5.4405 882	1.4358 109	3.0938 607	6.6644 437	1.0851 8927
2.97	1.7233 688	5.4497 706	1.4374 260	3.0968 403	6.6719 403	1.0885 6195
2.98	1.7262 677	5.4589 376	1.4390 374	3.1003 121	6.6794 200	1.0919 2330
2.99	1.7291 616	5.4680 892	1.4406 453	3.1037 762	6.6868 831	1.0952 7339
3.00	1.7320 508	5.4772 256	1.4422 496	3.1072 325	6.6943 295	1.0986 1229
$\begin{matrix}E_{ii}\\E_{iii}\end{matrix}$	.0000 007	.0000 022	.0000 005	.0000 011	.0000 024	.0000 0165

TABLE VII. SQUARE AND CUBE ROOTS AND NATURAL LOGARITHMS

N	$\sqrt{N}$	$\sqrt{10N}$	$\sqrt[3]{N}$	$\sqrt[3]{10N}$	$\sqrt[3]{100N}$	$\log_e N$
3.00	1.7320 508	5.4772 256	1.4422 496	3.1072 325	6.6943 295	1.0986 1229
3.01	1.7349 352	5.4863 467	1.4438 503	3.1106 812	6.7017 594	1.1019 4008
3.02	1.7378 147	5.4954 527	1.4454 475	3.1141 222	6.7091 729	1.1052 5683
3.03	1.7406 895	5.5045 436	1.4470 411	3.1175 556	6.7165 700	1.1085 6262
3.04	1.7435 596	5.5136 195	1.4486 313	3.1209 815	6.7239 508	1.1118 5752
3.05	1.7464 249	5.5226 805	1.4502 180	3.1243 999	6.7313 155	1.1151 4159
3.06	1.7492 856	5.5317 267	1.4518 012	3.1278 108	6.7386 641	1.1184 1492
3.07	1.7521 415	5.5407 581	1.4533 809	3.1312 143	6.7459 967	1.1216 7756
3.08	1.7549 929	5.5497 748	1.4549 573	3.1346 104	6.7533 134	1.1249 2960
3.09	1.7578 396	5.5587 768	1.4565 302	3.1379 992	6.7606 143	1.1281 7109
3.10	1.7606 817	5.5677 644	1.4580 997	3.1413 807	6.7678 995	1.1314 0211
3.11	1.7635 192	5.5767 374	1.4596 659	3.1447 549	6.7751 690	1.1346 2273
3.12	1.7663 522	5.5856 960	1.4612 287	3.1481 218	6.7824 229	1.1378 3300
3.13	1.7691 806	5.5946 403	1.4627 882	3.1514 816	6.7896 613	1.1410 3300
3.14	1.7720 045	5.6035 703	1.4643 444	3.1548 343	6.7968 844	1.1442 2280
3.15	1.7748 239	5.6124 861	1.4658 972	3.1581 798	6.8040 921	1.1474 0245
3.16	1.7776 389	5.6213 877	1.4674 468	3.1615 183	6.8112 846	1.1505 7203
3.17	1.7804 494	5.6302 753	1.4689 931	3.1648 497	6.8184 619	1.1537 3159
3.18	1.7832 555	5.6391 489	1.4705 362	3.1681 741	6.8256 242	1.1568 8120
3.19	1.7860 571	5.6480 085	1.4720 760	3.1714 916	6.8327 715	1.1600 2092
3.20	1.7888 544	5.6568 542	1.4736 126	3.1748 021	6.8399 038	1.1631 5081
3.21	1.7916 473	5.6656 862	1.4751 460	3.1781 058	6.8470 213	1.1662 7094
3.22	1.7944 358	5.6745 044	1.4766 763	3.1814 025	6.8541 240	1.1693 8136
3.23	1.7972 201	5.6833 089	1.4782 033	3.1846 925	6.8612 120	1.1724 8214
3.24	1.8000 000	5.6920 998	1.4797 272	3.1879 757	6.8682 855	1.1755 7333
3.25	1.8027 756	5.7008 771	1.4812 480	3.1912 521	6.8753 443	1.1786 5500
3.26	1.8055 470	5.7096 410	1.4827 657	3.1945 219	6.8823 888	1.1817 2720
3.27	1.8083 141	5.7183 914	1.4842 803	3.1977 849	6.8894 188	1.1847 8998
3.28	1.8110 770	5.7271 284	1.4857 918	3.2010 413	6.8964 345	1.1878 4342
3.29	1.8138 357	5.7358 522	1.4873 002	3.2042 911	6.9034 359	1.1908 8756
3.30	1.8165 902	5.7445 626	1.4888 056	3.2075 343	6.9104 232	1.1939 2247
3.31	1.8193 405	5.7532 599	1.4903 079	3.2107 710	6.9173 964	1.1969 4819
3.32	1.8220 867	5.7619 441	1.4918 072	3.2140 012	6.9243 556	1.1999 6478
3.33	1.8248 288	5.7706 152	1.4933 035	3.2172 248	6.9313 008	1.2029 7230
3.34	1.8275 667	5.7792 733	1.4947 968	3.2204 421	6.9382 321	1.2059 7081
3.35	1.8303 005	5.7879 185	1.4962 871	3.2236 529	6.9451 496	1.2089 6035
3.36	1.8330 303	5.7965 507	1.4977 745	3.2268 573	6.9520 533	1.2119 4097
3.37	1.8357 560	5.8051 701	1.4992 589	3.2300 554	6.9589 433	1.2149 1274
3.38	1.8384 776	5.8137 767	1.5007 404	3.2332 471	6.9658 198	1.2178 7571
3.39	1.8411 953	5.8223 707	1.5022 189	3.2364 326	6.9726 826	1.2208 2992
3.40	1.8439 089	5.8309 519	1.5036 946	3.2396 118	6.9795 320	1.2237 7543
3.41	1.8466 185	5.8395 205	1.5051 674	3.2427 848	6.9863 680	1.2267 1229
3.42	1.8493 242	5.8480 766	1.5066 373	3.2459 516	6.9931 907	1.2296 4055
3.43	1.8520 259	5.8566 202	1.5081 043	3.2491 122	7.0000 000	1.2325 6026
3.44	1.8547 237	5.8651 513	1.5095 685	3.2522 667	7.0067 961	1.2354 7147
3.45	1.8574 176	5.8736 701	1.5110 298	3.2554 150	7.0135 791	1.2383 7423
3.46	1.8601 075	5.8821 765	1.5124 883	3.2585 573	7.0203 490	1.2412 6859
3.47	1.8627 936	5.8906 706	1.5139 440	3.2616 936	7.0271 058	1.2441 5459
3.48	1.8654 758	5.8991 525	1.5153 970	3.2648 238	7.0338 497	1.2470 3229
3.49	1.8681 542	5.9076 222	1.5168 471	3.2679 480	7.0405 806	1.2499 0174
3.50	1.8708 287	5.9160 798	1.5182 945	3.2710 663	7.0472 987	1.2527 6297
$\begin{matrix}\mathbf{E_{iii}}\\\mathbf{E_{iii}}\end{matrix} =$	.0000 005	.0000 017	.0000 004	.0000 008	.0000 018	.0000 0118

TABLE VII. SQUARE AND CUBE ROOTS AND NATURAL LOGARITHMS

N	$\sqrt{N}$	$\sqrt{10N}$	$\sqrt[3]{N}$	$\sqrt[3]{10N}$	$\sqrt[3]{100N}$	$\log_e N$
3.50	1.8708 287	5.9160 798	1.5182 945	3.2710 663	7.0472 987	1.2527 6297
3.51	1.8734 994	5.9245 253	1.5197 391	3.2741 786	7.0540 041	1.2556 1604
3.52	1.8761 663	5.9329 588	1.5211 810	3.2772 851	7.0606 967	1.2584 6099
3.53	1.8788 294	5.9413 803	1.5226 201	3.2803 856	7.0673 766	1.2612 9787
3.54	1.8814 888	5.9497 899	1.5240 566	3.2834 803	7.0740 440	1.2641 2673
3.55	1.8841 444	5.9581 876	1.5254 903	3.2865 692	7.0806 988	1.2669 4760
3.56	1.8867 962	5.9665 736	1.5269 213	3.2896 523	7.0873 411	1.2697 6054
3.57	1.8894 444	5.9749 477	1.5283 497	3.2927 296	7.0939 709	1.2725 6560
3.58	1.8920 888	5.9833 101	1.5297 754	3.2958 012	7.1005 885	1.2753 6280
3.59	1.8947 295	5.9916 609	1.5311 985	3.2988 671	7.1071 937	1.2781 5220
3.60	1.8973 666	6.0000 000	1.5326 189	3.3019 272	7.1137 866	1.2809 3385
3.61	1.9000 000	6.0083 276	1.5340 366	3.3049 818	7.1203 674	1.2837 0777
3.62	1.9026 298	6.0166 436	1.5354 518	3.3080 306	7.1269 360	1.2864 7403
3.63	1.9052 559	6.0249 481	1.5368 644	3.3110 739	7.1334 925	1.2892 3265
3.64	1.9078 784	6.0332 413	1.5382 743	3.3141 116	7.1400 370	1.2919 8368
3.65	1.9104 973	6.0415 230	1.5396 817	3.3171 437	7.1465 695	1.2947 2717
3.66	1.9131 126	6.0497 934	1.5410 865	3.3201 703	7.1530 901	1.2974 6315
3.67	1.9157 244	6.0580 525	1.5424 888	3.3231 914	7.1595 988	1.3001 9166
3.68	1.9183 326	6.0663 004	1.5438 885	3.3262 070	7.1660 957	1.3029 1275
3.69	1.9209 373	6.0745 370	1.5452 857	3.3292 171	7.1725 809	1.3056 2646
3.70	1.9235 384	6.0827 625	1.5466 804	3.3322 219	7.1790 544	1.3083 3282
3.71	1.9261 360	6.0909 769	1.5480 725	3.3352 212	7.1855 162	1.3110 3188
3.72	1.9287 302	6.0991 803	1.5494 622	3.3382 151	7.1919 663	1.3137 2367
3.73	1.9313 208	6.1073 726	1.5508 493	3.3412 036	7.1984 050	1.3164 0823
3.74	1.9339 080	6.1155 539	1.5522 340	3.3441 868	7.2048 321	1.3190 8561
3.75	1.9364 917	6.1237 244	1.5536 163	3.3471 648	7.2112 479	1.3217 5584
3.76	1.9390 719	6.1318 839	1.5549 960	3.3501 374	7.2176 522	1.3244 1896
3.77	1.9416 488	6.1400 326	1.5563 733	3.3531 047	7.2240 451	1.3270 7500
3.78	1.9442 222	6.1481 705	1.5577 482	3.3560 668	7.2304 268	1.3297 2401
3.79	1.9467 922	6.1562 976	1.5591 207	3.3590 237	7.2367 972	1.3323 6602
3.80	1.9493 589	6.1644 140	1.5604 908	3.3619 754	7.2431 564	1.3350 0107
3.81	1.9519 221	6.1725 197	1.5618 584	3.3649 219	7.2495 045	1.3376 2919
3.82	1.9544 820	6.1806 149	1.5632 237	3.3678 633	7.2558 415	1.3402 5042
3.83	1.9570 386	6.1886 994	1.5645 865	3.3707 995	7.2621 674	1.3428 6480
3.84	1.9595 918	6.1967 734	1.5659 471	3.3737 307	7.2684 824	1.3454 7237
3.85	1.9621 417	6.2048 368	1.5673 052	3.3766 567	7.2747 863	1.3480 7315
3.86	1.9646 883	6.2128 898	1.5686 610	3.3795 777	7.2810 794	1.3506 6718
3.87	1.9672 316	6.2209 324	1.5700 145	3.3824 936	7.2873 616	1.3532 5451
3.88	1.9697 716	6.2289 646	1.5713 656	3.3854 046	7.2936 330	1.3558 3515
3.89	1.9723 083	6.2369 865	1.5727 144	3.3883 105	7.2998 937	1.3584 0916
3.90	1.9748 418	6.2449 980	1.5740 609	3.3912 114	7.3061 436	1.3609 7655
3.91	1.9773 720	6.2529 993	1.5754 051	3.3941 074	7.3123 828	1.3635 3737
3.92	1.9798 990	6.2609 903	1.5767 470	3.3969 985	7.3186 114	1.3660 9165
3.93	1.9824 228	6.2689 712	1.5780 867	3.3998 847	7.3248 294	1.3686 3943
3.94	1.9849 433	6.2769 419	1.5794 240	3.4027 659	7.3310 369	1.3711 8072
3.95	1.9874 607	6.2849 025	1.5807 591	3.4056 423	7.3372 339	1.3737 1558
3.96	1.9899 749	6.2928 531	1.5820 920	3.4085 138	7.3434 205	1.3762 4403
3.97	1.9924 859	6.3007 936	1.5834 226	3.4113 805	7.3495 966	1.3787 6609
3.98	1.9949 937	6.3087 241	1.5847 510	3.4142 424	7.3557 624	1.3812 8182
3.99	1.9974 984	6.3166 447	1.5860 771	3.4170 996	7.3619 178	1.3837 9123
4.00	2.0000 000	6.3245 553	1.5874 011	3.4199 519	7.3680 630	1.3862 9436
E_{ii} E_{iii}	.0000 004	.0000 014	.0000 003	.0000 007	.0000 014	.0000 0089

TABLE VII. SQUARE AND CUBE ROOTS AND NATURAL LOGARITHMS

N	$\sqrt{N}$	$\sqrt{10N}$	$\sqrt[3]{N}$	$\sqrt[3]{10N}$	$\sqrt[3]{100N}$	$\log_e N$
4.00	2.0000 000	6.3245 553	1.5874 011	3.4199 519	7.3680 630	1.3862 9436
4.01	2.0024 984	6.3324 561	1.5887 228	3.4227 995	7.3741 979	1.3887 9124
4.02	2.0049 938	6.3403 470	1.5900 423	3.4256 423	7.3803 227	1.3912 8190
4.03	2.0074 860	6.3482 281	1.5913 597	3.4284 805	7.3864 373	1.3937 6638
4.04	2.0099 751	6.3560 994	1.5926 748	3.4313 139	7.3925 418	1.3962 4469
4.05	2.0124 612	6.3639 610	1.5939 879	3.4341 427	7.3986 362	1.3987 1688
4.06	2.0149 442	6.3718 129	1.5952 987	3.4369 669	7.4047 206	1.4011 8297
4.07	2.0174 241	6.3796 552	1.5966 074	3.4397 864	7.4107 951	1.4036 4300
4.08	2.0199 010	6.3874 878	1.5979 139	3.4426 012	7.4168 595	1.4060 9699
4.09	2.0223 748	6.3953 108	1.5992 184	3.4454 115	7.4229 141	1.4085 4497
4.10	2.0248 457	6.4031 242	1.6005 207	3.4482 172	7.4289 588	1.4109 8697
4.11	2.0273 135	6.4109 282	1.6018 208	3.4510 184	7.4349 937	1.4134 2303
4.12	2.0297 783	6.4187 226	1.6031 189	3.4538 150	7.4410 189	1.4158 5316
4.13	2.0322 401	6.4265 076	1.6044 149	3.4566 071	7.4470 342	1.4182 7741
4.14	2.0346 990	6.4342 832	1.6057 088	3.4593 947	7.4530 399	1.4206 9579
4.15	2.0371 549	6.4420 494	1.6070 006	3.4621 778	7.4590 359	1.4231 0833
4.16	2.0396 078	6.4498 062	1.6082 903	3.4649 564	7.4650 223	1.4255 1507
4.17	2.0420 578	6.4575 537	1.6095 780	3.4677 306	7.4709 991	1.4279 1604
4.18	2.0445 048	6.4652 920	1.6108 636	3.4705 004	7.4769 664	1.4303 1125
4.19	2.0469 489	6.4730 209	1.6121 471	3.4732 657	7.4829 241	1.4327 0073
4.20	2.0493 902	6.4807 407	1.6134 286	3.4760 266	7.4888 724	1.4350 8453
4.21	2.0518 285	6.4884 513	1.6147 081	3.4787 832	7.4948 112	1.4374 6265
4.22	2.0542 639	6.4961 527	1.6159 856	3.4815 354	7.5007 407	1.4398 3513
4.23	2.0566 964	6.5038 450	1.6172 610	3.4842 833	7.5066 607	1.4422 0199
4.24	2.0591 260	6.5115 282	1.6185 345	3.4870 268	7.5125 715	1.4445 6327
4.25	2.0615 528	6.5192 024	1.6198 059	3.4897 660	7.5184 730	1.4469 1898
4.26	2.0639 767	6.5268 675	1.6210 753	3.4925 010	7.5243 652	1.4492 6916
4.27	2.0663 978	6.5345 237	1.6223 428	3.4952 316	7.5302 482	1.4516 1383
4.28	2.0688 161	6.5421 709	1.6236 083	3.4979 580	7.5361 220	1.4539 5301
4.29	2.0712 315	6.5498 092	1.6248 718	3.5006 801	7.5419 867	1.4562 8673
4.30	2.0736 441	6.5574 385	1.6261 333	3.5033 981	7.5478 423	1.4586 1502
4.31	2.0760 539	6.5650 590	1.6273 929	3.5061 118	7.5536 888	1.4609 3790
4.32	2.0784 610	6.5726 707	1.6286 506	3.5088 213	7.5595 263	1.4632 5540
4.33	2.0808 652	6.5802 736	1.6299 063	3.5115 266	7.5653 548	1.4655 6754
4.34	2.0832 667	6.5878 676	1.6311 601	3.5142 278	7.5711 743	1.4678 7435
4.35	2.0856 654	6.5954 530	1.6324 119	3.5169 248	7.5769 849	1.4701 7585
4.36	2.0880 613	6.6030 296	1.6336 618	3.5196 177	7.5827 865	1.4724 7206
4.37	2.0904 545	6.6105 976	1.6349 099	3.5223 065	7.5886 793	1.4747 6301
4.38	2.0928 450	6.6181 568	1.6361 560	3.5249 912	7.5943 633	1.4770 4872
4.39	2.0952 327	6.6257 075	1.6374 002	3.5276 718	7.6001 385	1.4793 2923
4.40	2.0976 177	6.6332 496	1.6386 425	3.5303 483	7.6059 049	1.4816 0454
4.41	2.1000 000	6.6407 831	1.6398 830	3.5330 208	7.6116 626	1.4838 7469
4.42	2.1023 796	6.6483 081	1.6411 216	3.5356 893	7.6174 116	1.4861 3970
4.43	2.1047 565	6.6558 245	1.6423 583	3.5383 537	7.6231 519	1.4883 9958
4.44	2.1071 308	6.6633 325	1.6435 932	3.5410 141	7.6288 836	1.4906 5438
4.45	2.1095 023	6.6708 320	1.6448 262	3.5436 705	7.6346 067	1.4929 0410
4.46	2.1118 712	6.6783 231	1.6460 573	3.5463 230	7.6403 212	1.4951 4877
4.47	2.1142 375	6.6858 059	1.6472 866	3.5489 715	7.6460 272	1.4973 8841
4.48	2.1166 010	6.6932 802	1.6485 141	3.5516 160	7.6517 247	1.4996 2305
4.49	2.1189 620	6.7007 462	1.6497 398	3.5542 566	7.6574 137	1.5018 5270
4.50	2.1213 203	6.7082 039	1.6509 636	3.5568 933	7.6630 943	1.5040 7740
$\underset{\varepsilon iii}{\varepsilon ii} =$	.0000 004	.0000 011	.0000 002	.0000 005	.0000 012	.0000 0069

TABLE VII. SQUARE AND CUBE ROOTS AND NATURAL LOGARITHMS

N	$\sqrt{N}$	$\sqrt{10N}$	$\sqrt[3]{N}$	$\sqrt[3]{10N}$	$\sqrt[3]{100N}$	$\log_e N$
4.50	2.1213 203	6.7082 039	1.6509 636	3.5568 933	7.6630 943	1.5040 7740
4.51	2.1236 761	6.7156 534	1.6521 857	3.5595 261	7.6687 665	1.5062 9715
4.52	2.1260 292	6.7230 945	1.6534 059	3.5621 550	7.6744 303	1.5085 1199
4.53	2.1283 797	6.7305 275	1.6546 243	3.5647 800	7.6800 857	1.5107 2194
4.54	2.1307 276	6.7379 522	1.6558 409	3.5674 012	7.6857 328	1.5129 2701
4.55	2.1330 729	6.7453 688	1.6570 558	3.5700 185	7.6913 717	1.5151 2723
4.56	2.1354 157	6.7527 772	1.6582 689	3.5726 320	7.6970 023	1.5173 2262
4.57	2.1377 558	6.7601 775	1.6594 802	3.5752 416	7.7026 246	1.5195 1320
4.58	2.1400 935	6.7675 697	1.6606 897	3.5778 475	7.7082 388	1.5216 9900
4.59	2.1424 285	6.7749 539	1.6618 975	3.5804 496	7.7138 448	1.5238 8002
4.60	2.1447 611	6.7823 300	1.6631 035	3.5830 479	7.7194 426	1.5260 5630
4.61	2.1470 911	6.7896 981	1.6643 078	3.5856 424	7.7250 324	1.5282 2786
4.62	2.1494 185	6.7970 582	1.6655 103	3.5882 332	7.7306 141	1.5303 9471
4.63	2.1517 435	6.8044 103	1.6667 111	3.5908 202	7.7361 877	1.5325 5687
4.64	2.1540 659	6.8117 545	1.6679 102	3.5934 036	7.7417 533	1.5347 1437
4.65	2.1563 859	6.8190 908	1.6691 075	3.5959 832	7.7473 109	1.5368 6722
4.66	2.1587 033	6.8264 193	1.6703 032	3.5985 591	7.7528 605	1.5390 1545
4.67	2.1610 183	6.8337 398	1.6714 971	3.6011 313	7.7584 023	1.5411 5907
4.68	2.1633 308	6.8410 526	1.6726 893	3.6036 999	7.7639 361	1.5432 9811
4.69	2.1656 408	6.8483 575	1.6738 798	3.6062 648	7.7694 620	1.5454 3258
4.70	2.1679 483	6.8556 546	1.6750 687	3.6088 261	7.7749 801	1.5475 6251
4.71	2.1702 534	6.8629 440	1.6762 558	3.6113 837	7.7804 904	1.5496 8791
4.72	2.1725 561	6.8702 256	1.6774 413	3.6139 377	7.7859 928	1.5518 0880
4.73	2.1748 563	6.8774 995	1.6786 251	3.6164 882	7.7914 875	1.5539 2520
4.74	2.1771 541	6.8847 658	1.6798 072	3.6190 350	7.7969 745	1.5560 3714
4.75	2.1794 495	6.8920 244	1.6809 877	3.6215 782	7.8024 538	1.5581 4462
4.76	2.1817 424	6.8992 753	1.6821 665	3.6241 179	7.8079 253	1.5602 4767
4.77	2.1840 330	6.9065 187	1.6833 437	3.6266 540	7.8133 892	1.5623 4630
4.78	2.1863 211	6.9137 544	1.6845 192	3.6291 866	7.8188 455	1.5644 4055
4.79	2.1886 069	6.9209 826	1.6856 931	3.6317 157	7.8242 942	1.5665 3041
4.80	2.1908 902	6.9282 032	1.6868 653	3.6342 412	7.8297 353	1.5686 1592
4.81	2.1931 712	6.9354 164	1.6880 360	3.6367 632	7.8351 688	1.5706 9708
4.82	2.1954 498	6.9426 220	1.6892 050	3.6392 817	7.8405 948	1.5727 7393
4.83	2.1977 261	6.9498 201	1.6903 723	3.6417 968	7.8460 134	1.5748 4647
4.84	2.2000 000	6.9570 109	1.6915 381	3.6443 084	7.8514 244	1.5769 1472
4.85	2.2022 716	6.9641 941	1.6927 023	3.6468 165	7.8568 280	1.5789 7870
4.86	2.2045 408	6.9713 700	1.6938 649	3.6493 212	7.8622 242	1.5810 3844
4.87	2.2068 076	6.9785 385	1.6950 258	3.6518 224	7.8676 130	1.5830 9394
4.88	2.2090 722	6.9856 997	1.6961 852	3.6543 203	7.8729 944	1.5851 4522
4.89	2.2113 344	6.9928 535	1.6973 430	3.6568 147	7.8783 684	1.5871 9230
4.90	2.2135 944	7.0000 000	1.6984 993	3.6593 057	7.8837 352	1.5892 3521
4.91	2.2158 520	7.0071 392	1.6996 539	3.6617 933	7.8890 946	1.5912 7394
4.92	2.2181 073	7.0142 712	1.7008 070	3.6642 776	7.8944 468	1.5933 0853
4.93	2.2203 603	7.0213 959	1.7019 585	3.6667 585	7.8997 917	1.5953 3899
4.94	2.2226 111	7.0285 134	1.7031 085	3.6692 360	7.9051 294	1.5973 6533
4.95	2.2248 595	7.0356 236	1.7042 569	3.6717 102	7.9104 599	1.5993 8758
4.96	2.2271 057	7.0427 267	1.7054 038	3.6741 811	7.9157 832	1.6014 0574
4.97	2.2293 497	7.0498 227	1.7065 491	3.6766 487	7.9210 994	1.6034 1984
4.98	2.2315 914	7.0569 115	1.7076 929	3.6791 129	7.9264 084	1.6054 2989
4.99	2.2338 308	7.0639 932	1.7088 352	3.6815 738	7.9317 104	1.6074 3591
5.00	2.2360 680	7.0710 678	1.7099 759	3.6840 315	7.9370 053	1.6094 3791
E_{ii} =	.0000 003	.0000 010	.0000 002	.0000 004	.0000 010	.0000 0055
E_{iii} =						

TABLE VII. SQUARE AND CUBE ROOTS AND NATURAL LOGARITHMS

N	$\sqrt{N}$	$\sqrt{10N}$	$\sqrt[3]{N}$	$\sqrt[3]{10N}$	$\sqrt[3]{100N}$	$\log_e N$
5.00	2.2360 680	7.0710 678	1.7099 759	3.6840 315	7.9370 053	1.6094 3791
5.01	2.2383 029	7.0781 353	1.7111 152	3.6864 859	7.9422 931	1.6114 3592
5.02	2.2405 357	7.0851 958	1.7122 529	3.6889 370	7.9475 739	1.6134 2993
5.03	2.2427 661	7.0922 493	1.7133 891	3.6913 849	7.9528 476	1.6154 1998
5.04	2.2449 944	7.0992 957	1.7145 238	3.6938 295	7.9581 144	1.6174 0608
5.05	2.2472 205	7.1063 352	1.7156 570	3.6962 709	7.9633 742	1.6193 8824
5.06	2.2494 444	7.1133 677	1.7167 887	3.6987 091	7.9686 271	1.6213 6648
5.07	2.2516 660	7.1203 932	1.7179 189	3.7011 440	7.9738 731	1.6233 4082
5.08	2.2538 855	7.1274 119	1.7190 476	3.7035 758	7.9791 122	1.6253 1126
5.09	2.2561 028	7.1344 236	1.7201 749	3.7060 044	7.9843 444	1.6272 7783
5.10	2.2583 180	7.1414 284	1.7213 006	3.7084 298	7.9895 697	1.6292 4054
5.11	2.2605 309	7.1484 264	1.7224 249	3.7108 520	7.9947 883	1.6311 9940
5.12	2.2627 417	7.1554 175	1.7235 478	3.7132 711	8.0000 000	1.6331 5444
5.13	2.2649 503	7.1624 018	1.7246 691	3.7156 870	8.0052 049	1.6351 0566
5.14	2.2671 568	7.1693 793	1.7257 890	3.7180 998	8.0104 031	1.6370 5308
5.15	2.2693 611	7.1763 500	1.7269 075	3.7205 094	8.0155 946	1.6389 9671
5.16	2.2715 633	7.1833 140	1.7280 245	3.7229 160	8.0207 793	1.6409 3658
5.17	2.2737 634	7.1902 712	1.7291 401	3.7253 194	8.0259 574	1.6428 7269
5.18	2.2759 613	7.1972 217	1.7302 542	3.7277 197	8.0311 287	1.6448 0506
5.19	2.2781 571	7.2041 655	1.7313 669	3.7301 170	8.0362 934	1.6467 3370
5.20	2.2803 509	7.2111 026	1.7324 782	3.7325 112	8.0414 515	1.6486 5863
5.21	2.2825 424	7.2180 330	1.7335 881	3.7349 023	8.0466 030	1.6505 7986
5.22	2.2847 319	7.2249 567	1.7346 965	3.7372 903	8.0517 479	1.6524 9740
5.23	2.2869 193	7.2318 739	1.7358 035	3.7396 753	8.0568 862	1.6544 1128
5.24	2.2891 046	7.2387 844	1.7369 091	3.7420 573	8.0620 180	1.6563 2150
5.25	2.2912 878	7.2456 884	1.7380 133	3.7444 362	8.0671 432	1.6582 2808
5.26	2.2934 690	7.2525 857	1.7391 161	3.7468 121	8.0722 620	1.6601 3103
5.27	2.2956 481	7.2594 766	1.7402 175	3.7491 850	8.0773 742	1.6620 3036
5.28	2.2978 251	7.2663 608	1.7413 175	3.7515 549	8.0824 800	1.6639 2610
5.29	2.3000 000	7.2732 386	1.7424 162	3.7539 218	8.0875 794	1.6658 1825
5.30	2.3021 729	7.2801 099	1.7435 134	3.7562 858	8.0926 723	1.6677 0682
5.31	2.3043 437	7.2869 747	1.7446 093	3.7586 467	8.0977 589	1.6695 9184
5.32	2.3065 125	7.2938 330	1.7457 037	3.7610 047	8.1028 390	1.6714 7330
5.33	2.3086 793	7.3006 849	1.7467 969	3.7633 598	8.1079 128	1.6733 5124
5.34	2.3108 440	7.3075 304	1.7478 886	3.7657 119	8.1129 803	1.6752 2565
5.35	2.3130 067	7.3143 694	1.7489 790	3.7680 610	8.1180 414	1.6770 9656
5.36	2.3151 674	7.3212 021	1.7500 680	3.7704 073	8.1230 962	1.6789 6398
5.37	2.3173 260	7.3280 284	1.7511 557	3.7727 506	8.1281 447	1.6808 2791
5.38	2.3194 827	7.3348 483	1.7522 420	3.7750 910	8.1331 870	1.6826 8837
5.39	2.3216 374	7.3416 619	1.7533 270	3.7774 285	8.1382 230	1.6845 4538
5.40	2.3237 900	7.3484 692	1.7544 106	3.7797 631	8.1432 528	1.6863 9895
5.41	2.3259 407	7.3552 702	1.7554 929	3.7820 949	8.1482 764	1.6882 4909
5.42	2.3280 893	7.3620 649	1.7565 739	3.7844 238	8.1532 939	1.6900 9581
5.43	2.3302 360	7.3688 534	1.7576 536	3.7867 498	8.1583 051	1.6919 3913
5.44	2.3323 808	7.3756 356	1.7587 319	3.7890 729	8.1633 102	1.6937 7906
5.45	2.3345 235	7.3824 115	1.7598 089	3.7913 933	8.1683 092	1.6956 1561
5.46	2.3366 643	7.3891 813	1.7608 845	3.7937 107	8.1733 020	1.6974 4879
5.47	2.3388 031	7.3959 448	1.7619 589	3.7960 254	8.1782 888	1.6992 7862
5.48	2.3409 400	7.4027 022	1.7630 320	3.7983 372	8.1832 695	1.7011 0510
5.49	2.3430 749	7.4094 534	1.7641 037	3.8006 462	8.1882 441	1.7029 2826
5.50	2.3452 079	7.4161 985	1.7651 742	3.8029 525	8.1932 127	1.7047 4809
$\epsilon^{II}=$ $\epsilon^{III}=$	.0000 003	.0000 008	.0000 002	.0000 004	.0000 008	.0000 0045

TABLE VII. SQUARE AND CUBE ROOTS AND NATURAL LOGARITHMS

N	$\sqrt{N}$	$\sqrt{10N}$	$\sqrt[3]{N}$	$\sqrt[3]{10N}$	$\sqrt[3]{100N}$	$\log_e N$
5.50	2.3452 079	7.4161 985	1.7651 742	3.8029 525	8.1932 127	1.7047 4809
5.51	2.3473 389	7.4229 374	1.7662 433	3.8052 559	8.1981 753	1.7065 6462
5.52	2.3494 680	7.4296 702	1.7673 112	3.8075 565	8.2031 319	1.7083 7786
5.53	2.3515 952	7.4363 970	1.7683 778	3.8098 544	8.2080 825	1.7101 8782
5.54	2.3537 205	7.4431 176	1.7694 430	3.8121 495	8.2130 271	1.7119 9450
5.55	2.3558 438	7.4498 322	1.7705 071	3.8144 418	8.2179 657	1.7137 9793
5.56	2.3579 652	7.4565 408	1.7715 698	3.8167 314	8.2228 985	1.7155 9811
5.57	2.3600 847	7.4632 433	1.7726 312	3.8190 182	8.2278 254	1.7173 9505
5.58	2.3622 024	7.4699 398	1.7736 914	3.8213 023	8.2327 463	1.7191 8878
5.59	2.3643 181	7.4766 303	1.7747 503	3.8235 837	8.2376 614	1.7209 7929
5.60	2.3664 319	7.4833 148	1.7758 080	3.8258 624	8.2425 706	1.7227 6660
5.61	2.3685 439	7.4899 933	1.7768 644	3.8281 383	8.2474 740	1.7245 5072
5.62	2.3706 539	7.4966 659	1.7779 195	3.8304 116	8.2523 715	1.7263 3166
5.63	2.3727 621	7.5033 326	1.7789 734	3.8326 821	8.2572 633	1.7281 0944
5.64	2.3748 684	7.5099 933	1.7800 261	3.8349 500	8.2621 492	1.7298 8407
5.65	2.3769 729	7.5166 482	1.7810 775	3.8372 151	8.2670 294	1.7316 5555
5.66	2.3790 755	7.5232 971	1.7821 277	3.8394 776	8.2719 038	1.7334 2389
5.67	2.3811 762	7.5299 402	1.7831 766	3.8417 375	8.2767 725	1.7351 8912
5.68	2.3832 751	7.5365 775	1.7842 243	3.8439 947	8.2816 355	1.7369 5123
5.69	2.3853 721	7.5432 089	1.7852 707	3.8462 492	8.2864 928	1.7387 1025
5.70	2.3874 673	7.5498 344	1.7863 160	3.8485 011	8.2913 443	1.7404 6617
5.71	2.3895 606	7.5564 542	1.7873 600	3.8507 504	8.2961 902	1.7422 1902
5.72	2.3916 521	7.5630 682	1.7884 028	3.8529 970	8.3010 305	1.7439 6881
5.73	2.3937 418	7.5696 763	1.7894 444	3.8552 411	8.3058 651	1.7457 1553
5.74	2.3958 297	7.5762 788	1.7904 848	3.8574 825	8.3106 941	1.7474 5921
5.75	2.3979 158	7.5828 754	1.7915 239	3.8597 213	8.3155 175	1.7491 9985
5.76	2.4000 000	7.5894 664	1.7925 619	3.8619 575	8.3203 353	1.7509 3747
5.77	2.4020 824	7.5960 516	1.7935 987	3.8641 912	8.3251 475	1.7526 7208
5.78	2.4041 631	7.6026 311	1.7946 342	3.8664 222	8.3299 542	1.7544 0368
5.79	2.4062 419	7.6092 050	1.7956 686	3.8686 507	8.3347 553	1.7561 3229
5.80	2.4083 189	7.6157 731	1.7967 018	3.8708 766	8.3395 509	1.7578 5792
5.81	2.4103 942	7.6223 356	1.7977 338	3.8731 000	8.3443 410	1.7595 8057
5.82	2.4124 676	7.6288 924	1.7987 646	3.8753 208	8.3491 256	1.7613 0026
5.83	2.4145 393	7.6354 437	1.7997 942	3.8775 391	8.3539 047	1.7630 1700
5.84	2.4166 092	7.6419 893	1.8008 227	3.8797 548	8.3586 784	1.7647 3080
5.85	2.4186 773	7.6485 293	1.8018 499	3.8819 680	8.3634 466	1.7664 4166
5.86	2.4207 437	7.6550 637	1.8028 761	3.8841 787	8.3682 094	1.7681 4960
5.87	2.4228 083	7.6615 925	1.8039 010	3.8863 869	8.3729 668	1.7698 5463
5.88	2.4248 711	7.6681 158	1.8049 248	3.8885 926	8.3777 187	1.7715 5676
5.89	2.4269 322	7.6746 335	1.8059 474	3.8907 957	8.3824 653	1.7732 5600
5.90	2.4289 916	7.6811 457	1.8069 689	3.8929 964	8.3872 065	1.7749 5235
5.91	2.4310 492	7.6876 524	1.8079 892	3.8951 946	8.3919 424	1.7766 4583
5.92	2.4331 050	7.6941 536	1.8090 083	3.8973 903	8.3966 729	1.7783 3645
5.93	2.4351 591	7.7006 493	1.8100 264	3.8995 836	8.4013 981	1.7800 2421
5.94	2.4372 115	7.7071 395	1.8110 432	3.9017 743	8.4061 180	1.7817 0913
5.95	2.4392 622	7.7136 243	1.8120 589	3.9039 627	8.4108 326	1.7833 9122
5.96	2.4413 111	7.7201 036	1.8130 735	3.9061 485	8.4155 419	1.7850 7048
5.97	2.4433 583	7.7265 775	1.8140 870	3.9083 319	8.4202 459	1.7867 4693
5.98	2.4454 039	7.7330 460	1.8150 993	3.9105 129	8.4249 447	1.7884 2057
5.99	2.4474 477	7.7395 090	1.8161 105	3.9126 915	8.4296 383	1.7900 9141
6.00	2.4494 897	7.7459 667	1.8171 206	3.9148 676	8.4343 267	1.7917 5947
$\begin{matrix}\text{E ii} \\ \text{E iii}\end{matrix}=$	.0000 002	.0000 007	.0000 001	.0000 003	.0000 007	.0000 0038

TABLE VII. SQUARE AND CUBE ROOTS AND NATURAL LOGARITHMS

N	$\sqrt{N}$	$\sqrt{10N}$	$\sqrt[3]{N}$	$\sqrt[3]{10N}$	$\sqrt[3]{100N}$	$\log_e N$
6.00	2.4494 897	7.7459 667	1.8171 206	3.9148 676	8.4343 267	1.7917 5947
6.01	2.4515 301	7.7524 190	1.8181 295	3.9170 414	8.4390 098	1.7934 2475
6.02	2.4535 688	7.7588 659	1.8191 374	3.9192 127	8.4436 877	1.7950 8726
6.03	2.4556 058	7.7653 075	1.8201 441	3.9213 816	8.4483 605	1.7967 4701
6.04	2.4576 411	7.7717 437	1.8211 497	3.9235 481	8.4530 281	1.7984 0401
6.05	2.4596 748	7.7781 746	1.8221 542	3.9257 122	8.4576 906	1.8000 5827
6.06	2.4617 067	7.7846 002	1.8231 576	3.9278 739	8.4623 479	1.8017 0980
6.07	2.4637 370	7.7910 205	1.8241 599	3.9300 333	8.4670 001	1.8033 5861
6.08	2.4657 656	7.7974 355	1.8251 611	3.9321 903	8.4716 472	1.8050 0470
6.09	2.4677 925	7.8038 452	1.8261 611	3.9343 449	8.4762 892	1.8066 4808
6.10	2.4698 178	7.8102 497	1.8271 601	3.9364 972	8.4809 261	1.8082 8877
6.11	2.4718 414	7.8166 489	1.8281 580	3.9386 471	8.4855 579	1.8099 2677
6.12	2.4738 634	7.8230 429	1.8291 549	3.9407 947	8.4901 847	1.8115 6210
6.13	2.4758 837	7.8294 317	1.8301 506	3.9429 399	8.4948 065	1.8131 9475
6.14	2.4779 023	7.8358 152	1.8311 452	3.9450 828	8.4994 233	1.8148 2474
6.15	2.4799 194	7.8421 936	1.8321 388	3.9472 234	8.5040 350	1.8164 5208
6.16	2.4819 347	7.8485 667	1.8331 313	3.9493 616	8.5086 417	1.8180 7678
6.17	2.4839 485	7.8549 348	1.8341 227	3.9514 976	8.5132 435	1.8196 9884
6.18	2.4859 606	7.8612 976	1.8351 131	3.9536 313	8.5178 403	1.8213 1827
6.19	2.4879 711	7.8676 553	1.8361 023	3.9557 626	8.5224 321	1.8229 3509
6.20	2.4899 799	7.8740 079	1.8370 906	3.9578 916	8.5270 190	1.8245 4929
6.21	2.4919 872	7.8803 553	1.8380 777	3.9600 184	8.5316 009	1.8261 6090
6.22	2.4929 902	7.8835 271	1.8390 638	3.9621 428	8.5361 780	1.8277 6991
6.23	2.4959 968	7.8930 349	1.8400 488	3.9642 650	8.5407 501	1.8293 7633
6.24	2.4979 992	7.8993 671	1.8410 328	3.9663 850	8.5453 174	1.8309 8018
6.25	2.5000 000	7.9056 942	1.8420 157	3.9685 026	8.5498 797	1.8325 8146
6.26	2.5019 992	7.9120 162	1.8429 976	3.9706 180	8.5544 372	1.8341 8019
6.27	2.5039 968	7.9183 332	1.8439 785	3.9727 312	8.5589 899	1.8357 7635
6.28	2.5059 928	7.9246 451	1.8449 583	3.9748 421	8.5635 377	1.8373 6998
6.29	2.5079 872	7.9309 520	1.8459 370	3.9769 508	8.5680 807	1.8389 6107
6.30	2.5099 801	7.9372 539	1.8469 148	3.9790 572	8.5726 189	1.8405 4963
6.31	2.5119 713	7.9435 508	1.8478 914	3.9811 614	8.5771 523	1.8421 3568
6.32	2.5139 610	7.9498 428	1.8488 671	3.9832 634	8.5816 809	1.8437 1921
6.33	2.5159 491	7.9561 297	1.8498 417	3.9853 632	8.5862 047	1.8453 0024
6.34	2.5179 357	7.9624 117	1.8508 153	3.9874 607	8.5907 237	1.8468 7877
6.35	2.5199 206	7.9686 887	1.8517 879	3.9895 561	8.5952 380	1.8484 5481
6.36	2.5219 040	7.9749 608	1.8527 595	3.9916 492	8.5997 476	1.8500 2838
6.37	2.5238 859	7.9812 280	1.8537 300	3.9937 402	8.6042 524	1.8515 9947
6.38	2.5258 662	7.9874 902	1.8546 995	3.9958 290	8.6087 526	1.8531 6810
6.39	2.5278 449	7.9937 476	1.8556 680	3.9979 156	8.6132 480	1.8547 3427
6.40	2.5298 221	8.0000 000	1.8566 355	4.0000 000	8.6177 388	1.8562 9799
6.41	2.5317 978	8.0062 476	1.8576 020	4.0020 822	8.6222 248	1.8578 5927
6.42	2.5337 719	8.0124 902	1.8585 675	4.0041 623	8.6267 062	1.8594 1812
6.43	2.5357 445	8.0187 281	1.8595 320	4.0062 403	8.6311 830	1.8609 7454
6.44	2.5377 155	8.0249 611	1.8604 955	4.0083 160	8.6356 551	1.8625 2854
6.45	2.5396 850	8.0311 892	1.8614 580	4.0103 897	8.6401 226	1.8640 8013
6.46	2.5416 530	8.0374 125	1.8624 195	4.0124 611	8.6445 855	1.8656 2932
6.47	2.5436 195	8.0436 310	1.8633 800	4.0145 305	8.6490 437	1.8671 7611
6.48	2.5455 844	8.0498 447	1.8643 395	4.0165 977	8.6534 974	1.8687 2051
6.49	2.5475 478	8.0560 536	1.8652 980	4.0186 628	8.6579 465	1.8702 6253
6.50	2.5495 098	8.0622 577	1.8662 556	4.0207 258	8.6623 911	1.8718 0218
Eii Eiii	.0000 002	.0000 006	.0000 001	.0000 003	.0000 006	.0000 0032

TABLE VII. SQUARE AND CUBE ROOTS AND NATURAL LOGARITHMS

N	$\sqrt{N}$	$\sqrt{10N}$	$\sqrt[3]{N}$	$\sqrt[3]{10N}$	$\sqrt[3]{100N}$	$\log_e N$
6.50	2.5495 098	8.0622 577	1.8662 556	4.0207 258	8.6623 911	1.8718 0218
6.51	2.5514 702	8.0684 571	1.8672 121	4.0227 866	8.6668 310	1.8733 3946
6.52	2.5534 291	8.0746 517	1.8681 677	4.0248 454	8.6712 665	1.8748 7438
6.53	2.5553 865	8.0808 415	1.8691 223	4.0269 020	8.6756 974	1.8764 0694
6.54	2.5573 424	8.0870 266	1.8700 760	4.0289 565	8.6801 237	1.8779 3717
6.55	2.5592 968	8.0932 070	1.8710 286	4.0310 090	8.6845 456	1.8794 6505
6.56	2.5612 497	8.0993 827	1.8719 803	4.0330 594	8.6889 630	1.8809 9060
6.57	2.5632 011	8.1055 537	1.8729 311	4.0351 076	8.6933 759	1.8825 1383
6.58	2.5651 511	8.1117 199	1.8738 808	4.0371 538	8.6977 843	1.8840 3475
6.59	2.5670 995	8.1178 815	1.8748 296	4.0391 980	8.7021 882	1.8855 5335
6.60	2.5690 465	8.1240 384	1.8757 775	4.0412 400	8.7065 877	1.8870 6965
6.61	2.5709 920	8.1301 906	1.8767 243	4.0432 800	8.7109 827	1.8885 8365
6.62	2.5729 361	8.1363 382	1.8776 703	4.0453 180	8.7153 734	1.8900 9537
6.63	2.5748 786	8.1424 812	1.8786 152	4.0473 539	8.7197 596	1.8916 0480
6.64	2.5768 197	8.1486 195	1.8795 593	4.0493 877	8.7241 413	1.8931 1196
6.65	2.5787 594	8.1547 532	1.8805 024	4.0514 195	8.7285 187	1.8946 1685
6.66	2.5806 976	8.1608 823	1.8814 445	4.0534 493	8.7328 917	1.8961 1948
6.67	2.5826 343	8.1670 068	1.8823 857	4.0554 770	8.7372 604	1.8976 1986
6.68	2.5845 696	8.1731 267	1.8833 259	4.0575 027	8.7416 246	1.8991 1799
6.69	2.5865 034	8.1792 420	1.8842 653	4.0595 264	8.7459 846	1.9006 1387
6.70	2.5884 358	8.1853 528	1.8852 036	4.0615 481	8.7503 401	1.9021 0753
6.71	2.5903 668	8.1914 590	1.8861 411	4.0635 678	8.7546 914	1.9035 9895
6.72	2.5922 963	8.1975 606	1.8870 776	4.0655 854	8.7590 383	1.9050 8815
6.73	2.5942 244	8.2036 577	1.8880 132	4.0676 011	8.7633 809	1.9065 7514
6.74	2.5961 510	8.2097 503	1.8889 478	4.0696 148	8.7677 192	1.9080 5992
6.75	2.5980 762	8.2158 384	1.8898 816	4.0716 264	8.7720 532	1.9095 4250
6.76	2.6000 000	8.2219 219	1.8908 144	4.0736 361	8.7763 830	1.9110 2289
6.77	2.6019 224	8.2280 010	1.8917 463	4.0756 438	8.7807 084	1.9125 0109
6.78	2.6038 433	8.2340 755	1.8926 773	4.0776 496	8.7850 296	1.9139 7710
6.79	2.6057 628	8.2401 456	1.8936 073	4.0796 533	8.7893 466	1.9154 5094
6.80	2.6076 810	8.2462 113	1.8945 365	4.0816 551	8.7936 593	1.9169 2261
6.81	2.6095 977	8.2522 724	1.8954 647	4.0836 549	8.7979 679	1.9183 9212
6.82	2.6115 130	8.2583 291	1.8963 920	4.0856 528	8.8022 721	1.9198 5947
6.83	2.6134 269	8.2643 814	1.8973 185	4.0876 487	8.8065 722	1.9213 2467
6.84	2.6153 394	8.2704 293	1.8982 440	4.0896 427	8.8108 681	1.9227 8773
6.85	2.6172 505	8.2764 727	1.8991 686	4.0916 347	8.8151 598	1.9242 4865
6.86	2.6191 602	8.2825 117	1.9000 923	4.0936 248	8.8194 473	1.9257 0744
6.87	2.6210 685	8.2885 463	1.9010 152	4.0956 130	8.8237 307	1.9271 6411
6.88	2.6229 754	8.2945 765	1.9019 371	4.0975 992	8.8280 099	1.9286 1865
6.89	2.6248 809	8.3006 024	1.9028 581	4.0995 835	8.8322 850	1.9300 7109
6.90	2.6267 851	8.3066 239	1.9037 783	4.1015 659	8.8365 559	1.9315 2141
6.91	2.6286 879	8.3126 410	1.9046 975	4.1035 464	8.8408 227	1.9329 6964
6.92	2.6305 893	8.3186 537	1.9056 159	4.1055 250	8.8450 854	1.9344 1577
6.93	2.6324 893	8.3246 622	1.9065 334	4.1075 016	8.8493 440	1.9358 5981
6.94	2.6343 880	8.3306 662	1.9074 500	4.1094 764	8.8535 985	1.9373 0177
6.95	2.6362 853	8.3366 660	1.9083 657	4.1114 493	8.8578 489	1.9387 4166
6.96	2.6381 812	8.3426 614	1.9092 805	4.1134 202	8.8620 952	1.9401 7947
6.97	2.6400 758	8.3486 526	1.9101 945	4.1153 893	8.8663 375	1.9416 1522
6.98	2.6419 690	8.3546 394	1.9111 076	4.1173 565	8.8705 757	1.9430 4892
6.99	2.6438 608	8.3606 220	1.9120 198	4.1193 218	8.8748 099	1.9444 8056
7.00	2.6457 513	8.3666 003	1.9129 312	4.1212 853	8.8790 400	1.9459 1015
$E_{ii}=$ $E_{iii}=$	.0000 002	.0000 006	.0000 001	.0000 002	.0000 005	.0000 0027

TABLE VII. SQUARE AND CUBE ROOTS AND NATURAL LOGARITHMS

N	$\sqrt{N}$	$\sqrt{10N}$	$\sqrt[3]{N}$	$\sqrt[3]{10N}$	$\sqrt[3]{100N}$	$\log_e N$
7.00	2.6457 513	8.3666 003	1.9129 312	4.1212 853	8.8790 400	1.9459 1015
7.01	2.6476 405	8.3725 743	1.9138 417	4.1232 469	8.8832 661	1.9473 3770
7.02	2.6495 283	8.3785 440	1.9147 513	4.1252 066	8.8874 882	1.9487 6322
7.03	2.6514 147	8.3845 095	1.9156 600	4.1271 645	8.8917 063	1.9501 8671
7.04	2.6532 998	8.3904 708	1.9165 679	4.1291 205	8.8959 204	1.9516 0817
7.05	2.6551 836	8.3964 278	1.9174 750	4.1310 746	8.9001 305	1.9530 2762
7.06	2.6570 661	8.4023 806	1.9183 812	4.1330 269	8.9043 366	1.9544 4505
7.07	2.6589 472	8.4083 292	1.9192 865	4.1349 774	8.9085 387	1.9558 6048
7.08	2.6608 269	8.4142 736	1.9201 910	4.1369 260	8.9127 369	1.9572 7391
7.09	2.6627 054	8.4202 138	1.9210 946	4.1388 728	8.9169 311	1.9586 8534
7.10	2.6645 825	8.4261 498	1.9219 973	4.1408 177	8.9211 214	1.9600 9478
7.11	2.6664 583	8.4320 816	1.9228 993	4.1427 609	8.9253 078	1.9615 0224
7.12	2.6683 328	8.4380 092	1.9238 003	4.1447 022	8.9294 902	1.9629 0773
7.13	2.6702 060	8.4439 327	1.9247 006	4.1466 417	8.9336 687	1.9643 1123
7.14	2.6720 778	8.4498 521	1.9256 000	4.1485 794	8.9378 433	1.9657 1278
7.15	2.6739 484	8.4557 673	1.9264 985	4.1505 153	8.9420 140	1.9671 1236
7.16	2.6758 176	8.4616 783	1.9273 962	4.1524 493	8.9461 809	1.9685 0998
7.17	2.6776 856	8.4675 853	1.9282 931	4.1543 816	8.9503 438	1.9699 0565
7.18	2.6795 522	8.4734 881	1.9291 892	4.1563 121	8.9545 029	1.9712 9938
7.19	2.6814 175	8.4793 868	1.9300 844	4.1582 407	8.9586 581	1.9726 9117
7.20	2.6832 816	8.4852 814	1.9309 788	4.1601 676	8.9628 095	1.9740 8103
7.21	2.6851 443	8.4911 719	1.9318 723	4.1620 928	8.9669 570	1.9754 6895
7.22	2.6870 058	8.4970 583	1.9327 651	4.1640 161	8.9711 007	1.9768 5495
7.23	2.6888 659	8.5029 407	1.9336 570	4.1659 377	8.9752 406	1.9782 3904
7.24	2.6907 248	8.5088 190	1.9345 481	4.1678 574	8.9793 766	1.9796 2121
7.25	2.6925 824	8.5146 932	1.9354 383	4.1697 755	8.9835 089	1.9810 0147
7.26	2.6944 387	8.5205 634	1.9363 278	4.1716 917	8.9876 373	1.9823 7983
7.27	2.6962 938	8.5264 295	1.9372 164	4.1736 062	8.9917 620	1.9837 5629
7.28	2.6981 475	8.5322 916	1.9381 042	4.1755 190	8.9958 829	1.9851 3086
7.29	2.7000 000	8.5381 497	1.9389 912	4.1774 300	9.0000 000	1.9865 0355
7.30	2.7018 512	8.5440 037	1.9398 774	4.1793 392	9.0041 133	1.9878 7435
7.31	2.7037 012	8.5498 538	1.9407 628	4.1812 467	9.0082 229	1.9892 4327
7.32	2.7055 499	8.5556 999	1.9416 474	4.1831 525	9.0123 288	1.9906 1033
7.33	2.7073 973	8.5615 419	1.9425 311	4.1850 565	9.0164 309	1.9919 7552
7.34	2.7092 434	8.5673 800	1.9434 141	4.1869 588	9.0205 293	1.9933 3884
7.35	2.7110 883	8.5732 141	1.9442 963	4.1888 594	9.0246 239	1.9947 0031
7.36	2.7129 320	8.5790 442	1.9451 777	4.1907 582	9.0287 149	1.9960 5993
7.37	2.7147 744	8.5848 704	1.9460 582	4.1926 553	9.0328 021	1.9974 1771
7.38	2.7166 155	8.5906 926	1.9469 380	4.1945 508	9.0368 857	1.9987 7364
7.39	2.7184 554	8.5965 109	1.9478 170	4.1964 445	9.0409 655	2.0001 2773
7.40	2.7202 941	8.6023 253	1.9486 952	4.1983 365	9.0450 417	2.0014 8000
7.41	2.7221 315	8.6081 357	1.9495 726	4.2002 267	9.0491 142	2.0028 3044
7.42	2.7239 677	8.6139 422	1.9504 492	4.2021 153	9.0531 831	2.0041 7906
7.43	2.7258 026	8.6197 448	1.9513 250	4.2040 022	9.0572 482	2.0055 2586
7.44	2.7276 363	8.6255 435	1.9522 000	4.2058 874	9.0613 098	2.0068 7085
7.45	2.7294 688	8.6313 383	1.9530 743	4.2077 709	9.0653 677	2.0082 1403
7.46	2.7313 001	8.6371 292	1.9539 477	4.2096 528	9.0694 220	2.0095 5541
7.47	2.7331 301	8.6429 162	1.9548 204	4.2115 329	9.0734 726	2.0108 9500
7.48	2.7349 589	8.6486 993	1.9556 923	4.2134 114	9.0775 197	2.0122 3279
7.49	2.7367 864	8.6544 786	1.9565 635	4.2152 882	9.0815 631	2.0135 6880
7.50	2.7386 128	8.6602 540	1.9574 338	4.2171 633	9.0856 030	2.0149 0302
$\begin{smallmatrix}\text{EII} \\ \text{EIII}\end{smallmatrix} \equiv$	.0000 002	.0000 005	.0000 001	.0000 002	.0000 005	.0000 0024

TABLE VII. SQUARE AND CUBE ROOTS AND NATURAL LOGARITHMS

N	$\sqrt{N}$	$\sqrt{10N}$	$\sqrt[3]{N}$	$\sqrt[3]{10N}$	$\sqrt[3]{100N}$	$\log_e N$
7.50	2.7386 128	8.6602 540	1.9574 338	4.2171 633	9.0856 030	2.0149 0302
7.51	2.7404 379	8.6660 256	1.9583 034	4.2190 368	9.0896 392	2.0162 3547
7.52	2.7422 618	8.6717 934	1.9591 722	4.2209 086	9.0936 719	2.0175 6614
7.53	2.7440 845	8.6775 573	1.9600 403	4.2227 787	9.0977 010	2.0188 9504
7.54	2.7459 060	8.6833 173	1.9609 075	4.2246 472	9.1017 265	2.0202 2218
7.55	2.7477 263	8.6890 736	1.9617 740	4.2265 141	9.1057 485	2.0215 4756
7.56	2.7495 454	8.6948 260	1.9626 398	4.2283 792	9.1097 669	2.0228 7119
7.57	2.7513 633	8.7005 747	1.9635 048	4.2302 428	9.1137 818	2.0241 9307
7.58	2.7531 800	8.7063 195	1.9643 690	4.2321 047	9.1177 931	2.0255 1320
7.59	2.7549 955	8.7120 606	1.9652 324	4.2339 650	9.1218 010	2.0268 3159
7.60	2.7568 098	8.7177 979	1.9660 951	4.2358 236	9.1258 053	2.0281 4825
7.61	2.7586 228	8.7235 314	1.9669 571	4.2376 806	9.1298 061	2.0294 6317
7.62	2.7604 347	8.7292 611	1.9678 183	4.2395 360	9,1338 034	2.0307 7637
7.63	2.7622 455	8.7349 871	1.9686 787	4.2413 897	9.1377 971	2.0320 8785
7.64	2.7640 550	8.7407 094	1.9695 384	4.2432 419	9.1417 874	2.0333 9760
7.65	2.7658 633	8.7464 278	1.9703 973	4.2450 924	9.1457 743	2.0347 0565
7.66	2.7676 705	8.7521 426	1.9712 555	4.2469 413	9.1497 576	2.0360 1198
7.67	2.7694 765	8.7578 536	1.9721 130	4.2487 886	9.1537 375	2.0373 1662
7.68	2.7712 813	8.7635 609	1.9729 697	4.2506 343	9.1577 139	2.0386 1955
7.69	2.7730 849	8.7692 645	1.9738 256	4.2524 784	9.1616 869	2.0399 2078
7.70	2.7748 874	8.7749 644	1.9746 808	4.2543 209	9.1656 565	2.0412 2033
7.71	2.7766 887	8.7806 606	1.9755 353	4.2561 618	9.1696 226	2.0425 1819
7.72	2.7784 888	8.7863 531	1.9763 890	4.2580 011	9.1735 852	2.0438 1436
7.73	2.7802 878	8.7920 419	1.9772 420	4.2598 388	9.1775 445	2.0451 0886
7.74	2.7820 855	8.7977 270	1.9780 943	4.2616 749	9.1815 003	2.0464 0169
7.75	2.7838 822	8.8034 084	1.9789 458	4.2635 095	9.1854 528	2.0476 9284
7.76	2.7856 777	8.8090 862	1.9797 966	4.2653 425	9.1894 018	2.0489 8233
7.77	2.7874 720	8.8147 603	1.9806 467	4.2671 739	9.1933 474	2.0502 7016
7.78	2.7892 651	8.8204 308	1.9814 960	4.2690 037	9.1972 897	2.0515 5634
7.79	2.7910 571	8.8260 977	1.9823 446	4.2708 320	9.2012 286	2.0528 4086
7.80	2.7928 480	8.8317 609	1.9831 925	4.2726 587	9.2051 641	2.0541 2373
7.81	2.7946 377	8.8374 204	1.9840 396	4.2744 838	9.2090 962	2.0554 0496
7.82	2.7964 263	8.8430 764	1.9848 861	4.2763 074	9.2130 250	2.0566 8455
7.83	2.7982 137	8.8487 287	1.9857 318	4.2781 294	9.2169 505	2.0579 6251
7.84	2.8000 000	8.8543 774	1.9865 768	4.2799 499	9.2208 726	2.0592 3883
7.85	2.8017 851	8.8600 226	1.9874 211	4.2817 689	9.2247 914	2.0605 1353
7.86	2.8035 692	8.8656 641	1.9882 646	4.2835 862	9.2287 068	2.0617 8661
7.87	2.8053 520	8.8713 020	1.9891 075	4.2854 021	9.2326 189	2.0630 5806
7.88	2.8071 338	8.8769 364	1.9899 496	4.2872 164	9.2365 277	2.0643 2790
7.89	2.8089 144	8.8825 672	1.9907 910	4.2890 292	9.2404 333	2.0655 9613
7.90	2.8106 939	8.8881 944	1.9916 317	4.2908 404	9.2443 355	2.0668 6276
7.91	2.8124 722	8.8938 181	1.9924 717	4.2926 501	9.2482 344	2.0681 2778
7.92	2.8142 495	8.8994 382	1.9933 110	4.2944 583	9.2521 300	2.0693 9121
7.93	2.8160 256	8.9050 547	1.9941 496	4.2962 650	9.2560 224	2.0706 5304
7.94	2.8178 006	8.9106 678	1.9949 874	4.2980 702	9.2599 115	2.0719 1328
7.95	2.8195 744	8.9162 773	1.9958 246	4.2998 738	9.2637 973	2.0731 7193
7.96	2.8213 472	8.9218 832	1.9966 611	4.3016 759	9.2676 798	2.0744 2900
7.97	2.8231 188	8.9274 856	1.9974 969	4.3034 765	9.2715 592	2.0756 8449
7.98	2.8248 894	8.9330 846	1.9983 319	4.3052 757	9.2754 352	2.0769 3841
7.99	2.8266 588	8.9386 800	1.9991 663	4.3070 733	9.2793 081	2.0781 9076
8.00	2.8284 271	8.9442 719	2.0000 000	4.3088 694	9.2831 777	2.0794 4154
$\begin{smallmatrix}Ell\\Elll\end{smallmatrix}$	.0000 002	.0000 005	.0000 001	.0000 002	.0000 004	.0000 0021

TABLE VII. SQUARE AND CUBE ROOTS AND NATURAL LOGARITHMS

N	$\sqrt{N}$	$\sqrt{10N}$	$\sqrt[3]{N}$	$\sqrt[3]{10N}$	$\sqrt[3]{100N}$	$\log_e N$
8.00	2.8284 271	8.9442 719	2.0000 000	4.3088 694	9.2831 777	2.0794 4154
8.01	2.8301 943	8.9498 603	2.0008 330	4.3106 640	9.2870 440	2.0806 9076
8.02	2.8319 605	8.9554 453	2.0016 653	4.3124 571	9.2909 072	2.0819 3842
8.03	2.8337 255	8.9610 267	2.0024 969	4.3142 487	9.2947 672	2.0831 8453
8.04	2.8354 894	8.9666 047	2.0033 278	4.3160 389	9.2986 239	2.0844 2908
8.05	2.8372 522	8.9721 792	2.0041 580	4.3178 276	9.3024 775	2.0856 7209
8.06	2.8390 139	8.9777 503	2.0049 876	4.3196 147	9.3063 278	2.0869 1356
8.07	2.8407 745	8.9833 179	2.0058 164	4.3214 004	9.3101 750	2.0881 5348
8.08	2.8425 341	8.9888 820	2.0066 446	4.3231 847	9.3140 190	2.0893 9187
8.09	2.8442 925	8.9944 427	2.0074 720	4.3249 674	9.3178 598	2.0906 2873
8.10	2.8460 499	9.0000 000	2.0082 989	4.3267 487	9.3216 975	2.0918 6406
8.11	2.8478 062	9.0055 538	2.0091 250	4.3285 285	9.3255 320	2.0930 9787
8.12	2.8495 614	9.0111 043	2.0099 504	4.3303 069	9.3293 634	2.0943 3015
8.13	2.8513 155	9.0166 513	2.0107 752	4.3320 838	9.3331 916	2.0955 6092
8.14	2.8530 685	9.0221 949	2.0115 993	4.3338 592	9.3370 167	2.0967 9018
8.15	2.8548 205	9.0277 350	2.0124 227	4.3356 332	9.3408 386	2.0980 1793
8.16	2.8565 714	9.0332 718	2.0132 454	4.3374 058	9.3446 575	2.0992 4417
8.17	2.8583 212	9.0388 052	2.0140 675	4.3391 769	9.3484 732	2.1004 6891
8.18	2.8600 699	9.0443 352	2.0148 889	4.3409 465	9.3522 858	2.1016 9215
8.19	2.8618 176	9.0498 619	2.0157 096	4.3427 147	9.3560 952	2.1029 1390
8.20	2.8635 642	9.0553 851	2.0165 297	4.3444 815	9.3599 016	2.1041 3415
8.21	2.8653 098	9.0609 050	2.0173 491	4.3462 468	9.3637 049	2.1053 5292
8.22	2.8670 542	9.0664 216	2.0181 678	4.3480 107	9.3675 051	2.1065 7021
8.23	2.8687 977	9.0719 347	2.0189 859	4.3497 732	9.3713 022	2.1077 8601
8.24	2.8705 400	9.0774 446	2.0198 033	4.3515 342	9.3750 963	2.1090 0034
8.25	2.8722 813	9.0829 511	2.0206 200	4.3532 938	9.3788 873	2.1102 1320
8.26	2.8740 216	9.0884 542	2.0214 361	4.3550 520	9.3826 752	2.1114 2459
8.27	2.8757 608	9.0939 540	2.0222 515	4.3568 088	9.3864 601	2.1126 3451
8.28	2.8774 989	9.0994 505	2.0230 663	4.3585 642	9.3902 419	2.1138 4297
8.29	2.8792 360	9.1049 437	2.0238 804	4.3603 181	9.3940 206	2.1150 4997
8.30	2.8809 721	9.1104 336	2.0246 939	4.3620 707	9.3977 964	2.1162 5551
8.31	2.8827 071	9.1159 201	2.0255 067	4.3638 218	9.4015 691	2.1174 5961
8.32	2.8844 410	9.1214 034	2.0263 188	4.3655 715	9.4053 388	2.1186 6225
8.33	2.8861 739	9.1268 834	2.0271 303	4.3673 199	9.4091 054	2.1198 6346
8.34	2.8879 058	9.1323 600	2.0279 412	4.3690 668	9.4128 690	2.1210 6322
8.35	2.8896 367	9.1378 334	2.0287 514	4.3708 123	9.4166 297	2.1222 6154
8.36	2.8913 665	9.1433 036	2.0295 609	4.3725 565	9.4203 873	2.1234 5843
8.37	2.8930 952	9.1487 704	2.0303 698	4.3742 992	9.4241 420	2.1246 5388
8.38	2.8948 230	9.1542 340	2.0311 781	4.3760 406	9.4278 936	2.1258 4791
8.39	2.8965 497	9.1596 943	2.0319 857	4.3777 805	9.4316 423	2.1270 4052
8.40	2.8982 753	9.1651 514	2.0327 927	4.3795 191	9.4353 880	2.1282 3171
8.41	2.9000 000	9.1706 052	2.0335 991	4.3812 564	9.4391 307	2.1294 2147
8.42	2.9017 236	9.1760 558	2.0344 048	4.3829 922	9.4428 704	2.1306 0983
8.43	2.9034 462	9.1815 031	2.0352 098	4.3847 267	9.4466 072	2.1317 9677
8.44	2.9051 678	9.1869 473	2.0360 143	4.3864 598	9.4503 411	2.1329 8231
8.45	2.9068 884	9.1923 882	2.0368 181	4.3881 915	9.4540 719	2.1341 6644
8.46	2.9086 079	9.1978 258	2.0376 212	4.3899 218	9.4577 999	2.1353 4917
8.47	2.9103 264	9.2032 603	2.0384 237	4.3916 508	9.4615 249	2.1365 3051
8.48	2.9120 440	9.2086 915	2.0392 256	4.3933 785	9.4652 470	2.1377 1045
8 49	2.9137 605	9.2141 196	2.0400 269	4.3951 047	9.4689 661	2.1388 8900
8.50	2.9154 759	9.2195 445	2.0408 276	4.3968 297	9.4726 824	2.1400 6616
$\begin{matrix}\mathrm{Eii} \\ \mathrm{Eiii}\end{matrix}$ =	.0000 001	.0000 004	.0000 001	.0000 002	.0000 004	.0000 0018

TABLE VII. SQUARE AND CUBE ROOTS AND NATURAL LOGARITHMS

N	$\sqrt{N}$	$\sqrt{10N}$	$\sqrt[3]{N}$	$\sqrt[3]{10N}$	$\sqrt[3]{100N}$	$\log_e N$
8.50	2.9154 759	9.2195 445	2.0408 276	4.3968 297	9.4726 824	2.1400 6616
8.51	2.9171 904	9.2249 661	2.0416 276	4.3985 532	9.4763 957	2.1412 4194
8.52	2.9189 039	9.2303 846	2.0424 269	4.4002 755	9.4801 061	2.1424 1634
8.53	2.9206 164	9.2357 999	2.0432 257	4.4019 963	9.4838 136	2.1435 8936
8.54	2.9223 278	9.2412 120	2.0440 238	4.4037 159	9.4875 182	2.1447 6101
8.55	2.9240 383	9.2466 210	2.0448 214	4.4054 341	9.4912 200	2.1459 3128
8.56	2.9257 478	9.2520 268	2.0456 182	4.4071 509	9.4949 188	2.1471 0019
8.57	2.9274 562	9.2574 294	2.0464 145	4.4088 664	9.4986 148	2.1482 6773
8.58	2.9291 637	9.2628 289	2.0472 102	4.4105 806	9.5023 078	2.1494 3391
8.59	2.9308 702	9.2682 253	2.0480 052	4.4122 934	9.5059 981	2.1505 9874
8.60	2.9325 757	9.2736 185	2.0487 996	4.4140 050	9.5096 854	2.1517 6220
8.61	2.9342 802	9.2790 086	2.0495 934	4.4157 152	9.5133 699	2.1529 2432
8.62	2.9359 837	9.2843 955	2.0503 866	4.4174 240	9.5170 516	2.1540 8508
8.63	2.9376 862	9.2897 793	2.0511 792	4.4191 316	9.5207 304	2.1552 4451
8.64	2.9393 877	9.2951 600	2.0519 711	4.4208 378	9.5244 063	2.1564 0258
8.65	2.9410 882	9.3005 376	2.0527 625	4.4225 427	9.5280 794	2.1575 5932
8.66	2.9427 878	9.3059 121	2.0535 532	4.4242 463	9.5317 497	2.1587 1472
8.67	2.9444 864	9.3112 835	2.0543 434	4.4259 486	9.5354 172	2.1598 6879
8.68	2.9461 840	9.3166 518	2.0551 329	4.4276 496	9.5390 818	2.1610 2153
8.69	2.9478 806	9.3220 169	2.0559 218	4.4293 493	9.5427 437	2.1621 7294
8.70	2.9495 762	9.3273 791	2.0567 101	4.4310 476	9.5464 027	2.1633 2303
8.71	2.9512 709	9.3327 381	2.0574 978	4.4327 447	9.5500 589	2.1644 7179
8.72	2.9529 646	9.3380 940	2.0582 849	4.4344 405	9.5537 124	2.1656 1924
8.73	2.9546 573	9.3434 469	2.0590 714	4.4361 349	9.5573 630	2.1667 6537
8.74	2.9563 491	9.3487 967	2.0598 573	4.4378 281	9.5610 108	2.1679 1019
8.75	2.9580 399	9.3541 435	2.0606 426	4.4395 200	9.5646 559	2.1690 5370
8.76	2.9597 297	9.3594 872	2.0614 274	4.4412 106	9.5682 982	2.1701 9590
8.77	2.9614 186	9.3648 278	2.0622 115	4.4428 999	9.5719 377	2.1713 3681
8.78	2.9631 065	9.3701 654	2.0629 950	4.4445 880	9.5755 745	2.1724 7641
8.79	2.9647 934	9.3755 000	2.0637 779	4.4462 747	9.5792 085	2.1736 1471
8.80	2.9664 794	9.3808 315	2.0645 602	4.4479 602	9.5828 397	2.1747 5172
8.81	2.9681 644	9.3861 600	2.0653 420	4.4496 444	9.5864 682	2.1758 8744
8.82	2.9698 485	9.3914 855	2.0661 231	4.4513 273	9.5900 939	2.1770 2187
8.83	2.9715 316	9.3968 080	2.0669 037	4.4530 089	9.5937 170	2.1781 5501
8.84	2.9732 137	9.4021 274	2.0676 836	4.4546 893	9.5973 372	2.1792 8688
8.85	2.9748 950	9.4074 439	2.0684 630	4.4563 684	9.6009 548	2.1804 1746
8.86	2.9765 752	9.4127 573	2.0692 418	4.4580 463	9.6045 696	2.1815 4676
8.87	2.9782 545	9.4180 677	2.0700 200	4.4597 229	9.6081 817	2.1826 7480
8.88	2.9799 329	9.4233 752	2.0707 976	4.4613 982	9.6117 911	2.1838 0156
8.89	2.9816 103	9.4286 797	2.0715 746	4.4630 723	9.6153 977	2.1849 2705
8.90	2.9832 868	9.4339 811	2.0723 511	4.4647 451	9.6190 017	2.1860 5128
8.91	2.9849 623	9.4392 796	2.0731 270	4.4664 167	9.6226 030	2.1871 7424
8.92	2.9866 369	9.4445 752	2.0739 023	4.4680 870	9.6262 016	2.1882 9595
8.93	2.9883 106	9.4498 677	2.0746 770	4.4697 560	9.6297 975	2.1894 1639
8.94	2.9899 833	9.4551 573	2.0754 511	4.4714 239	9.6333 907	2.1905 3559
8.95	2.9916 551	9.4604 440	2.0762 247	4.4730 904	9.6369 812	2.1916 5353
8.96	2.9933 259	9.4657 277	2.0769 976	4.4747 558	9.6405 691	2.1927 7023
8.97	2.9949 958	9.4710 084	2.0777 700	4.4764 199	9.6441 542	2.1938 8568
8.98	2.9966 648	9.4762 862	2.0785 419	4.4780 827	9.6477 368	2.1949 9988
8.99	2.9983 329	9.4815 611	2.0793 131	4.4797 444	9.6513 166	2.1961 1285
9.00	3.0000 000	9.4868 330	2.0800 838	4.4814 047	9.6548 938	2.1972 2458
E±i E±ii	.0000 001	.0000 004	.0000 001	.0000 002	.0000 004	.0000 0016

TABLE VII. SQUARE AND CUBE ROOTS AND NATURAL LOGARITHMS

N	$\sqrt{N}$	$\sqrt{10N}$	$\sqrt[3]{N}$	$\sqrt[3]{10N}$	$\sqrt[3]{100N}$	$\log_e N$
9.00	3.0000 000	9.4868 330	2.0800 838	4.4814 047	9.6548 938	2.1972 2458
9.01	3.0016 662	9.4921 020	2.0808 539	4.4830 639	9.6584 684	2.1983 3507
9.02	3.0033 315	9.4973 681	2.0816 235	4.4847 218	9.6620 403	2.1994 4433
9.03	3.0049 958	9.5026 312	2.0823 925	4.4863 786	9.6656 096	2.2005 5237
9.04	3.0066 593	9.5078 915	2.0831 609	4.4880 341	9.6691 763	2.2016 5917
9.05	3.0083 218	9.5131 488	2.0839 287	4.4896 883	9.6727 403	2.2027 6476
9.06	3.0099 834	9.5184 032	2.0846 960	4.4913 414	9.6763 017	2.2038 6912
9.07	3.0116 441	9.5236 548	2.0854 627	4.4929 932	9.6798 604	2.2049 7226
9.08	3.0133 038	9.5289 034	2.0862 289	4.4946 438	9.6834 166	2.2060 7419
9.09	3.0149 627	9.5341 491	2.0869 945	4.4962 932	9.6869 701	2.2071 7491
9.10	3.0166 206	9.5393 920	2.0877 595	4.4979 414	9.6905 211	2.2082 7441
9.11	3.0182 777	9.5446 320	2.0885 239	4.4995 884	9.6940 694	2.2093 7271
9.12	3.0199 338	9.5498 691	2.0892 879	4.5012 342	9.6976 152	2.2104 6980
9.13	3.0215 890	9.5551 033	2.0900 512	4.5028 788	9.7011 583	2.2115 6569
9.14	3.0232 433	9.5603 347	2.0908 140	4.5045 222	9.7046 989	2.2126 6039
9.15	3.0248 967	9.5655 632	2.0915 762	4.5061 644	9.7082 369	2.2137 5388
9.16	3.0265 492	9.5707 889	2.0923 379	4.5078 054	9.7117 723	2.2148 4618
9.17	3.0282 008	9.5760 117	2.0930 990	4.5094 452	9.7153 051	2.2159 3729
9.18	3.0298 515	9.5812 317	2.0938 596	4.5110 838	9.7188 354	2.2170 2720
9.19	3.0315 013	9.5864 488	2.0946 196	4.5127 212	9.7223 631	2.2181 1594
9.20	3.0331 502	9.5916 630	2.0953 791	4.5143 574	9.7258 883	2.2192 0348
9.21	3.0347 982	9.5968 745	2.0961 380	4.5159 925	9.7294 109	2.2202 8985
9.22	3.0364 453	9.6020 831	2.0968 964	4.5176 263	9.7329 309	2.2213 7504
9.23	3.0380 915	9.6072 889	2.0976 542	4.5192 590	9.7364 484	2.2224 5905
9.24	3.0397 368	9.6124 919	2.0984 115	4.5208 905	9.7399 634	2.2235 4189
9.25	3.0413 813	9.6176 920	2.0991 682	4.5225 208	9.7434 758	2.2246 2355
9.26	3.0430 248	9.6228 894	2.0999 244	4.5241 500	9.7469 857	2.2257 0405
9.27	3.0446 675	9.6280 839	2.1006 801	4.5257 780	9.7504 931	2.2267 8338
9.28	3.0463 092	9.6332 757	2.1014 351	4.5274 048	9.7539 979	2.2278 6155
9.29	3.0479 501	9.6384 646	2.1021 897	4.5290 304	9.7575 003	2.2289 3855
9.30	3.0495 901	9.6436 508	2.1029 437	4.5306 549	9.7610 001	2.2300 1440
9.31	3.0512 293	9.6488 341	2.1036 972	4.5322 782	9.7644 974	2.2310 8909
9.32	3.0528 675	9.6540 147	2.1044 501	4.5339 004	9.7679 922	2.2321 6263
9.33	3.0545 049	9.6591 925	2.1052 025	4.5355 213	9.7714 845	2.2332 3501
9.34	3.0561 414	9.6643 675	2.1059 544	4.5371 412	9.7749 743	2.2343 0625
9.35	3.0577 770	9.6695 398	2.1067 057	4.5387 598	9.7784 617	2.2353 7634
9.36	3.0594 117	9.6747 093	2.1074 565	4.5403 774	9.7819 465	2.2364 4529
9.37	3.0610 456	9.6798 760	2.1082 067	4.5419 937	9.7854 289	2.2375 1310
9.38	3.0626 786	9.6850 400	2.1089 565	4.5436 089	9.7889 087	2.2385 7976
9.39	3.0643 107	9.6902 012	2.1097 056	4.5452 230	9.7923 861	2.2396 4529
9.40	3.0659 419	9.6953 597	2.1104 543	4.5468 359	9.7958 611	2.2407 0969
9.41	3.0675 723	9.7005 155	2.1112 024	4.5484 477	9.7993 336	2.2417 7295
9.42	3.0692 019	9.7056 684	2.1119 500	4.5500 584	9.8028 036	2.2428 3509
9.43	3.0708 305	9.7108 187	2.1126 971	4.5516 679	9.8062 711	2.2438 9610
9.44	3.0724 583	9.7159 662	2.1134 436	4.5532 762	9.8097 363	2.2449 5598
9.45	3.0740 852	9.7211 110	2.1141 896	4.5548 835	9.8131 989	2.2460 1474
9.46	3.0757 113	9.7262 531	2.1149 351	4.5564 896	9.8166 592	2.2470 7238
9.47	3.0773 365	9.7313 925	2.1156 801	4.5580 945	9.8201 169	2.2481 2891
9.48	3.0789 609	9.7365 292	2.1164 245	4.5596 983	9.8235 723	2.2491 8432
9.49	3.0805 844	9.7416 631	2.1171 684	4.5613 011	9.8270 252	2.2502 3861
9.50	3.0822 070	9.7467 943	2.1179 118	4.5629 026	9.8304 757	2.2512 9180
$\varepsilon_{II} =$ $\varepsilon_{III} =$	.0000 001	.0000 003	.0000 001	.0000 001	.0000 003	.0000 0015

TABLE VII. SQUARE AND CUBE ROOTS AND NATURAL LOGARITHMS

N	$\sqrt{N}$	$\sqrt{10N}$	$\sqrt[3]{N}$	$\sqrt[3]{10N}$	$\sqrt[3]{100N}$	$\log_e N$
9.50	3.0822 070	9.7467 943	2.1179 118	4.5629 026	9.8304 757	2.2512 9180
9.51	3.0838 288	9.7519 229	2.1186 547	4.5645 031	9.8339 238	2.2523 4388
9.52	3.0854 497	9.7570 487	2.1193 970	4.5661 024	9.8373 695	2.2533 9485
9.53	3.0870 698	9.7621 719	2.1201 388	4.5677 006	9.8408 127	2.2544 4472
9.54	3.0886 890	9.7672 924	2.1208 801	4.5692 977	9.8442 536	2.2554 9349
9.55	3.0903 074	9.7724 101	2.1216 209	4.5708 937	9.8476 920	2.2565 4115
9.56	3.0919 250	9.7775 252	2.1223 612	4.5724 886	9.8511 280	2.2575 8773
9.57	3.0935 417	9.7826 377	2.1231 010	4.5740 823	9.8545 617	2.2586 3321
9.58	3.0951 575	9.7877 474	2.1238 402	4.5756 750	9.8579 929	2.2596 7759
9.59	3.0967 725	9.7928 545	2.1245 789	4.5772 665	9.8614 218	2.2607 2089
9.60	3.0983 867	9.7979 590	2.1253 171	4.5788 570	9.8648 483	2.2617 6310
9.61	3.1000 000	9.8030 607	2.1260 548	4.5804 463	9.8682 724	2.2628 0422
9.62	3.1016 125	9.8081 599	2.1267 920	4.5820 345	9.8716 941	2.2638 4426
9.63	3.1032 241	9.8132 563	2.1275 287	4.5836 217	9.8751 135	2.2648 8323
9.64	3.1048 349	9.8183 502	2.1282 649	4.5852 077	9.8785 305	2.2659 2111
9.65	3.1064 449	9.8234 414	2.1290 005	4.5867 926	9.8819 451	2.2669 5792
9.66	3.1080 541	9.8285 299	2.1297 357	4.5883 765	9.8853 574	2.2679 9365
9.67	3.1096 624	9.8336 158	2.1304 703	4.5899 592	9.8887 673	2.2690 2831
9.68	3.1112 698	9.8386 991	2.1312 045	4.5912 408	9.8921 749	2.2700 6190
9.69	3.1128 765	9.8437 798	2.1319 381	4.5931 214	9.8955 801	2.2710 9443
9.70	3.1144 823	9.8488 578	2.1326 712	4.5947 009	9.8989 830	2.2721 2589
9.71	3.1160 873	9.8539 332	2.1334 039	4.5962 793	9.9023 835	2.2731 5628
9.72	3.1176 915	9.8590 060	2.1341 360	4.5978 566	9.9057 817	2.2741 8562
9.73	3.1192 948	9.8640 762	2.1348 676	4.5994 328	9.9091 776	2.2752 1390
9.74	3.1208 973	9.8691 438	2.1355 987	4.6010 080	9.9125 712	2.2762 4112
9.75	3.1224 990	9.8742 088	2.1363 293	4.6025 320	9.9159 624	2.2772 6729
9.76	3.1240 999	9.8792 712	2.1370 595	4.6041 550	9.9193 513	2.2782 9240
9.77	3.1256 999	9.8843 310	2.1377 891	4.6057 270	9.9227 379	2.2793 1647
9.78	3.1272 992	9.8893 883	2.1385 182	4.6072 978	9.9261 222	2.2803 3948
9.79	3.1288 976	9.8944 429	2.1392 468	4.6088 676	9.9295 042	2.2813 6146
9.80	3.1304 952	9.8994 949	2.1399 750	4.6104 363	9.9328 839	2.2823 8239
9.81	3.1320 920	9.9045 444	2.1407 026	4.6120 039	9.9362 613	2.2834 0227
9.82	3.1336 879	9.9095 913	2.1414 297	4.6135 705	9.9396 364	2.2844 2112
9.83	3.1352 831	9.9146 356	2.1421 564	4.6151 360	9.9430 092	2.2854 3893
9.84	3.1368 774	9.9196 774	2.1428 825	4.6167 005	9.9463 797	2.2864 5571
9.85	3.1384 710	9.9247 166	2.1436 082	4.6182 639	9.9497 479	2.2874 7146
9.86	3.1400 637	9.9297 533	2.1443 334	4.6198 262	9.9531 138	2.2884 8617
9.87	3.1416 556	9.9347 874	2.1450 581	4.6213 875	9.9564 775	2.2894 9985
9.88	3.1432 467	9.9398 189	2.1457 822	4.6229 477	9.9598 389	2.2905 1251
9.89	3.1448 370	9.9448 479	2.1465 060	4.6245 069	9.9631 981	2.2915 2415
9.90	3.1464 265	9.9498 744	2.1472 292	4.6260 650	9.9665 549	2.2925 3476
9.91	3.1480 152	9.9548 983	2.1479 519	4.6276 221	9.9699 095	2.2935 4435
9.92	3.1496 031	9.9599 197	2.1486 741	4.6291 781	9.9732 619	2.2945 5292
9.93	3.1511 903	9.9649 385	2.1493 959	4.6307 331	9.9766 120	2.2955 6048
9.94	3.1527 766	9.9699 549	2.1501 172	4.6322 870	9.9799 599	2.2965 6702
9.95	3.1543 621	9.9749 687	2.1508 380	4.6338 399	9.9833 055	2.2975 7255
9.96	3.1559 468	9.9799 800	2.1515 583	4.6353 918	9.9866 489	2.2985 7707
9.97	3.1575 307	9.9849 887	2.1522 781	4.6369 426	9.9899 900	2.2995 8058
9.98	3.1591 138	9.9899 950	2.1529 974	4.6384 924	9.9933 289	2.3005 8309
9.99	3.1606 961	9.9949 987	2.1537 163	4.6400 411	9.9966 656	2.3015 8459
10.00	3.1622 777	10.0000 000	2.1544 347	4.6415 888	10.0000 000	2.3025 8509
$\begin{smallmatrix} Ell \\ Elll \end{smallmatrix}$	.0000 001	.0000 003	.0000 001	.0000 001	.0000 003	.0000 0013

TABLE VIII

Table of θ Values

i	θ_i	i	θ_i	i	θ_i
1	1.6499158	43	2.1103574	85	2.1157744
2	1.8856181	44	2.1106066	86	2.1158389
3	1.9641855	45	2.1108447	87	2.1159019
4	2.0034692	46	2.1110724	88	2.1159635
5	2.0270394	47	2.1112905	89	2.1160237
6	2.0427529	48	2.1114994	90	2.1160825
7	2.0539768	49	2.1116998	91	2.1161401
8	2.0623948	50	2.1118923	92	2.1161964
9	2.0689421	51	2.1120771	93	2.1162515
10	2.0741799	52	2.1122549	94	2.1163054
11	2.0784654	53	2.1124259	95	2.1163582
12	2.0820366	54	2.1125906	96	2.1164099
13	2.0850585	55	2.1127494	97	2.1164605
14	2.0876488	56	2.1129024	98	2.1165101
15	2.0898934	57	2.1130501	99	2.1165587
16	2.0918576	38	2.1131927	100	2.1166063
17	2.0935907	59	2.1133304	110	2.1170348
18	2.0951312	60	2.1134636	120	2.1173920
19	2.0965096	61	2.1135924	130	2.1176942
20	2.0977481	62	2.1137170	140	2.1179532
21	2.0988725	63	2.1138377	150	2.1181776
22	2.0998929	64	2.1139546	160	2.1183741
23	2.1008245	65	2.1140680	170	2.1185474
24	2.1016785	66	2.1141779	180	2.1187014
25	2.1024642	67	2.1142845	190	2.1188393
26	2.1031894	68	2.1143879	200	2.1189633
27	2.1038609	69	2.1144884	220	2.1191776
28	2.1044845	70	2.1145860	240	2.1193562
29	2.1050650	71	2.1146808	260	2.1195072
30	2.1056069	72	2.1147731	280	2.1196368
31	2.1061137	73	2.1148627	300	2.1197490
32	2.1065890	74	2.1149500	350	2.1199735
33	2.1070354	75	2.1150350	400	2.1201418
34	2.1074555	76	2.1151177	450	2.1202728
35	2.1078516	77	2.1151982	500	2.1203775
36	2.1082258	78	2.1152767	600	2.1205347
37	2.1085797	79	2.1153532	700	2.1206469
38	2.1089150	80	2.1154278	800	2.1207311
39	2.1092330	81	2.1155005	900	2.1207966
40	2.1095352	82	2.1155715	1000	2.1208489
41	2.1098227	83	2.1156408	10000	2.1212732
42	2.1100964	84	2.1157084	∞	2.1213203

TABLE IX
Constants Frequently Needed

		Logarithms to base 10	Logarithms to base e
2	2.00000 00000 00000	0.30102 99956 63981	.69314 71805 59945
$\sqrt{2}$	1.41421 35623 73095	0.15051 49978 31991	.34657 35902 79973
$1/\sqrt{2}$	.70710 67811 86548		
$\sqrt[3]{2}$	1.25992 10498 94873	.10034 33318 87994	.23104 90601 86648
10	10.00000 00000 00000	1.00000 00000 00000	2.30258 50929 94046
$\sqrt{10}$	3.16227 76601 68379	.50000 00000 00000	1.15129 25464 97023
$\sqrt[3]{10}$	2.15443 46900 31884	.33333 33333 33333	.76752 83643 31349
100	100.00000 00000 00000	2.00000 00000 00000	4.60517 01859 88091
$\sqrt[3]{100}$	4.64158 88336 12779	.66666 66666 66667	1.53505 67286 62697
π	3.14159 26535 89793	0.49714 98726 94134	1.14472 98858 49400
$1/\pi$	.31830 98861 83791		
$\sqrt{\pi}$	1.77245 38509 05516	0.24857 49363 47067	.57236 49429 24700
$1/\sqrt{\pi}$	.56418 95835 47756		
$\sqrt{2\pi}$	2.50662 82746 31001	0.39908 99341 79058	.91893 85332 04673
$1/\sqrt{2\pi}$	.39894 22804 01433		
π^2	9.86960 44010 89359	.99429 97453 88268	2.28945 97716 98800
180	180.00000 00000 00000	2.25527 25051 03306	5.19295 68508 90210
$\dfrac{\pi}{180}$	0.̅01745 32925 19943	$\overline{2}$.24187 73675 90828	−4.04822 69650 40810
$\dfrac{180}{\pi}$	57.°29577 95130 82321		
e	2.71828 18284 59045	0.43429 44819 03252 the modulus	1.00000 00000 00000
1/e	.36787 94411 71442		
$\sqrt{e}$	1.64872 12707 00128	0.21714 72409 51626	.50000 00000 00000
$1/\sqrt{e}$	.60653 06597 12633		
e^2	7.38905 60989 30650	0.86858 89638 06504	2.00000 00000 00000
$1/e^2$	.13533 52832 36613		
the modulus	.43429 44819 03252	$\overline{1}$.63778 43113 00537	
Euler's	.57721 56649 01533	$\overline{1}$.76133 81087 83168	−0.54953 93129 81645